Ethics, Politics, and Whistleblowing in Engineering

Ethics, Politics, and Whistleblowing in Engineering

Edited by
Nicholas Sakellariou
Rania Milleron

CRC Press
Taylor & Francis Group
Boca Raton London New York

CRC Press is an imprint of the
Taylor & Francis Group, an **informa** business

CRC Press
Taylor & Francis Group
6000 Broken Sound Parkway NW, Suite 300
Boca Raton, FL 33487-2742

© 2019 by Taylor & Francis Group, LLC
CRC Press is an imprint of Taylor & Francis Group, an Informa business

No claim to original U.S. Government works

Printed on acid-free paper

International Standard Book Number-13: 978-1-138-56265-3 (Paperback)
International Standard Book Number-13: 978-0-815-37434-3 (Hardback)

Library of Congress Cataloging-in-Publication Data

Names: Sakellariou, Nicholas, author. | Milleron, Rania, author.
Title: Ethics, politics, and whistleblowing in engineering / Nicholas
Sakellariou and Rania Milleron.
Description: Boca Raton, FL : CRC Press, 2018. | Includes bibliographical
references and index.
Identifiers: LCCN 2018019730| ISBN 9781138562653 (pbk. : alk. paper) | ISBN
9781351242417 (e-book)
Subjects: LCSH: Engineering ethics. | Whistle blowing.
Classification: LCC TA157 .S275 2018 | DDC 174/.962--dc23
LC record available at https://lccn.loc.gov/2018019730

Visit the Taylor & Francis Web site at
http://www.taylorandfrancis.com

and the CRC Press Web site at
http://www.crcpress.com

To the Ethical Engineers

Contents

PART I Engineering Leadership

PART II   Daily Practice

PART III   Raising the Bar

Foreword

Ethics, Politics, and Whistleblowing in Engineering is about broader levels of reality and thought than most engineers are permitted or expected to connect with in their daily work. These frames of reference and insight relate to what engineers can and cannot do to apply their expert knowledge in ethical contexts. The structures, principles, rights, and actions described herein are designed to make it easier for them to apply their conscience to their labors when the going gets tough.

Although belonging to a largely employed profession, engineers need to value more the importance of securing more independence in their profession from their commercial employers or clients, including their work for governments and universities. Professions embody more than specialized skills for sale; they are given, by licensure and public trust, a privileged position which underlines the critical features of a profession. These are a reasonable degree of *independence*, a *learned tradition* that is self-critical and noncensorious, and an obligation for *public service*.

Taken together, these attributes constitute a responsibility to place the interests of the public—which are uniquely in the knowledgeable embrace of the profession—above the interests of commercial gains and individual/group ambitions—when the two conflict. Toward these ends, professional engineering societies—which are numerous—need to defend and uphold these roles by their conscientious members whether as employees, consultants, or independent practitioners. In short, such services go beyond just being commerce; otherwise, engineering is merely a trade, not a profession, and a more dangerous trade at that.

Starting with the education of the student engineer, these wider responsibilities are usually not reified so as to be reflected throughout the central core of the curriculum. Because engineering knowledge is proliferating so rapidly, professors feel that more teaching and learning have to be compressed in the available hours, leaving little time to study the larger realities that will envelop their students' life work. This argues for the integration of these nontechnical dimensions into the regular courses, whenever possible, in addition to some special course work in these fields.

Simply having student engineers read and discuss the seminal article "Why the Future Doesn't Need Us" (*Wired Magazine*, 2000), by Bill Joy, would motivate broader intellectual and normative thinking with sound concerns about their own future roles. Engineering students could also read and discuss the more recent 2015 letter to the world, "Open Letter on Artificial Intelligence," signed by Stephen Hawking and other leading scientists and technologists, warning us about the catastrophic perils of unleashing autonomous robotic systems. Even something as straightforward as the long-recognized duty, within the ethical code of the professional engineer, to report covered-up hazards by one's intransigent superior to the proper outside authorities is not impressed very deeply in the minds of engineering students. Thus, the case is made for the study of engineering history, as was exemplified so marvelously by the popular courses of Princeton University engineering professor David P. Billington.

When and how to blow the whistle invoking legal safeguards, how to appeal to outside professional engineering societies whose ethical codes you are invoking, and how to participate in strengthening these societies' fortitude and their work on product standards are all central to the courageous stands of stalwart professionals. Yet, as the articles in this book reveal, the structures of professional independence and defense of aggrieved members are sorely lacking, to the ultimate detriment of the trusting public, consumers, and taxpayers. Big negligent or willful technological disasters may be coming in our intricate, dependent, vulnerable world.

In the fast-growing areas of government procurement—such as ballistic missile defense and new weapon systems—there are taboos on public discussion, both technical and policy. So, too, with emerging technologies—surveillance, detection, medical, mining, and chemical products—whose benefits are touted by their vendors as much as their risks are obscured.

Freedom of information laws cannot keep up with impositions of government secrecy and corporate trade secrets that enable these black ops. Openness in technological deliberations is hindered when engineering societies are so indentured to commercially driven or dominating interests that they fail in their proper watchdog role to sustain engineering integrity. Before the advent of the motor vehicle safety law in the mid-1960s, the role of the industry-controlled Society of Automotive Engineers was to provide a cover for the technological stagnation of the industry. That spelled the loss of many lives; these and many more injuries could have been prevented by seat belts, collapsible steering columns, padded dashboards, airbags, and many other devices invented long ago by forgotten engineers.

Having to remain silent about ways to prevent hazards to public safety or to avert other conditions that are wasteful and disruptive of the ostensible mission, writ large or writ small, is the antithesis of professional behavior. Yet we cannot expect many engineers to risk their careers or advancement, unless there are legal protections and professional society support for their forthright declarations. A climate of fearful self-censorship of one's pertinent but ignored knowledge feeds the overall censorious cultures of the employing organization. Such silence prevails within the gigantic (over $10 billion a year) ballistic missile defense program, criticized by the American Physical Society and Massachusetts Institute of Technology professor Theodore Postol as unworkable, covered with deception, and easily decoyed. The program's promoters in business, Congress, and the Executive branch refuse to allow such technical doubts to be addressed in their many fora. Taboos also permeated the history of atomic energy—a technology early on garnished with secrecy, even including area evacuation plans at one time, as to preclude the exposed citizenry's elementary right to know until, that is, the citizenry became more demanding and stimulated some leading nuclear engineers and physicists to speak out.

Now, there is still too little openness regarding recognized consequences which can flow from the spread, mostly unregulated, fields of artificial intelligence (robotics), biotechnology, and nanotechnology into the Internet, the environment, and the national security, medical, consumer products, and workplace arenas. These technologies are often hyped and driven by profits and careerism at the expense of sober caveats. Taboos fueled by secrecy continue to drag on fuller, timely exposition of broad environmental problems flowing from technologically corner-cutting diseconomies— a fancy phrase for describing the silent violence of pollution and other disruptions upon humankind and the biosphere.

Engineering education that ignores such obstacles to conscientious free speech is not preparing its students to resolutely handle with knowledge and character the deplorable pressures on the job between public duty and private ambitions.

Does the engineer have a role in resisting the onset of gross but profitable waste, fraud, and adverse mission creep, or confronting the failed logistics of postdisaster efforts, or highlighting risks and backlogs in public infrastructure (think Boston's Big Dig)? Are not engineers often well positioned to show the availability of products and devices capable of saving lives and preventing injuries and illnesses, even though they perturb the National Institutes of Health (not-invented-here) syndrome? Who knows earliest about readily available or adaptable innovations or techniques for diminishing the limitations of communications, transportation, and construction technologies as if people users matter first? Indeed, the engineer is often the first alerter, or the first sentinel for the unsuspecting public.

There is also the matter of restraint incumbent upon engineers when their technologies are widely deployed too far ahead of the science that should be their governing discipline. The internal combustion engine and its emissions are such an example. Motor vehicle-sourced photochemical smog over cities and suburbs lethally filled many lungs and corroded many properties for decades before the connection to motor vehicles was indisputably made by Professor Arlie Haagen-Smit at the California Institute of Technology in the 1950s. Once the science caught up, stronger regulations finally were issued to rework tens of millions of later cleaner vehicles. Indeed, the foundational need for legal and ethical frameworks for rapidly deployed new technologies is not being remotely met.

Broader awareness of consequences from corporate–governmental decisions determining engineering priorities has come from the research of outspoken professors. Industrial engineering professor Seymour Melman of Columbia University was a lonely voice, meticulously pointing out the economic and employment costs of diverting massive scientific and engineering human resources from the civilian economy to excessive military weapons programs.

Civic pursuits by engineers are involved in the work of Ralf Hotchkiss. Rendered a paraplegic before college, this creative and utterly courageous engineer has revolutionized the practical and economical design of superior wheelchairs, including using local materials in poor countries. He helped break the virtual wheelchair monopoly by a British multinational company in the process. (Find out more about Ralf Hotchkiss in my article "Ralf Hotchkiss & His Whirlwind Wheelchairs.")

Widespread media attention is given every year to the annual report by the American Society of Civil Engineers, which details large backlogs of deferred maintenance and the investment required for the repair and modernization of our nation's critical public works and public buildings.

This role of defining unmet needs is carried into hands on practice abroad by the non-profit organization Engineers without Borders—involving students as well—to provide engineering assistance for clean water systems and ways to foresee coming breakdowns and their remedies in impoverished communities around the world.

Prevention is one of the hallmarks of any profession. It should work to make some of its services unnecessary. So, too, should professionals keep "options open for revision," in Alfred North Whitehead's wise words. For example, electrical engineers should heed these principles as they follow their corporate superiors' orders and implement the arrogant algorithms pouring out of Silicon Valley to tighten their grips on humanity. Those engineered controlling systems by Google, Facebook, Apple, Amazon, and Microsoft, so far, have escaped the restraints of the law, the rights of privacy, timely disclosures, two-way negotiated terms and conditions, viable competition, and other basic accountabilities to their various publics.

As the reader moves through this collection of writings, he or she will meet some of the profession's critical idealists, operating realists, setters of standards for products, educational innovators, protectors, evokers of conscience, heroic stories of the valiant successes overcoming tall odds, and futurists depicting disturbing scenarios. These authors are coalescing around a central theme—namely, the light of a moral and technical imagination that should penetrate the fog of able minds too subordinated to the strictures which chain them. They should have a higher estimate of their own significance.

In 1966, in an address before the Middle Atlantic Section meeting of the American Society of Engineering Education (reprinted in this book), I said: "the profession must assert itself toward its most magnificent aspirations—for so much of our future is in your trust."

This liberating direction accelerates with one engineer at a time, one engineering society or one whistleblowing protection advocacy group at a time, and one engineering professor or student at a time driven, we can hope, conterminously by informing the public toward higher expectations for the requisite culture change. As Nassim Taleb, author of *The Black Swan*, has written: "mental clarity is the child of courage, not the other way around."

Editors Rania Milleron and Nicholas Sakellariou deserve much praise for persevering to compile such an important stimulation of the working engineer's mind and conscience. May this descriptive book generate wide public media and professional attention that their labors deserve and that our future needs.

Ralph Nader

Preface

This book is a selection of documents covering issues of interest to all engineers and a short introduction to guide readers. The selections cover engineering as more than a machine-producing profession or occupation. The articles explore engineering as a liberal art—going beyond the technical to be able to interpret impact—pollution, ugliness, dangerous defects, and waste of resources. It can be read by young engineers as well as consumers, communities, workers, and government officials, all those who have an interest in engineers with broader frameworks of knowledge and thinking. But first we might ask, what is engineering? A traditional definition of engineering reads "the planning, designing, construction or management of machinery, roads, bridges, buildings, fortifications, and waterways, etc.; science, or work of an engineer" (*Webster's English Dictionary*). But what is left out is the engineering audience. By comparison, out of six definitions of law, the people to be served, "society," are operative in the definition.[1] Engineering is rarely defined by its dedication to serving society. The readings encourage engineers to view their profession in terms of who they serve—humanizing the engineering profession. Keeping in mind the actual people, the designs will impact ethics and professional responsibility intuitively.

The purpose of *Ethics, Politics, and Whistleblowing in Engineering* is twofold: to recognize sustainable engineering and to inform current and future generations of engineers and even nonengineers about the practice and promise of their own profession. It brings together in a single book materials about engineers' noteworthy accomplishments against the background of daily struggles for professional autonomy. The selections in this book urge engineers to take an active part in designing their profession, to stretch their concerns and involvement beyond the narrow confines constructed by the profession and corporate employers and to expand public expectation. Such a rethinking of the relevance of engineering might recruit graduating engineers from Princeton to Cal Tech into public interest occupations and encourage older engineers to broaden their frame of reference to include the Commons. American engineers have made history as social crusaders,[2] and the readings chronicle the history of the engineering imagination.

While other anthologies explore the ethics of engineering,[3] this book stresses engineers' critiques of their own profession. Such self-criticism can lead to stronger performance and technologies that are safer and more responsive to people's needs. The book stresses engineering reflexivity; namely, critiques from within the profession, as well as solutions for the future. Our approach is intentionally holistic and brings perspectives from a variety of venues and interdisciplinary perspectives. The mission is both broad and immediate.

The selections are divided into three parts. Part I, "Engineering Leadership," includes chapters that exalt engineers, commentaries on engineering education by engineers, and a discussion of the status of engineering as a liberal art, rather than as simply a technical discipline. In Part II, "Daily Practice," the selections analyze facets of the work place and focus on what engineers worry about design, defects, levels of allowable risk, determining standards, and whistleblowing. Exactly how do engineers apply their conscience when it conflicts with corporate or employer imperatives? To whom can engineers appeal when they find themselves in conflicts of interest and what can the public learn from such cases? Part III, "Raising the Bar," discusses leaps of understanding associated with diversity in the engineering workforce and points to future directions for an engineering profession that demands more.

Ethics, Politics, and Whistleblowing in Engineering is neither a complete nor a comprehensive survey of all engineering issues that merit attention: engineering is among the most broad reaching of all the traditional professions, spanning fields as diverse as aeronautical and biological engineering. The more modest aim of this book is to make insider critiques and review of the engineering profession accessible—by means of this cross section of writing, beginning in over 50 years ago. Our research revealed a number of recurrent themes in engineering journals over the past several decades. Throughout this period, engineers have often expressed themselves on philosophical topics such as whether engineering is a trade or a profession, an art or a science. These voices have addressed the definition and identity of the profession.

Although engineering-related health and safety assumptions fluctuate, definitive changes have occurred since the 1950s. There is evidence that the engineering profession takes external criticism. Ralph Nader (1965) effected a fundamentally new awareness in engineering by demonstrating that the consumer who operates a car or other machine is not at fault for design inadequacies. The public should demand more from engineering. The airbag generation may take statements about safety for granted, because we have come to expect safe designs. However, in the 1950s and 1960s, the use of the leading dogma resulted in blaming the driver and in calling for more rigorous driver training as a way to bring down the automobile death toll. This book will contribute to the "critical awareness" of their public required of any dedicated engineer who must contend with the daily work of preserving and improving the quality of life in a society where new technologies have a pervasive distribution.[4]

Expressing viewpoints on engineering is crucial for all engineers. *Ethics, Politics, and Whistleblowing in Engineering* is an approbation of the engineer who sends in letters to the editor or writes articles in engineering society journals voicing passionate comments and critiques. Critiques of engineering need not be relegated only to historians of science and science and technology studies specialists. To emphasize the importance of critique for all engineers, the book encourages engineering students, before they begin a career in engineering, to publicly share comments and critiques that may lead to innovation. These engineering viewpoints are chosen in particular as inspiration.

The book serves to remind the engineering profession that however distantly related the lives of any two engineers may be, the gap is bridged by the commitment of the profession to public service—a commitment which means realizing designs, the impact of which is taken into account. The selections reflect a broad variety of subjects within engineering, but point to the same conclusion: a need for less idle talk and more action.

ENDNOTES

1. Bonsignore, John J. et al., *Before the Law: An Introduction to Legal Process*. Boston, MA: Houghton Mifflin, 1984, p. xi.
2. See Layton Jr., Edwin T. *The Revolt of the Engineers: Social Responsibility and the American Engineering Profession*. Cleveland, OH: Case Western Reserve University Press, 1971—2nd ed., Baltimore, MD: Johns Hopkins University Press, 1986 for a historical account of the engineering profession in the first half of the twentieth century.
3. These important collections have been edited primarily by engineers and philosophers of engineering. A sample of characteristic volumes may include some of the following: Baum, Robert, and Flores, Albert, eds. *Ethical Problems in Engineering*. Vols 1 and 2. Troy, NY: Center for the Study of the Human Dimensions of Science and Technology, Rensselaer Polytechnic Institute, 1978; Unger, Stephen H. *Controlling Technology: Ethics and the Responsible Engineer*. Holt, Rinehart and Winston, 1982— 2nd ed. New York: Wiley, 1994; Harris Jr., Charles E., Pritchard, Michael S., and Rabins, Michael J., eds. *Engineering Ethics: Concepts and Cases:* Wadsworth, 1995—2nd ed., 2000. 3rd ed., 2005. 4th ed., 2008; Whitbeck, Caroline. *Ethics in Engineering Practice and Research*: Cambridge, UK: Cambridge University Press, 1998; Herkert, Joseph R., ed., *Social, Ethical, and Policy Implications of Engineering: Selected Readings*. New York: Wiley/IEEE Press, 2000; Davis, Michael, ed. *Engineering Ethics*. Aldershot, UK: Ashgate, 2005.
4. The claim that engineers "shall hold paramount the safety, health and welfare of the public" is characteristic of most engineering codes of ethics.

Acknowledgments

Without Ralph Nader, this book would never have come to fruition. The vision for this book developed from his writing of *Unsafe at Any Speed* as a way to further champion the work done by engineers. He inspired this book project when he invited Rania Milleron to work at the Center for the Study of Responsive Law in Washington, DC. Dr. Claire Nader and Dr. Laura Nader were also instrumental in bringing this book to you. They facilitated aspects of the project, never gave up on the vision, and made time for vigorous discussions on the engineering profession. Thanks are also due to professors David Billington of Princeton University for his sage, if brief advice, to a young author, and to Roberto Gonzalez of San Jose State University for his insightful comments on practicing as a young engineer. Finally, thank you to Ricardo Cruz for his honest commentary.

Nicholas Sakellariou joined the project, and he acknowledges with gratitude the inspiration and resources provided by Professors Gary Downey and Matt Wisnioski at Virginia Tech, with whom he studied. Downey and Wisnioski early on recognized the need to reconsider engineering identity, practice, and education in their scholarship and teaching. In addition, he thanks faculty and fellow students, the people he met in Greece and later in the United States, who led him from "doing engineering" to "interpreting engineering." University of California Berkeley Professor Alastair Iles has been continuously supportive and encouraging.

Once we joined together as coeditors, there are several people we both acknowledge with great appreciation. We are indebted to Hunter Jones who worked tirelessly, with the help of Henry Trotter, until the copyright permissions were obtained. John Richard was a stalwart supporter of this book project. He supervised the copyright permission acquisition and understood the value of our project for engineers. Laura Nader provided guidance on composing and compiling an edited book and on the valuable role of a public outsider. Betina Lewis was our very patient editor. Librarians at the George Washington University Library in Washington, DC, the University of California Berkeley Library, especially the Bechtel Engineering Library, were helpful, as was the Applied Science and Technology Index, which was used extensively. During the final stages of our project, we were grateful to have worked with Joe Clements, senior acquisitions editor at Taylor & Francis/CRC Press. Our book benefited from his involvement and from the guidance provided by CRC editorial assistants.

Most of all, we thank the engineers and public authors who are included in the book. They are professionals who see beyond their own narrow interest and who lend their informed zeal to innovative and often holistic thinking in the engineering profession. Both editors are grateful to the authors and publishers who granted permission to reprint these papers.

Editors

Rania Milleron is a microbiologist at the Texas Department of State Health Services, Austin, Texas. She is both a biomedical scientist and a public health professional who always brings a multidisciplinary perspective to her work. She finds that being able to look at problems from many perspectives helps her find innovative solutions. Dr. Milleron received a master of science degree in public health from Harvard University (global health and infectious disease epidemiology) and a PhD degree from the University of Texas Medical Branch in Galveston (molecular genetics of infectious disease vectors). She subsequently completed a postdoctoral fellowship at the University of Texas Austin (molecular and cellular biology). Dr. Milleron has published numerous experimental research articles. She is not an engineer. But because of her outsider perspective, Ralph Nader urged her to unify in a book voices that champion good works in engineering. After creating the methodological framework for the project, carrying out literature reviews spanning five decades, and writing the preface and introduction to the book, she combined forces with Nicholas Sakellariou, an engineering insider.

Nicholas Sakellariou is a lecturer at California Polytechnic State University (San Luis Obispo, California), College of Engineering, Department of Computer Science, where he teaches engineering ethics/professional responsibilities classes. He received his PhD degree in environmental science policy and management from the University of California at Berkeley in 2015. He holds postgraduate degrees from the National Technical University of Athens (history and philosophy of science and technology) and Virginia Polytechnic Institute and State University (science and technology studies). He is the author of *Life Cycle Assessment of Energy Systems: Closing the Ethical Loophole of Social Sustainability* (Wiley, 2018).

Contributors

Alison Adam
Professor of Science
Technology and Society Sheffield Hallam
    University
Sheffield, UK

Albert L. Batik
Chemist (deceased)
Denver, Colorado (CO)

David P. Billington
Gordon Y.S. Wu Professor of Engineering
    Emeritus (deceased March 2018)
Princeton University
Princeton, NJ

Kate Blackwell
Author
http://www.kateblackwell.com/

Edward Conlon
Structured Lecturer
School of Multidisciplinary Technology
College of Engineering and Built Environment
Dublin Institute of Technology
Dublin, Ireland

Michael Davis
Professor of Philosophy
Center for the Study of Ethics in the
    Professions
Illinois Institute of Technology
Chicago, Illinois

Gary Lee Downey
Alumni Distinguished Professor of Science
    and Technology Studies and Affiliated
    Faculty Member in Women's and
    Gender Studies
Virginia Polytechnic Institute and State
    University (Virginia Tech)
Blacksburg, VA

Ron Eglash
Professor of Science and Technology Studies
    Rensselaer Polytechnic Institute (RPI)
Troy, NY

Karen Fitzgerald
Editor (retired)
IEEE Spectrum

Domenico Grasso
Chancellor
University of Michigan-Dearborn
Dearborn, MI

Susan Karlin
Freelance Journalist and Contributor
Fast Company, National Public Radio
    and *IEEE Spectrum*
Los Angeles, CA

Jean Kumagai
Senior Editor and Contributor
IEEE Spectrum
New York, NY

Jon A. Leydens
Associate Professor
Humanities, Arts, & Social Sciences
Colorado School of Mines
Golden, Colorado

Taylor Loy
Poet
Knoxville, TN

Juan C. Lucena
Professor
Engineering, Design, & Society
Colorado School of Mines
Golden, Colorado

Heinz C. Luegenbiehl
Professor Emeritus of Philosophy and
  Technology Studies
Rose-Hulman Institute of Technology
Terre Haute, Indiana

Peter Maier
Consumer Lawyer
Seattle, WA

Mark Manion
Professor (retired)
Department of Philosophy
Drexel University
Philadelphia, PA

Usman Mushtaq
Youth Development Coordinator
City Centre Community Centre (City of
  Richmond)
Vancouver, Canada

Ralph Nader
Founder
Center for Study of Responsive Law
Washington, D.C.

Tekla S. Perry
Senior Editor and Contributor
IEEE Spectrum
Palo Alto, CA

Peter Petkas
Lawyer
Houston, Texas

Jonathan D. J. VanderSteen
Water and Wastewater Project Engineer
Ontario, Canada

Introduction

Compiling *Ethics, Politics, and Whistleblowing in Engineering* has posed a number of interesting dilemmas. At the outset, this book was called *Rallying the Engineers: A Reader for Young Engineers and Their Public*. But the more the work progressed, the clearer it became that age per se and public did not accurately describe our book. There were issues with the quality of the literature about engineering, as well as the different types of engineers and different patrons. The age dimension of the audience was not simple either, especially if changing the status quo was a primary goal. Besides, the consumers of engineering artifacts vary with the products: bridges, dams, nuclear plants, iPhones, computer programs, gas pipelines, kitchen stoves, and the ever-present gadgets from toys to tools.

From the outset, a key issue was the "lack of much reflective literature either on engineering or by engineers."[1] Professor Billington of the School of Engineering at Princeton University advised that a reflective tradition was emerging, albeit slowly, in the Society for the History of Technology. He also suggested that the book might help by identifying quality writing and giving it some order: there were different types of engineering and different types of engineers. There are engineers in professional practice (as in architecture or law firms), those in public service (as in government, regulatory, and design agencies), and yet others in private industry, both large and small. As Thomas Kuhn (1962) noted in his book *The Structure of Scientific Revolutions*, and Billington also noted, in America, engineering has always been pioneered by people from outside the mainstream of society.[2]

Another set of observations was also helpful in grounding our methodology. Roberto Gonzalez had been an engineering major at the University of Texas in Austin and, as such, had experience as a student at both General Motors and Southwest Research Institute. For him, age was important in differentiating engineers:

> You describe a vision of engineering that only three years ago, as a mechanical engineering student, I was desperately trying to find in books like Pirsig's *Zen and the Art of Motorcycle Maintenance* and [Samuel Florman's] *The Existential Pleasures of Engineering*. We never covered anything like that in my engineering classes at the University of Texas....
>
> In my experience, both at General Motors and at Southwest Research, "middle-aged engineers" often came in two varieties. On the one hand, there were the engineers-turned-managers who supervised younger engineers. On the other hand, there were the engineers-stayed-engineers, who, because they lacked the "interpersonal skills" (or, sometimes the MBA), were never promoted very far up the chain of command. I do not think that the latter would argue for more specialized training; often, as I have noted, they are resentful of the younger engineers, many of whom are more knowledgeable in narrowly specialized fields....
>
> At the Oklahoma City plant where I worked, managers were often industrial engineers, originally trained to rationalize assembly line operations. Once in management positions, they applied the same principles to engineering work in the office. What I am trying to say is that based on my experience, engineering managers are not so much "profit-hungry" as they are "efficiency-hungry"; and they are not so much driven by "wage-labor mentalities" as they are by (ironically enough) "engineering mentalities"....
>
> In my opinion, engineering is not unlike law in that it is extremely competitive, and as a result, the quest for promotion often takes precedence over the quest for responsible designs. Another problem is that starting engineers are often given quite trivial, mundane tasks, and eventually, I think that they are unable to see the total results of what they are doing. The way engineers are trained to solve problems feeds right into this—the students draw sharp boundaries ("controls") around the system, make a number of simplifying assumptions, and proceed to solve the problem. In these sorts of environments, I think that "common candor" is still considered "courage," as Ralph Nader has said.

But this requires a sense of values, a moral frame of reference for the student to work from. Teaching ethics is pointless. When applying for a job at the placement office of the UT Austin engineering school, there is a box that you may check off if you do not wish to work for a company that develops weapons systems. This, to a large degree, is as far as ethical considerations go for many undergraduate engineers. I was so starved for more, I opted to take a philosophy course ("Contemporary Moral Problems") as my (only) required humanities elective....

Furthermore, the engineering students I knew spent a great deal of time on their lab reports and home-work assignments (you had to, if you wanted to succeed) and in general took little interest in the world around them. At UT Austin, the engineering buildings were a metaphor for this isolation—they con-sisted of four steel and concrete high rise buildings in the northeast corner of the university, separated from the rest of the campus by a busy four-lane road....

In short, I believe that there is a dilemma here. I think you are right in stating that the best hope for a new engineering resides in young engineers. The problem is that so many of them are being pulled in other directions that they never do get around to thinking about where they fit in the grand scheme of things. Even women and minorities, who one would think are in the position most likely to call their profession into question, are quickly absorbed by the biggest corporations (the highest bidders), who are desperate to meet Equal Opportunity requirements to secure government contract work. There are, of course, some fine exceptions.[3]

Gonzalez left engineering and is now a professor of anthropology at San Jose State University.

University of California–Berkeley's professor of mechanical engineering Alice Agogino, for example, summarized the situation which applies far beyond women and minorities in schools of engineering. She noted, "I think that too often women in engineering basically become 'male.' And like everyone else in the field to succeed they don't stand out as women. In fact, they may fight hard not to be labeled a 'woman engineer.'" She goes on to report that when her graduate students begin to integrate engineering with public health and sustainability, they are sometimes treated as second-class citizens as they are not working on some narrowly defined vision of engineering: "The story I would like to see is women changing the face of engineering by redefining it in human terms. Groups like 'Engineers for a Sustainable World' are predominantly female in membership and leadership. Women are very active in biomedical engineering, environmental engineering and human-centered design. It is these very students who hold my hope for radical change to the engi-neering profession—to transform it into a profession of engineering leaders who value engineering as a means to improve the human condition."[4]

That we encountered many of the dilemmas posed earlier justified a less orderly set of selections than Professor Billington might have thought possible or necessary. The twenty-first century is a time for deliberation in search of new frames of reference. It is difficult, for example, to draw a sharp line between engineering and management for the simple reason that a high percentage of manag-ers are engineers. Ethnographic, sociological, and historical work on engineers and engineering has always dealt with the issue of the alleged strict demarcation between engineering and manage-ment. Michael Davis (1998), for example, concludes that the relationship between engineering and management is similar to that between faculty and administrators: the faculty advise, and they more or less expect administrators to take their advice. In that case, Davis concludes that deference to management was not what was expected from engineers.[5]

Although we have many more engineers than lawyers and economists, it is principally the law-yers, economists, and a few other professionals who are the decision makers. This can be a problem especially when the future of technologies, with potentially dangerous consequences, is determined by people in the driver seat who do not have the capacity to understand the technologies. Historically in our country, few cabinet members in Washington have had technical backgrounds. One simply has to look at the problems facing the country—energy, transportation, infrastructure decay, and water and air pollution—and ask about the talents of those in charge. It is instructive that President Carter was trained as a nuclear engineer, and it was he who articulated an energy policy that got watered down, even though he put solar panels on the White House roof. Thus, history shows that when technical factors are the core of the argument, chances are that appropriate decision makers will be found lacking.

We do not educate our college populations in technical matters as a matter of liberal education, nor do we educate our engineers to know how to integrate their liberal arts courses with their technical courses. People limited to only narrow skills will tend to employ them to every problem they encounter. Engineers are not alone in this difficulty. The integration of technical and nontechnical factors must occur not through interdisciplinary teamwork, but through people whose education outlooks span disciplines. For coming generations, the world must not look like a nail to be hammered into place by technicians. While this is being recognized, the urgency is not yet apparent.

Cherry Murray, former dean of the Harvard School of Engineering and Applied Sciences, is one of the many who say solving "big problems will demand big thinking across the disciplines."[6] If this is to mean anything in this world of galloping changes, it has to mean more than new designs of robotic bees, or robotic systems for heart surgery, or nanoscale machines, interesting as they may be. Adapting humans to machines is business as usual. New thinking means envisaging an engineering profession prepared to lead in the direction of steady-state economics (Daly, 1977),[7] material scarcities, conservation, environmental protection, consumer product safety, and all the other issues addressed in the next three parts of this Introduction.

ENDNOTES

1. David Billington, personal communication, 1995.
2. Kuhn, Thomas S. *The Structure of Scientific Revolutions.* Chicago, IL: University of Chicago Press, 1962.
3. Personal communication, 2012.
4. Personal communication, 2012.
5. Davis, Michael. *Thinking Like an Engineer: Studies in the Ethics of a Profession.* New York: Oxford University Press, 1998.
6. Rutter, Michael P. "An Open Source Model for Innovation." Fall 2009, page 2.
7. Daly, Herman. *Steady-State Economics: The Economics of Biophysical Equilibrium.* San Francisco, CA: W.H. Freeman, 1977.

Part I

Engineering Leadership

> ... these men, like their counterparts in universities and government who knew of the suppression of safer automobile development, yet remained silent year after year, will look back with shame on the time when common candor was considered courage.
>
> **Ralph Nader (1965, p. 346)**

An engineer's conceptualization of his or her impact on the world *has* evolved since *Unsafe at Any Speed* hit the shelves.[1] More than ever, engineering practitioners are attentive to, and sometimes remarkably thoughtful about, the technologies they create and the public they create them for. At the same time, controversy still surrounds the environmental and social ramifications of technical work. Indeed, Larry Jacobson, former executive director of the National Society of Professional Engineers (NSPE), lamented the fact that "[o]nly about 10% [of American engineers] have an affirmative ethical duty to place their *professional* practice ahead of profit" (Jacobson, 2010; emphasis ours). Some engineers still do not realize that being socially responsible is not a choice but a professional obligation.[2]

Unsafe at Any Speed was published in 1965, whereas "the very first conference ever held dealing exclusively with problems of engineering ethics" was undertaken in 1976 (ASCE, 1976). More recently, the World Federation of Engineering Organizations (WFEO, 2001) revised its *Code of Ethics* and stated that engineers "shall report to their association and/or appropriate agencies any illegal or unethical engineering decisions or practices of engineers or others." Unfortunately, engineers often ignore or consider casually general assertions called "environmental ethics," which urge engineers to "reject any kind of commitment that involves unfair damages for human surroundings and nature..." (WFEO, 2001).[3] Some practitioners, in light of what engineering commentators perceive as the "modern litigation crisis," place a higher priority on financial and legal considerations rather than on exceptional engineering work at home and abroad (Kemper, 1989).[4] A number of articles in Part I of *Ethics, Politics, and Whistleblowing in Engineering* demonstrate how engineers can and do maintain professional equanimity while being at once civic-minded citizens and exemplary technical workers. A broader frame of reference—inherently involving input from the communities for which engineering projects are designed—is needed to address the perceived tension between the individual engineer or corporate social responsibility versus their legal liability.

The book first features engineers who decided to honor, and not deviate from, what they perceived to be their professional duties. Communicating those engineers' legacy reconnects the public with its technical experts. Heroic engineers are those engineers whose frame of reference not only includes but also transcends engineering, as valuable reminders of the links between engineering practice and the nonengineering world.

It is often forgotten that in times past, engineers were considered heroic (not saints), and engineering itself, specialized artistry. These important qualities of engineers need reaffirmation. Although the theme of the "engineer-hero" has been a focus in literature, contemporary films or television programs are more likely to feature lawyers and doctors.[5] In those rare cases where engineering is the subject of a film or documentary, the directors have placed emphasis on the very technical feat (or failure), rather than on the individual engineers involved with the project at stake.[6,7] The engineer's low prestige, along with decreasing numbers in engineering enrollments, has led the Pentagon and the American Film Institute to fund teams of prominent engineers who are expected to write engineering-friendly movies for the American youth, an illustration of the current public devaluation of engineers who barely make their appearance in public affairs: they have become accustomed to being invisible to the public (Halbfinger, 2005).[8] Engineering projects are blatantly visible, as opposed to the individual engineer or engineering team operations behind the scenes. When the public does pause to think about the engineering profession, usually it is the result of a catastrophic event, such as Hurricane Katrina in New Orleans or the Gulf Oil spill, or the publicizing of innovations such as gadgets and social media apps. Thankfully, change is being made. The feature IMAX film *Dream Big: Engineering Our World* redefines what it means to be an engineer.

The engineer was once celebrated as "America's best symbol of the future": the years between 1850 and 1950 are usually referred to as "the golden age of engineering" (Tichi, 1987b). Pathbreaking inventions such as automobiles or telephones generated awe and admiration, while large engineering projects prompted celebrations to pay tribute to the engineer, the "exponent of efficiency and the slayer of that dragon, waste" (Tichi, 1987b, p. 105). Today, we continue to glorify the golden age of engineering. Indeed, two hundred thousand people celebrated the opening of the Golden Gate Bridge in 1937 and over one-half million people celebrated its 50th anniversary in 1987. Simultaneously, enormous advertising campaigns—such as the promotion of the so-called smartphones—draw our attention to contemporary engineering feats.

Included in Part I of *Ethics, Politics, and Whistleblowing in Engineering* are a variety of engineering viewpoints. Donald Schön demonstrates the positive implications of "reflective practice" for professionals, including engineers (Schön, 1984). It is critical that we listen carefully to what engineers have to say, since we hear disproportionately from marketing and chief executive officers. Some of the subjects that have prompted reflection by engineers include professional ethics, sustainability, and the use, likelihood and desirability of the future engineer as a "global citizen." Engineers may be "invisible," but their perceptions, whether on automobile safety, a bridge's durability, or a sanitation system's effectiveness, can be critical to our survival.

Why should consumers, workers, and government show interest in engineers? There are manifold answers as there are numerous, multidimensional and quite heterogeneous, "engineering cultures." We live in an engineered world: it is shaped by engineers and their projects. The status quo will remain as long as engineering managers engage in decision-making driven by profit and efficiency mind-sets. No new direction will result. If, on the contrary, this engineering environment is managed as a true "people-serving profession" rather than as a controlled trade, the flow of information beyond solely commerce will gain power, shaping, channeling, and filtering the engineer's imagination.[9]

Critiques are often most effectively communicated through humor. Rube Goldberg was a Pulitzer Prize-winning cartoonist whose brief work as an engineer informed his most acclaimed work where he satirizes man's capacity "[to exert] maximum effort to achieve minimal results." His name appears as a noun in the *Webster's New World Dictionary* and is defined as "a comical involved, complicated invention, laboriously contrived to perform a simple operation." Through his satire, he has inspired engineers around the world. These choices reflect our purpose to reveal widespread societal perceptions of engineers.

Making the engineer aware of how he or she is perceived by society will impact the way engineers are educated. Subsequent articles examine identified pathologies of teaching and practicing engineering as well as the values that underlie the behavior of students and educators. Other selections include recommendations of how engineering education, in general, and engineering curricula, more specifically, may be improved to transition from problem-solving in a flat world to problem definition in multinational environments.[10]

Discussions on engineering curricula or on teaching engineering ethics can be pervasively superficial. One generally has the impression that when engineering journals print articles dealing with "socially responsible" engineering they may do so as an altruistic gesture to the benefit of large engineering firms. Importantly, these publications perpetuate an ideology that serves to divert the genuine concern of a great deal of practitioners with maintaining their rightful position as socially minded technical experts. Samuel Florman (1987, p. 187), a prominent civil engineer and a writer with a master's degree in English literature from Columbia University, has noted the following: "A certain amount of this clamoring for well-rounded engineers is mere lip service. I once heard a highly placed executive say in a moment of candor, 'I want one engineer out of twenty to be eloquent and imaginative. I can't use a whole division of hotshots!'"

One of the goals of *Ethics, Politics, and Whistleblowing in Engineering* is to make technically inclined students realize at the very beginning of their careers that the best kind of engineering is a balanced marriage between the applied sciences and the humanities. The engineer is as much a technologist as an "applied humanist" or artist. Such an idea, though, has been met with considerable resistance for some engineers view liberal education as an "additional burden on students," which "might deter … [them] from pursuing engineering..." (Greenberg, 2005).[11] Could it be, as Florman suggests, that those in powerful positions often feel too much creativity threatens authority? To what degree, then, does engineering education remain a "controlling process" that limits the critical talents necessary for self-reflection and critique? How, for example, will the existence of "diametrically opposed forces trying to squeeze more content into the [engineering] baccalaureate curriculum" affect the efforts of technologists to "raise the education bar" for their discipline?[12]

The readings included in Part I of the *Ethics, Politics, and Whistleblowing in Engineering* go beyond mere lip service to examine the "traditional approach to teaching and learning" in the discipline and involves designing a liberally enriched engineering education. In a similar vein, by acknowledging the profession's own integrative essence, educators call attention to the need for holistic approaches in teaching students and professors of engineering alike.

On the whole, "Engineering Leadership" recognizes not only the individual engineers as heroes or advocates of social change but also the engineering education-related initiatives and institutions that encourage engineers to review the roots of their noble profession. Part I is meant to stimulate engineering educators to experiment broadly and open-mindedly in liberal education curricula, to promote unpopular but fact-based viewpoints, and to encourage students to learn about the heroic roots of engineering.

ENDNOTES

1. "We aspire to a future where engineers are prepared to adapt to changes in global forces and trends and to ethically assist the world in creating a balance in the standards of living for developing and developed countries alike" (NAE, 2004, p. 51).

2. It must be noted that the appropriation of the concept of social responsibility by American engineers is not devoid of historical context. See, for example, Paul T. Durbin's (1987) *Technology and Responsibility, Kelly Moore's (2008) Disrupting Science: Social Movements, American Scientists, and the Politics of the Military*, 1945–1975, and Matthew H. Wisnioski's (2012) *Engineers for Change: Competing Visions of Technology in 1960's America*. The classic reference, of course, to the relationship between engineers and social responsibility is Edwin Layton's (1971) *The Revolt of the Engineers: Social Responsibility and the American Engineering Profession*.

3. In a recent article on corporate ethics by two Australian engineers, the authors mention that "[t]he incidence of codes of ethics was also an interesting finding in that 67% of our respondents said that codes of ethics are seldom or never used in the industry to promote ethical behavior" (Tow and Loosemore, 2009).

4. In 1983, the Environmental Impact Analysis Research Council (EIARC), as part of a committee of theAmerican Society of Civil Engineers (ASCE), proposed an "eighth fundamental canon for the ASCE *Code of Ethics.*" This canon, requiring that engineers "shall perform service in such a manner as to husband the world's resources and the natural and cultured environment for the benefit of present and future generations," was disapproved by the leadership of ASCE, for the latter expressed concerns regarding the association of the canon with environmental liability claims. According to a survey designed for the ASCE EIARC, 93% of "the contracting firms that responded, . . . saw future problems with environmental consulting liability." The NSPE states that "[l]egal liability exposures may represent the greatest threat to the financial health of professional engineers and engineering firms." See the papers of Vesilind and Gunn (1998) and Nelson (1983). Regarding NSPE's statement on legal liability, see NSPE (2011).

5. In her *Shifting Gears: Technology, Literature, Culture in Modernist America*, Cecelia Tichi (1987a:, p. 98) remarks that "[a]s engineers made their presence felt in the American scene, they began to figure prominently in imaginative literature." To the best of our knowledge, the narrative theme of the engineer-hero appears in Richard Harding Davis's (1912 [1897]) *Soldiers of Fortune*, Bernhard Kellermann's (1913) *Der Tunnel*, and Ayn Rand's *Fountainhead* (Rayn, 1994 [1943]) and *Atlas Shrugged* (Rayn, 2005 [1957]). Engineer-heroes have made their way in Tom Wolfe's (1979) *The Right Stuff*, as well as to some science-fiction books such as Larry Niven's (1980) *The Ringworld Engineers* and Lois McMaster Bujold's (1988) *Falling Free*. Other popular novels that feature engineer-protagonists include Max Frisch's (1959) *Homo Faber* and John Richard Hersey's (1956) *A Single Pebble*, which begins with "I became an engineer."

6. See Appendix IV.

7. *Falling Down* (1993), directed by Joel Schumacher and written by Ebbe Roe Smith and starring Hollywood actor Michael Douglas as an engineer working for the industry, is the exception that proves the rule. Still, the inception of the protagonist of the film, whose license plates read "D-Fens," is based on the stereotypical and sometimes overly unfair depiction of the "engineer."

8. See, for example, Petroski's (1990) article.

9. This refers to the catchphrase commonly used by ASCE. The concept, however, of engineering professionalism, is fraught with ambiguity. See, for example, our introduction to Part II.

10. See "National Identities in Multinational Worlds: Engineers and 'Engineering Culture,'" by Gary Lee Downey and Juan C. Lucena, and "The Problem of Knowledge in Incorporating Humanitarian Ethics in Engineering Education: Barriers and Opportunities," by Jon A. Leydens and Juan C. Lucena, Chapter 5.

11. For a similar point of view that cautions against the liberal education of engineers, "at the expense of their technical education," see "The Engineer's Responsibility for Product Safety," by Herbert Egeder, which appears in Part II.

12. For example, see ASCE (2007). See also Russell et al.'s (2002) paper.

REFERENCES

ASCE (American Society of Civil Engineers). *Conference on Engineering Ethics.* New York: ASCE, 1976.

ASCE. Policy Statement 465 (approved by the Policy Review Committee on March 9, 2007). ASCE, 2009. http://www.asce.org/Content.aspx?id=26194. Accessed December 9, 2011.

Bujold, Lois McMaster. *Falling Free.* New York: Baen Books, 1988.

Davis, Richard Harding. *Soldiers of Fortune.* New York: Scribner's, 1912 [1897].

Durbin, Paul T. ed. *Technology and Responsibility.* Dordrecht: Kluwer Academic Publishers, 1987.

Florman, Samuel. *The Civilized Engineer.* New York: St. Martin Press, 1987.

Frisch, Max. *Homo Faber.* London: Abelard-Schuman, 1959.

Greenberg, Lisa. Engineering certification: License to engineer. ChemicalProcessing.com, 2005. http://www .chemicalprocessing.com/articles/2005/391.html. Accessed December 9, 2011.

Halbfinger, David M. Pentagon's new goal: Put science into scripts. *New York Times*, August 4, 2005.

Hersey, John Richard. *A Single Pebble.* New York: Knopf, 1956.

Jacobson, Larry. Statement on the Gulf Oil Spill and Licensed Professional Engineers. National Society of Professional Engineers, 2010. http://www.nspe.org/AboutNSPE/jun2310oilspill.html. Accessed October 21, 2011.

Kellermann, Bernhard. *Der Tunnel*. Berlin: Fischer, 1913.

Kemper, Robin A. Which comes first—Responsibility or liability? *Journal of Professional Issues in Engineering* 115, no. 4 (1989): 438–440.

Layton, Edwin. *The Revolt of the Engineers: Social Responsibility and the American Engineering Profession*. Cleveland, OH: Case Western Reserve University Press, 1971.

Moore, Kelly. *Disrupting Science: Social Movements, American Scientists, and the Politics of the Military, 1945–1975*. Princeton, NJ: Princeton University Press, 2008.

Nader, Ralph. *Unsafe at Any Speed: The Designed-In Danger of the American Automobile*. New York: Grossman Publishers, 1965.

NAE (National Academy of Engineering). *The Engineer of 2020: Visions of Engineering in the New Century*. Washington, DC: The National Academies Press, 2004.

Nelson, Roger E. An eighth canon for the practitioner. *Journal of Professional Issues in Engineering* 109, no. 1 (1983): 53–57.

Niven, Larry. *The Ringworld Engineers*. New York: Holt, Rinehart, and Winston, 1980.

NSPE (National Society of Professional Engineers). Licensure—Resources—Professional liability. NSPE, 2011. http://www.nspe.org/Licensure/Resources/liability/index.html. Accessed October 24, 2011.

Petroski, Henry. The invisible engineer. Civil Engineering—ASCE, *60*, no. 11 (1990): 46–49.

Rand, Ayn. *Fountainhead*. New York: Plume, 1994 [1943].

Rand, Ayn. *Atlas Shrugged*. New York: Dutton, 2005 [1957].

Russell, Jeffrey S. et al. Why raise the education bar for civil engineers? *Proceedings of the 2002 American Society for Engineering Education Annual Conference and Exposition*. 2002.

Schön, Donald. *The Reflective Practitioner: How Professionals Think in Action*. New York: Basic Books, 1984.

Tichi, Cecelia. *Shifting Gears: Technology, Literature, Culture in Modernist America*. Chapel Hill, NC: University of North Carolina Press, 1987a.

Tichi, Cecelia. *Technology, Literature, Culture in Modernist America*. Chapel Hill, NC: University of North Carolina Press, 1987b.

Tow, D. and Loosemore, Martin. Corporate ethics in the construction and engineering industry. *Journal of Legal Affairs and Dispute Resolution in Engineering and Construction* 1, no. 3 (2009): 122–129.

Vesilind, Aarne P. and Gunn, Alastair S. Sustainable development and the ASCE Code of Ethics. *Journal of Professional Issues in Engineering Education and Practice* 124, no. 3 (1998): 72–74.

WFEO (World Federation of Engineering Organizations). *Code of Ethics*, revised in 2001. http://wfeo.net /au_subrules.aspx. Accessed October 23, 2011.

Wisnioski, Matthew H. *Engineers for Change: Competing Visions of Technology in 1960's America*. Cambridge, MA: The MIT Press, 2012.

Wolfe, Tom. *The Right Stuff*. New York: Farrar, Straus, & Giroux, 1979.

1 Fred Lang

Ralph Nader

This is a celebration of the life of one engineer, a friend, Fred Lang, who lost his struggle against cancer last week in Florida. Engineers rarely receive much attention. By temperament, they are withdrawn; by profession they are often hidden behind their corporate or government employers.

But Fred Lang was different. He was up front, a pipeline safety crusader who took on the big boys, an inventor who tried to give you smooth, non-cracking highways, and a philanthropist whose work will go on.

I first met Lang at a conference in the late Sixties. He was then a DuPont engineer from Pennsylvania whose rural home was downhill to a new underground oil pipeline owned by the Colonial Pipeline company.

Since he did not relish the idea that highly pressurized, boiling hot oil could come down into his living room, from what he believed was an excessively brittle form of pipe, he asked me to help raise the larger issue of deficiencies in hundreds of thousands of miles of oil and natural gas pipelines.

We worked together to persuade Congress to enact the natural gas pipeline safety act of 1968. He then served on the advisory committee to the Department of Transportation's office of pipeline safety for six years before he quit in disgust. For the advisory committee and the office were dominated by corporate pipeline interests. There was no room for an independent engineer. The government's standards were to be written essentially by the very industry that was expected to obey them.

Lang left DuPont in the early Seventies to pursue his inventions that DuPont was not interested in developing. One was a way to build pothole free highways that do not buckle or crack in winter or summer. Such highways would need very little repair. Billions of dollars and large amounts of motorists' time and fuel, waiting or taking detours, could be saved every year.

Once again, Lang came up against a vested interest—the highway lobby composed of the cement, asphalt and state-federal bureaucrats who opposed another way of building highway pavement. Lang's pavement is made of two layers with a plastic sheet between them and uses less cement or asphalt. That means less money for those companies whose influence over the Federal Highway Administration—the agency that writes the standards for the states to meet—is all powerful.

Although strips of highway showed that Lang's invention would work and although companies stole his invention for use on vertical parking garages here and abroad—indicating its practical utility—, Lang could never break through. Bureaucrats and legislators would listen politely and then do nothing. His university engineering colleagues, with a few notable exceptions, were reluctant to speak out. They had consultantships, grants and contracts to worry about.

So Lang developed an alternative idea. Why not have the government simply rent the pavement under a competitive bidding process in accordance with smooth pavement standards? The winner would get the business (the government would still own the land under the highway) and receive rent payments as long as the pavement meets the standards. In this way, he reasoned, the best pavement design will win and corporate obstructionism would be bypassed.

His rent-a-pavement idea received some publicity but no adoptions. It would have made an excellent addition to the highway-mass transit law that passed Congress in 1992. But the highway lobby's dominance squashed it at the outset.

Lang's other crusade was on behalf of the lone inventor—still a prolific source of inventiveness in an era of big labs and big companies. Stealing the ideas of lone inventors or challenging these less than affluent people to sue them under the patent laws is a pastime of many corporations. Lang himself found that companies were purloining his invention without paying him royalties.

He retained a law firm in New Jersey whose lawyers take lone inventor cases on a contingent fee which they collect only if they win. The law firm won some cases for Lang and recovered several million dollars—only a fraction of what he was owed.

But Lang wanted to pursue numerous other ideas and ventures instead of wallowing in regret. Even when he was in his late Seventies, his youthful optimism and inventive mind never flagged. When the Colonial pipeline blew up outside of Washington over a year ago, he was on the phone with the press giving them his technical views. The same was true a few weeks ago when a pipeline blew up in Edison, New Jersey.

Lang won and lost many battles. In so doing he set an example for engineers to speak up and apply their code of ethics in courageous activity rather than hunker down and just grumble.

He leaves behind a foundation and active children who will further his causes. Fred Lang's example and legacy will continue his noble vision of an engineer's unique role in the betterment of society.

2 Peter Paltchinsky
The Hauntology of an Anthropological Perspective to Engineering

Nicholas Sakellariou

To Lefteris Larentzakis and Myrto Sakellariou-Larentzakis; two fellow engineers, whose hauntologies have sustained me.

Engineering projects are designed to endure. Yet these same characteristics—durability and effectiveness, for example—purposefully built into engineering design, often cause a project's performance to become a haunting issue of controversy. As the Modern Greek folk poem puts it: "Unless a person becomes haunted, the bridge shall not hold." For each engineering project to last and be appreciated by its users, a team of engineers must "connect the dots." For controversial engineered products, engineers themselves play an important role in sustaining the myths that haunt their profession.

Peter Paltchinsky connected the engineering dots: For this reason, his "hauntology" (the logic of a ghost, if you may) suggests a convincing example of a professional engineer who addressed his subject matter as a *whole*.[1] Some basic facts about Paltchinsky's life and, specifically, about his execution by Stalin's secret police in 1929, have been quoted by Aleksandr Solzhenitsyn in his *Gulag Archipelago* [2]. Nevertheless, to Loren Graham's persistence we owe the most detailed description of the prominent Russian mining engineer's Life and Times. Graham, also an engineer by training, became a professional historian and devoted his life to the study of Soviet science and technology.[2] Fortunately for us, on one of his trips to Moscow during Gorbachev's *Perestroika* he discovered a missing file of "P. A. Paltchinsky." As a result, he wrote *The Ghost of the Executed Engineer*, a highly readable book. According to my interpretation of both Paltchinsky and Graham, the *Ghost* serves to indicate that engineering must become more anthropological and, therefore, enter upon a thorough criticism of its intellectual foundations.[3]

Paltchinsky never received academic training in anthropology. Nevertheless, he came from "a large, complicated and troubled family," a fact that contributed greatly to his becoming a patient listener and someone who could easily read human relationships and complex political circumstances. Apart from that, he was a generalist by nature and nurture: in high school, he excelled in math and science as well as in literature, music and the foreign languages. Eventually, his dedication and skill won him admission to the prestigious Mining Institute where he soon became critical of the technically oriented and "academic-dilettantish" approach to engineering education. He was convinced that Russian engineering graduates would never be able to successfully compete with their fellow professionals from around the world, unless they became "realistic engineers who evaluate problems in all their aspects, particularly the economic [and social] ones." According to Paltchinsky, the most important element to inform engineering considerations "was human beings." Paltchinsky, then, *was* an engineer-anthropologist, if not by training, at least by intuition.

Paltchinsky approached engineering earnestly, as he refused to "give an evaluation of a problem before he had patiently collected all the relevant data." This mentality, which he himself contrasted to the "gusher psychology" of delivering the most technological innovations in the least time, made him a target under Stalin's regime where the engineer was expected to do "as he is ordered." Contrary to the formal accusation that was raised against him, Paltchinsky sought no personal benefit in supporting the Czar. According to Graham, Paltchinsky "considered himself a democratic socialist and he favored the overthrow of [both] the tsarist [and the Bolshevik] government for which he worked." Paltchinsky was *simultaneously* a sophisticated technical mind—his four volume reports of European seaports were widely read and translated in several languages—and someone who consistently insisted on "viewing engineering plans within their political, social and economic contexts."

Paltchinsky's tumultuous memoirs reflect his preoccupation with outstanding engineering work. His first professional assignment was to address the "worker question" in the mines of Don Basin, surmised at the time to be the source of 70 percent of the country's overall coal supply. But Paltchinsky's innovative approach to the issue he was then looking at—based on the collection of a vast amount of heterogeneous data—was hardly appreciated by his supervisors, who sentenced him to "administrative exile" in Siberia. Throughout his life, Paltchinsky was exiled and jailed a couple of times. He worked tirelessly for the government regardless of his country's political regime, while earlier in his career he had managed to pursue a highly successful and prolific career as an engineer-consultant in Eastern Europe.

For budding Russian engineers at the turn of the 20th century, being involved in radical politics was not uncommon, so the young Paltchinsky found inspiration in the works of the anarchist theoretician Peter Kropotkin. The latter, a proponent of technological utopianism and father of the evolutionary theory of "mutual aid," remained a continuous source of inspiration as well as dispute to Paltchinsky. The implementation of Kropotkin's ideas into his fellow countryman's vision of technical work preceded the efforts of American engineers in the nineteen-sixties to incorporate humanists' intellectual territory into their vision of a new paradigm of engineering.[4] In light of this phenomenon of "intellectual appropriation," we are better able to appreciate Paltchinsky's criticism of Soviet industrial practices as well as his positioning in technical controversies: between advocates of functional (i.e., centrally controlled and operated), versus regional planning, for instance.[5] Paltchinsky's preference for a mixture of these two planning methodologies clearly stems from Kropotkin's vision of a society in which sociotechnical components would be heterogeneous, namely a combination of some large projects with a great number of small-scale projects.[6]

Generally speaking, there is an important difference in point of view between anthropology and engineering. On the one hand, anthropology looks at its subject matter (the interaction between nature and culture) from a time perspective of thousands of years. On the other hand, when engineers look at an engineering project, they usually refer to an economic perspective. Consequently, the engineering perspective looks at its subject matter (the interaction between nature and culture!) from a short–term, business cycle-point of view. Anthropology may teach us how the ancient Persians first used windmills to pump water, or how the ancient Greeks employed passive solar orientation for their buildings and cities. A counter-example would be the argument that *a priori* devalues solar power strictly on economic grounds, ignoring, for instance, the huge risks and investments that were critical in cementing the oil infrastructure in the early 20th century United States.[7]

The anthropological perspective towards engineering recognizes that performing sophisticated technical work is only one element, albeit a critical one, in the overall process of connecting the engineering dots. In addition, the dots must be inclusive, holistic and comparative. Paltchinsky, for instance, himself a mining engineer, had a deep appreciation for comparing engineering projects' working conditions. As a consequence, he conceived of "humanitarian engineering" on the basis of advancing the workers' education and job satisfaction. Importantly, such an approach remains respectful of local culture and heterogeneity, while also challenging the allegedly straightforward boundaries between generalist and technical specialist. As the linguist, who, in the beginning of the 20th century, traced the genealogy of the word *engineer*, explained: "I suppose an engineer ought to be both *ingenious* and *ingenuous*, artful and artless, sophisticated and unsophisticated, bond and free."[8]

Hence, the anthropological perspective should combine technical sophistication with a liberating awareness of the underlying assumptions of our technological choices.

Paltchinsky used to say that "a good engineer could not perform miracles, only the maximum of what was possible." Indeed, his life provides an important correction to those who assume the impossibility of approaching engineering problems holistically. Paltchinsky's anthropological approach to engineering was not "soft," if "soft" is taken to mean "weak" or "irrelevant." On the contrary, his humanitarian and holistic rationality was, by definition, associated with an exhaustive, context-informed data collection. But as I have tried to show through Paltchinsky's example, it would be misleading and in fact unfortunate to conceptualize the anthropological perspective to engineering as another version of the recurring, but still problematic, argument to "broaden" engineering horizons by infusing engineering education with humanities courses.[9]

In what sense, then, does the logic of Paltchinsky's ghost inform the present, as well as the future of engineering? In part, Paltchinsky's hauntology supplements the explanation of the USSR's technological modernization pathologies via the limitations of a centrally-run economy. To that effect, Graham contends that modernization in the USSR did not scale up due to a "loss of faith" to the system. To this last point, I would simply add that faith in the system was lost, for the technologists who were behind the process of modernization failed in/were obstructed from connecting the engineering dots. The social studies of technology and, specifically, the history of engineering will always provide insights for practicing engineers as long as engineering elites around the world reinvent themselves by picturing (their style of) technological change as the sole guarantor of national progress.

Besides its historical lessons, Paltchinsky's hauntology stimulates reconsideration of some of the basic assumptions in engineering education and practice. How many of our young engineers, for example, have worked as laborers so as to appreciate the dangers and working conditions of technical labor? Should engineers follow in Paltchinsky's footsteps and simply withdraw from an engineering project when confronted with lack of heterogeneous data that incorporate humans into the big engineering picture? Shall post construction assessment be considered an integral part of engineering project design?[10] In that regard, recent attempts to revive or redefine "humanitarian engineering" have indeed led to more community-focused, human centric design approaches.

Two images emerge: A fairly dominant image views engineering practitioners as technical experts preoccupied exclusively with perfecting rational means; in other words, making sure that engineering projects will last, without questioning why they were built in the first place and for whom. On the other hand, Paltchinsky's hauntology implies that in engineering "the ends may play a more significant role than it is often assumed."[11] Ironically, the only instance in which Paltchinsky failed to contextualize a problem was in his own case of refusing to "confess to crimes he did not commit." This decision, one may argue, was quite "unreasonable," for it literally caused him his life.

As I have argued elsewhere, self-sacrifice—what Downey and Lucena label *the element of altruism inherent in engineering*—is one of the standard, even noble forces behind technological change.[12] Whereas engineering projects' permanence is increasingly a matter of social, political and environmental struggle, Paltchinsky's perspective would premise "a society where all human needs are fulfilled, a goal that cannot be achieved without heterogeneity in scale, style and organization." In the words of his wife, Nina Aleksadrovna, her husband was "'one of those people who wanted to take action up to the moment when there was nothing left to do, up to the last possibility.' I see you [the engineer!] in just this way."[13]

ENDNOTES

1. The idea, as well as the term "hauntology" belongs to French philosopher Jaques Derrida [1] and was introduced in his 1993 work called *Spectres of Marx*. The concept speaks to a philosophy of history according to which the present is haunted by images of both the past *and* the future.

2. In 2010 Anthony Heywood of King's College, University of Aberdeen, published a detailed history of the life and work of Iurii Vladimirovich Lomonosov [3]; a second, highly successful, and equally controversial Russian technical expert who happened to be Kropotkin's contemporary. Heywood's book constitutes another remarkable source of information as regards the interactive relationship between engineering and society in late tsarist and early Soviet Russia.

3. All quotations in the text were drawn from *The Ghost*.

4. Wisnioski, Matthew. *Engineers for Change: Competing Visions of Technology in 1960s America*. MIT Press, 2012.

5. On the idea of "intellectual appropriation," see Hård, Mikael and Jamison Andrew, eds. *The Intellectual Appropriation of Technology: Discourses on Modernity, 1900–1939*. MIT Press, 1998.

6. See Kropotkin's *Mutual Aid* [4].

7. The same shortsighted rationale undergirds claims like that of policy analyst Ben Lieberman's who argues that "[i]f wind power made sense, why would it need a government subsidy in the first place? It's a bubble which bursts as soon as the government subsidies end." See Walden, Andrew. "Wind Energy Ghosts." *American Thinker*, February 15, 2010. A notable exception to this mentality is engineer Steven Johnson's "My First Year with Solar." In this article published in the *IEEE Spectrum* in September 2010, Johnson explains how "[s]witching to solar wasn't just about dollars and cents."

8 ASCE Committee on History and Heritage of American Civil Engineering. *The civil engineer: His origins*. Historical Publication No. 1, 1970, ASCE, New York.

9. Downey, Gary Lee. "Are Engineers Losing Control of Technology? From 'Problem-Solving' to 'Problem-Definition and Solution' Engineering Education." *Chemical Engineering Research and Design*, 83 (A6) (2005): 583–595.

10. The case of dam-displacement being a major cause of community impoverishment comes to mind. See, for example, Richter, Brian D., Sandra Postel, Carmen Revenga, Thayer Scudder, Bernhard Lehner, Allegra Churchill, and Morgan Chow. "Lost in development's shadow: The downstream human consequences of dams." *Water Alternatives* 3, no. 2 (2010): 14–42.

11. Picon, 2004 [5].

12. Sakellariou, Nicholas. *Engineers and the Greek State: 1830–1878* M.Sc. Thesis, National Technical University of Athens, 2006 (in Greek.) Downey, Gary Lee, and C. Juan Lucena. "The Need for *Engineering Studies: Engineers in Development*." In Engineering: Issues and Challenges for Development, edited by Tony Marjoram, 167–171. Paris: United Nations Educational, Scientific, and Cultural Organization, 2010.

13. Paltchinsky's wife quoting Romain Rolland's novel *Jean Christophe* in Graham's *The Ghost of the Executed Engineer* [6].

REFERENCES

1. Derrida, Jacques. *Specters of Marx: The State of the Debt, the Work of Mourning, and the New International*. New York: Routledge, 1994.

2. Solzhenitsyn, Aleksandr. *The Gulag Archipelago: 1918–1956*. Harper Perennial Modern Classics, 2002.

3. Heywood, Anthony. *Engineer of Revolutionary Russia Iurii V. Lomonosov (1876–1952) and the Railways*. Burlington: Ashgate Press, 2010.

4. Kropotkin, Peter. *Mutual Aid: A Factor of Evolution*, 1902. <http://dwardmac.pitzer.edu/Anarchist _Archives/kropotkin/mutaidcontents.html> Anarchy Archives. Accessed October 15, 2015.

5. Picon, Antoine. "Engineers and Engineering History. Problems and Perspectives." *History and Technology* 4, (2004): 421–436.

6. Graham, Lauren. *The Ghost of the Executed Engineer: Technology and the Fall of the Soviet Union*. Harvard University Press, 1993.

3 Heroes or Sibyls?
Gender and Engineering Ethics

Alison Adam

CONTENTS

Interest in gender as an analytical category in the study of engineering has grown rapidly through continuing concerns about the low numbers of women in the profession. A simple world wide web search on "women and engineering" throws up hundreds of project sites spread throughout the world.

At the same time, disasters such as the Challenger accident have accelerated the need to develop professional engineering ethics, in particular through the technical curriculum and codes of conduct in professional societies.

Meanwhile, in other parts of the academy, there has been an upsurge of research on feminist ethics, broadly speaking, based on the position that traditional ethical theories may not serve men and women equally well, and that alternative moral theories need to be developed [1].

Yet, other than liberal calls for engaging in professional relationships "without bias because of race, religion, sex, age, national origin or handicap" [2, p. 407] and some references to Gilligan's *In a Different Voice* (still the best known work of feminist ethics), these three strands of thought almost never come together [3].

What would happen if they did connect? What would a gender informed engineering ethics look like?

In exploring this question my aim is not to look for ways to attract more women into engineering, which is how the "gender question" is often cast in the technical domain, even though this remains important. The reasons why women are not attracted to engineering in great numbers are subtle and complex, defying monocausal explanations and solutions. It may well be that, in the teeth of the apparently intractable problem of women's absence from engineering and technology, we need new, imaginative ways of thinking about the whole question. In that spirit, my aim is to think about how these concerns translate into engineering ethics, and thereby to demonstrate that certain constructs may not serve men and women equally well. Hence exploring such tensions and making novel connections between engineering ethics and feminist ethics can offer the beginnings of a gender-aware and ultimately more inclusive ethics for engineering.

To achieve this, I examine one aspect of contemporary thinking in engineering ethics in terms of the "moral hero." I do not intend to claim that this is the dominant position in engineering ethics thinking. Nevertheless the concept of hero is a strong theme, often explicit in popular discussions of engineering risks, and sometimes more implicit as in connection with "whistleblowing." More particularly, it has even been offered as a conscious role model for potentially immature engineering undergraduates [4,5]. Analyzing and making explicit the heroic theme in engineering ethics demonstrates that it is an overtly masculine concept, deriving initially from storytelling and folklore. It is woven through many epical classical tales, such as *The Iliad*, and extends into Enlightenment narratives of scientific exploration. I argue that the heroic theme is such a strong part of our traditions in folklore and legend that its influence extends into many traditional ethical theories, notably Kantian theory, where the moral agent can be cast as a kind of "moral hero." It is unlikely that authors, extolling the virtues of the moral hero, intend to exclude women in their definitions; some even explicitly include heroines and "sheroes" [4,5]. Nevertheless I shall argue that the moral hero, Kantian or otherwise, is not only strongly masculine but is also very individualistic. As recent writing on feminist ethics argues against hegemonic masculinity, and the inherent individualism it often entails [1], I offer two alternatives. The first is a consideration of the possibilities inherent in care ethics for a more collective and less overtly masculine theme for engineering and technology ethics. The second is a classical concept, in terms of the "Sibyl," the ancient prophetess or oracle, which may be particularly useful in understanding the historical exclusion of women from technical disciplines, and which also may inspire alternative moral decision heuristics, although she is not a symbol that emphasizes collectivity in ethics.

ORIGINS OF THE MORAL HERO

From classical, heroic epic tales such as Homer's *Odyssey* and *Iliad* and Virgil's *Aeneid* through to Bunyan's *The Pilgrim's Progress* and Enlightenment tales of scientific exploration and discovery, the hero triumphing against all odds exerts a powerful hold. The notion of the "hero" in fable, folklore, and fairytales is such a strong element of mythical stories in Western and other traditions that it is not surprising that it is also woven into our theories of morality. Indeed the link between myths and morality, where the myth explicitly offers moral guidance, must be emphasized. The initial function of the Homeric epic hero tale was to show the qualities that men are supposed to have—to be brave and skillful warriors, to know one's place in the moral order, to perform one's socially allocated function according to an idealized view of social life [6]. Hence myth, legend, and storytelling are inextricably linked to our moral thinking.

Terrall [7] points to the way that the hero myth is re-invented and reinterpreted in Enlightenment voyages of discovery. It is here that the hero becomes specifically associated with knowledge, progress, and technical and scientific disciplines. Through emulating the classical hero's extreme physical effort, these early scientific explorers took risks to bring back data, the raw material for scientific progress thus strengthening the link between heroic effort and the grand narrative of scientific progress. The notion of the hero was strongly gendered:

> Narratives of discovery, and in particular heroic discovery, promoted the value of science … they elevated the heroes of science above the rest of humanity, often in gendered terms. The Enlightenment grand narrative of progress, then, entailed a paradoxical conjunction of universal human reason with the special attributes reserved for a heroic masculine elite … [7, p. 225].

The "moral hero" is more than just a hero. In folklore genre he (and it is usually "he") risks everything including reputation, status, and even life for the sake of a moral principle. Searle [8] includes the following elements as part of the unwritten rules of the modern moral hero drama. First there is a protagonist driven by the need to restore moral order in society who is completely convinced and will stand up for his inner convictions. The hero once engaged in the moral crusade cannot be deterred and expects those close to him to adhere to the standards, cutting ties with them if they do not.

The moral hero steps out of the story or myth and into modern ethical and political life in various ways. For instance, Walton argues that, in professional practice, moral heroism may be required to stick to a potentially violated code of ethics: "someone who stands out from the rest of us because of extraordinary courage and dedication, willingness to suffer for the sake of the code and thence for all the rest of us within that profession, so as not to have all of our names besmirched as peers under our otherwise-violated code" [9].

THE MORAL HERO IN ENGINEERING

The preceding section has taken a broad view to argue not only that the notion of the moral hero is a theme woven through the history of ethical thinking but also that the moral hero has particularly masculine connotations and associations. In this section I consider ways in which the moral hero theme has been expressed more particularly in writing on technology and engineering ethics.

Popular accounts of engineering projects often take an heroic *leitmotiv* as a subtext. Notable examples in this genre include Levy's *Hackers* [10], Kidder's *The Soul of a New Machine* [11] and Wolfe's *The Right Stuff* [12]. Almost all the characters in Levy's densely told tale of early hacking days are men. It is notable that one of the few women programmers mentioned in the story is the butt of pranks and jokes from her male colleagues; she went on to be one of NASA's top programmers. Such is the devotion of the male characters to their cause that they go without sleep, proper food and other creature comforts; they shun friendship and love outside the hacker circle. Etched in silicon, they are undertaking heroic journeys to build better and faster machines. And as Wajcman points out, in relation to the thirteen women pilots, judged NASA's top astronauts in 1960, but who were subsequently left on the ground, one can be the right stuff but the wrong sex [13].

Broome's writing on heroic themes in engineering is particularly pertinent as it explicitly advocates the adoption of a heroic approach in engineering ethics to counter perceived problems of immaturity and "toxic mentoring." The toxic mentor may steal others' work (perhaps even that of the mentee) and/or falsify results, thus setting the worst sort of example of professional life for less experienced colleagues to follow [4]. In further writing Broome elaborates heroism more specifically with respect to engineering. He examines ways in which engineering students may be taught to achieve heroism through stories of heroism in the practice of engineering [5]. The tale of the "Concrete Sumo" (this refers to the central figure in the story, a somewhat large man delivering concrete to an engineering site) is used to elaborate the hero ethic in terms of a moral heuristic or "praxistic" derived from African rather than Western ethics [14].

The fault lies, Broome argues, partly with immaturity itself in young engineers, and partly with a romanticism with boyish behavior and outlook prevalent in American culture and universities. He emphasizes a fixation on and protracted tolerance for immaturity, especially adolescence in males, where the qualities of a love of fun, a disdain for women and other perceived encumbrances, and an unwillingness to give up boyish liberties are all part of the condition.

Traditional ethical theories may not serve men and women equally well.

Broome offers instead the notion of the "heroic mentor" which he elaborates through the heuristic of storytelling, very much in the tradition of the classical epic moral fable, the mythological hero's journey [4]. Following the writings of Campbell, Broome recasts an ideal of mentoring in terms of the mythological "hero helper," who can be an old witch, and the mentor as veteran mythological hero [15]. Broome's archetypal hero answers some kind of call to separate from family and loved ones, then embarks on a perilous journey undertaken for the sake of others, accomplishing the task essentially alone. It is clear that separation, i.e., separation from Mother, parents, and others is seen as part of the maturation of the hero. The hero's "inner journey" is rational and corresponds with the physical journey. Underlining the heroic aspects of Kantian theory, Broome [4, p. 417] claims that the hero must possess the decision making powers of deontological reasoning as only mature adults possess these. They bestow the power to make decisions according to principles or moral rules. The mature adult does not just make such a journey once; indeed a lifetime of heroic journeys is required as a prerequisite for gaining the maturity and wisdom necessary for successful heroic mentoring of others.

GENDER AND THE MORAL HERO IN ENGINEERING

It is not my intention to claim that Broome's view as to the value of the hero in the development of engineering ethics teaching is necessarily the dominant position in the field. Nevertheless, I have argued for the prevalence of the hero, albeit often implicitly, in moral theories, especially insofar as they are based on our myths and legends. I have also suggested that many popular accounts of engineering involve a strong heroic element. Hence, although the moral hero may not, at first, appear to be a predominant theme in engineering ethics, nevertheless it bubbles just beneath the surface as a meta-text in narratives of engineering accomplishment

If we accept the move away from liberal arguments which regard engineering as essentially neutral, and which see women's relative absence as something to be addressed by publicity campaigns, we begin to see the possibility that technology is deeply gendered and, indeed gendered masculine. If that is the case then we must ask if the models we develop in engineering ethics also have a gendered character. If engineering ethics serves to reinforce masculinity in engineering, then this may further reinforce women's marginality and exclusion from the subject. For these reasons we must ask if the hero is a concept that reinforces existing gender stereotypes and if so can it be modified or replaced by a model that is more inclusive of both genders?

The heroes of classical epic poems were male warriors. As Tong [1, p. 119] points out, Homer's *Iliad* offers a celebration of the warrior hero, coupling Greek rationality and moderation with irrationality and violence. The warrior hero is praised for virtues that are masculine values; boldness, energy, and valor. Epic poems may contain strong female figures, such as the tragic Queen Dido in Virgil's *Aeneid*, but women are not represented as heroic journeying warriors setting the standard for morality. They are often to be prizes, those that cause the war and the journey such as Helen of Troy, or instead, those that passively watch, the spectators. Ruddick points to the way that military combat is almost exclusively a male preserve [16]. This serves as a reminder that war and the military are definers of a type of masculinity, and that winning is the goal [17]. The link between the military and technology and thence to definitions of appropriate masculinity is further underlined. The birth and rapid development of the modern computer in the midst of World War II and the development of the Internet in defense communications are pertinent examples. Women take a marginal part in the military and are excluded from war except as those to be fought for or as the *mater dolorosa*. Women were also excluded from the heroic journey of the Enlightenment explorer, serving as an audience back home rather than taking an active role [7]. Therefore a model for engineering ethics that derives, at root, from a heroic conception of warrior or explorer masculinity appears to offer little, if any space, for feminine inspired moral precepts.

It is notable that Broome offers the heroic model to counter a sense of immaturity which is couched in terms of *boyishness* rather than *childishness*, in other words it is to a masculine rather than a more inclusive form of immaturity which he alludes [4,5]. But we must be wary that, in seeking to supplant problematic boyishness with something better, we are not unconsciously reproducing its problematic aspects in the model of the hero. For instance "a love of fun" and a "disdain for women and other perceived encumbrances" looks very much like on the one hand, the hero's thirst for adventure, and on the other hand, the hero's leaving behind those that he loves as he sets out on his travels. To "protect his childhood freedoms" then may look little different to protecting his adult freedoms.

Two further concepts serve to mark out the moral hero as a concept which struggles to include convincing feminine moral values. The first is its inherent individualism, the second connected aspect its separation from emotion.

In order to undertake his journey, physical/and or spiritual, the hero must separate himself from those around him and journey alone to pursue his cause. This sense of individualism is often present in classical moral theories, from the Stoics to Kant, where the ideal moral agent carries out his or her duty without regard to the material things of the world [9]. This highlights the individualism inherent in many traditional themes in ethical theory and emphasizes the link with the Homeric epic hero seen as the precursor of modern Western ethics. This further signals a long standing detachment of reason from emotion, prevalent in Western philosophy in general, and moral theory in particular, adding further to the emergence of the moral hero as a detached, heroic individual. The Kantian moral agent rationally decides on his/her duty as an individual without recourse to emotions either his/hers or those of others. Individualism and moral autonomy go hand in hand with reason and they are achieved, not in cultivating relationships, but in weakening their hold, much as the epic hero does in embarking on his journey.

In fact, Broome appears ambivalent on this point [5]. He agrees that a proper adult separation from family is required for moral maturity, and the beginning of the heroic journey; yet, at the same time, he emphasizes influences of family members in his own route to moral maturity [14]. Rationalist, unemotional, rugged heroic individualism in ethics cuts against the grain of writings on ethics by feminist authors who emphasize both connectedness and emotion in moral reasoning.

FEMINIST ETHICS

Feminist writing is, of course, not the only source of a more collective approach toward more connected ways of casting moral imperatives [18]. However, feminist ethics has been developed over the last twenty or so years, both to criticize existing moral theories for the ways in which they do not include women's ways of reasoning morally, and also to offer alternative approaches to ethics. Generally, these attempt both to avoid the individualism of traditional ethical theories, and at the same time to be more inclusive of both men and women. The canonical work of feminist ethics is Gilligan's much quoted text, *In a Different Voice*, which criticized Kohlberg's views on the stages of moral reasoning, which had argued that women rarely reach the more advanced stages of moral decision making [3]. Gilligan offered a different way of thinking about woman's moral reasoning, which emphasizes its caring role rather than individual rational decision making. Although this is by no means the only strand within feminist ethics, since then, a considerable literature has been developed on the "ethics of care" [19].

The ethics of care emphasizes the emotional as well as the rational. One cares for one's children, for instance, not just out of a sense of duty, but also because of love. Relationships and responsibilities are emphasized over an ethics of justice which accentuates rules and rights. Gilligan initially connected such an approach with women's ways of thinking and talking, but was criticized for this, and later conceded that men often adhere to an ethics of care. However, her empirical evidence suggests that women are more likely to hold to this view than the alternative justice- and rights-based view. The fact that Gilligan's models are already found in both men's and women's reasoning should broaden rather than weaken its appeal as this indicates that it is already a more inclusive approach than traditional individualistic ethics.

ALTERNATIVES TO THE "MORAL HERO"—TWO SUGGESTIONS

ETHICS OF CARE

What alternative models may be drawn, particularly with respect to feminist views of ethics, to counter the problematic aspects of the moral hero in engineering ethics? A first thought might be that rather than setting off on an epic journey the moral hero should stay home and look after the kids! But this is more than just a frivolous detail. Those who raise and nurture children cannot set off on epic journeys, they transmit culture and moral values in ways other than the hero's journey. Broome's own moving story of his grandmother and great uncles' attempts to better themselves and make their ways in the world after World War I illustrates this point [4]. While his great uncles set off on their epic journeys to college, grandmother stays at home to raise a large brood of children. We would all regard her as just as much a hero as the others, perhaps more so, yet she has undertaken no epic journey.

In offering the moral hero as an alternative to counter boyish immaturity among engineering students we are in danger of replacing one unsatisfactory masculine model of morality with another. The more collective approach implied by an ethics of care can translate into practical action in several ways, in ways that are already recognized by some authors in the engineering ethics literature.

What would a gender informed engineering ethics look like?

For instance, Walton argues that, rather than expecting moral heroism when potential violations of codes of ethics occur, professional organizations need to set up extensive support systems for their members involving activities such as web sites, phone trees, chapter committees and roving delegates [9]. Clearly some organizations already undertake activities such as these. There are dangers, of course, in a profession's abilities to present a closed front against justified criticism from the public. Yet at the same time such networks could begin to alleviate the sacrifices of individuals in activities such as whistleblowing. This "support network" approach may seem rather obvious, and hardly novel, yet we have barely explored its possibilities and potential against the ethics of care. This, then, is the next step although beyond the scope of the present paper. We need to explore more fully how an ethics of care, derived largely from feminist ethics, can be translated more thoroughly and more inclusively for men and women, into practical action in engineering practice. There is a need to include alternative ethical theories such as the ethics of care in the engineering ethics curriculum. Through discussions of both theoretical aspects and by building up a pedagogical library of engineering cases that can be analyzed in terms of their correspondences with an ethics of care (rather than their more heroic aspects), we may build up alternative approaches to ethics teaching which are more inclusive of both genders.

The moral hero, Kantian or otherwise, is not only strongly masculine, but also very individualistic.

Much of Broome's argument is directed toward seeing the hero in a mentoring role [4]. It is tempting to suggest that the alternative of the wise woman, or Sibyl, which I discuss below, offers a model for mentoring women in technical domains. It may well be that future elaborations of the Sibyl can offer such scope. However, I am wary of offering too simplistic a view of mentoring, especially as it may seem to attempt an overly facile solution to the more deep-seated problem of women's marginality in technical domains. Instead, I argue that the image of the Sibyl may conjure up something much more useful. First, she may offer up a contrast to the heroic as a recipe for moral action. Second, considering the broken threads of her history, as opposed to the continuous historical lineage of the hero, may give us important clues to the historical depth of the exclusion of women from science and technology. Third, her story may offer the beginnings of alternative practical moral heuristics or praxistics. It must be emphasized that the concept of the Sibyl is one which juxtaposes a feminist slant against the traditional heroic trope, and is offered as a separate alternative to the much more developed notion of the feminist ethics of care, for which she appears, at first sight, to have little to say. This suggests that not all feminist alternatives must necessarily be directed toward collectivist ethics, an approach that could push women toward a stereotype where they are always seen in a caring role.

THE SIBYL

One clue toward a more gender-balanced approach to the hero, lies in Campbell's "hero helper," and this concept serves to underline the way in which the concept of hero is neither as independent in his epic journey nor necessarily as morally mature as one might expect. The helper is often needed to point out the true path to the hero [15]. As Broome points out, the helper may be male or female and is often old, ugly, and shabby [4]. In masculine guise the hero helper is ambiguous—he may lure the hero into trials of character; in female form the helper is usually benign, the Cosmic Mother or Mother Nature and can even be an old witch. In Virgil's *Aeneid*, Aeneas seeks direction from the Sibyl of Cumae. As a hero helper this Sibyl is ambivalent. She is a wise old woman, the archetypal old witch, but she is frightening rather than motherly, indeed as a virgin prophetess she never assumes a maternal role. She gives her prophecies to a grateful Aeneas but they are ambiguous; they do not tell him exactly what to do; he must assume the moral maturity to work it out for himself.

The Sibyl, then, may be a sort of hero helper, as in Broome's and Campbell's writing but she is much more than this. As Warner [20] describes, the concept of the Sibyl has been extremely powerful in classical mythology as a bridge between classical and Christian times. The sibylline oracles, *Oracula sibillina* are an alternative to the heroic epic poem. She is a strong part of the storytelling tradition, perhaps the origin of the concept of "old wives' tales." Her visionary gifts transcend historical time; her messages are moral as well as prophetic.

> In their very identity as truth-tellers, the Sibyls of tradition cancel connections to history ... They speak their verses, or sing their messages, and though they are always communicating a prior, universal wisdom, they are seen as actively shaping it—their voices are the instruments of the knowledge they pass on in order to prepare for life ahead [20, p. 71].

It is instructive to make a brief exploration of the reasons why the concept of the Sibyl, the wise woman, has not survived into modern times as a mythical symbol, by way of contrast to the unbroken historical journey of the classical hero. Warner [20] points out that adventurers, inspired by the oracles of the Sibyl, still searched for her grotto until well on into the middle of the fifteenth century. Yet something happened after this to exclude and marginalize the wise woman, with her herbal potions and amulets, from her prior central role, in the making of technical and moral knowledge. This exclusion took place through the extraordinarily ferocious witch-hunts in Europe in the sixteenth and seventeenth centuries where thousands of peasant women were put to death as witches in Germany, France, and Britain. As Easlea [21] points out, the witch hunts ended only with the triumph of the new mechanical philosophy, confirmed as the new "establishment" philosophy, which upheld religion and the social order against a perceived threat from natural magic. At the same time it signaled the exclusion of women from the making of technical knowledge, from the scientific and technical elites, which, to some extent, continues to this day.

Understanding the origins of the wise woman, the Sibyl, and her demise, especially in relation to the technical arts, does not offer us easy recipes for action, e.g., such as in mentoring programs. However, although only briefly discussed here, it can offer us a better understanding of the roots of present day problems in the exclusion of women from technical disciplines and ideas as to how the Sibyl may be reinstated alongside the moral hero as a positive role model.

A Sibylline Ethical Praxistic?

In addition to thinking of the roots of women's exclusion from science and technology, the concept of the Sibyl can be pressed into service, in other ways, for a feminine inspired ethical model in technical disciplines.

Just as Broome [14] develops a moral praxistic from his "Concrete Sumo" tale, might the story of the Sibyl generate an alternative moral praxistic? The classic Sibyl delivers an oracle, a divine communication, often a prophecy of some future event. The words of the oracle required considerable interpretation. This is an early form of hermeneutics, the interpreting and analyzing of texts. The business of looking for hidden meanings and interpretation has long been of considerable fascination to scholars, and cannot be regarded as somehow separate from scientific progress. For instance, in addition to his scientific laws and discoveries, Newton expended considerable effort on searching for hidden meanings in the Bible. One could even say that the computer-based techniques of data mining and knowledge discovery are but modern forms of the ancient art of interpreting the oracles. But the element of prophecy is also important as it yields the idea that we should heed the prophetic pronouncements of those who are judged venerable and wise, whether it be the classical Sibyl for her prophecies of war, or the engineers who predicted the loss of Challenger and tried to stop its launch.

One of the methods of receiving the oracle [22] was through "incubation" where the applicant was required to sleep in a prescribed place, e.g., invalids slept in the hall of Aesclepius, the god of medicine. They were reputed to receive the oracle through their dreams. We often talk of incubating an idea in the hope that it may hatch into something useful, as an appropriately incubated egg hatches into a chick. The modern message from the ancient Sibylline and oracular notion can be interpreted thus. Hasty moral decisions are to be avoided (although this is clearly not always possible as in Broome's "Concrete Sumo.") Moral decision making should be made slowly and thoughtfully. We sometimes, quite literally, need to "sleep on it." This suggests a contrast between the swift adventurousness of the masculine hero and the slower, more mediated, feminine form of moral decision making in the tradition of the wise woman, or Sibyl. Space does not permit an elaboration of this idea here, but I suggest that it would be a fruitful path to explore in feminist ethics for technical domains as an alternative to an ethics of care, particularly as an antidote to the more masculine inspired conceptions of the hero.

> From classical, heroic epic tales … to … Enlightenment tales of scientific exploration and discovery, the hero, triumphing against all odds, exerts a powerful hold.

IMAGINATIVE APPROACH REQUIRED

At least some of the notions currently promulgated in engineering ethics, and predicated on concepts from traditional ethical theories, are unlikely to be inclusive to both genders. Indeed they may be actively hostile toward women. Recognizing that many of our moral intuitions have roots in story telling, myths, and legends with a lengthy history, I have picked the theme of the "moral hero." Deriving from classical times, it can be seen as a thread which is woven into ethical theories from classical times onwards, implicitly or explicitly, and which also has been advocated in engineering ethics education as an exemplar to counter possible moral immaturity in engineering students. Having argued that the hero is essentially a masculine concept, I have shown that at least one of its aspects, the individualism of the hero is problematic. Feminist ethics emphasizes connectedness over individualism and this has been developed into a substantial literature in the ethics of care. I make some very preliminary suggestions as to how an ethics of care may be developed both in professional life and for engineering education. More tentatively, as an alternative to the hero and not specifically connected to the ethics of care, I discuss the alternative concept of the Sibyl as a kind of antidote, also deriving from classical roots, to the masculinity of the epic moral hero. This is offered as an imaginative beginning to new ways of thinking through problems in gender and technology ethics, underlining the need to look to the "wise women" of our societies for alternative ethical praxistics on our heroic journeys.

It is clear that separation—from Mother, parents, and others—is seen as part of the maturation of the hero.

While researching this article I have been struck by the deep-rooted nature of many of our concepts of masculinity and femininity which extend right back into classical times. Heroes and Sibyls are both such concepts but they do not remain fixed. From classical, to medieval, to modern times they are reinterpreted. They may be pressed into service in a modern setting far from the imaginings of their originators. As the gender inequalities existing in engineering and technical professions appear similarly deep-rooted, reinforced by the systems of ethics we develop, and difficult to change, it may well be that a more imaginative approach, such as offered here, is required.

ENDNOTE

1. The author is with the Information Systems Research Centre, University of Salford, Salford, Manchester M5 4WT U.K. Email: a.adam@salford.ac.uk.

REFERENCES

1. R. Tong, *Feminine and Feminist Ethics*. Belmont, CA: Wadsworth, 1993.
2. M.W. Martin and R. Schinzinger, *Ethics in Engineering*, 3rd ed. New York, NY: McGraw-Hill, 1996.
3. C. Gilligan, In a Different Voice: Psychological Theory and Women's Development. Cambridge, MA: Harvard Univ. Press, 1982.
4. T.H. Broome, "The heroic mentorship," *Science Communication*, vol. 17, no. 4, pp. 398–429, June 1996.
5. T.H. Broome, "The heroic engineer," *J. Engineering Education*, pp. 51–55, Jan. 1997.
6. A. MacIntyre, *A Short History of Ethics*, 2nd ed. London, UK: Routledge, 1998, p. 6.
7. M. Terrall, "Heroic narratives of quest and discovery," *Configurations*, vol. 6, pp. 223–242, 1998.
8. J. Searle, Story Genres and Enneagram Types. http://members.aol.com/jsearle479/3.html, Dec. 15, 2000.
9. C. Walton, "When the code meets the road: Professional ethics and the need for sanctions," http://www.unlv.edu/Colleges/Liberal_Arts/Ethics_and_Policy/paper2.html, Dec. 15, 2000.
10. S. Levy, *Hackers*, Harmondsworth, UK: Penguin, 1994.
11. T. Kidder's *The Soul of a New Machine*. Boston, MA: Back Bay, 2000.
12. T. Wolfe, *The Right Stuff*. Picador, 1991.
13. J. Wajcman, "Reflections on gender and technology studies: In what state is the art?" *Social Studies of Science*, vol. 30, no. 3, pp. 447–464, 2000.
14. T. Broome, V. Weil, M. Pritchard, J. Herkert, M. Davies, "'The Concrete Sumo' Exigent decision-making in engineering," *Science and Engineering Ethics*, vol. 5, pp. 541–567, 1999.
15. J. Campbell, *The Hero with a Thousand Faces*. Princeton, NJ: Princeton Univ. Press, 1973.
16. S. Ruddick, Maternal Thinking: Toward a Politics of Peace, Boston, MA: Beacon, 1989.
17. N.C.M. Hartsock, "Masculinity, heroism, and the making of war," in *Rocking the Ship of State: Toward a Feminist Peace Politics*, A. Harris and Y. King, Eds., Boulder, CO: Westview, pp. 133–152, 1989.
18. L. May and S. Hoffman, *Collective Responsibility*. Rowman and Littlefield, 1991.
19. M. Larrabee, Ed., *An Ethic of Care*. New York and London: Routledge, 1993.
20. M. Warner, From the Beast to the Blonde: On Fairy Tales and Their Tellers. London, UK: Chatto and Windus, 1994.
21. B. Easlea, Witch Hunting, Magic and the New Philosophy: An Introduction to the Debates of the Scientific Revolution 1450–1750. Sussex, UK: Harvester, 1980.
22. "oracle" (on-line Encyclopaedia Britannica entry) http://www.britannica.com/bcom/eb/article/printable/9/0,05722,58679,00.html, Mar. 22, 2001.

4 Engineer on a Mission

A Jesuit Priest EE Finds a New Calling in Africa

Susan Karlin

As both an electrical engineer and a Jesuit priest, Lammert B. "Bert" Otten can lead a spiritual retreat just as easily as a dam-building project in Zambia. "As an engineer," he says, "you're cocreating with God to make life better for people."

Zambia has been Otten's adopted African home since 2005, when he retired from the University of Seattle. As a consultant in appropriate technology for the Diocesan Promoters Office in Monze, about 180 kilometers southwest of the capital city of Lusaka, he has installed solar-powered water pumps for irrigation, solar vegetable dryers, solar lighting, and refrigeration. He has turned the oil from the seeds of local plants, such as jatropha, into biofuel, soap, and candles. And, yes, he's built a dam—on the Ngwerere River.

As if his retirement weren't busy enough, Otten also uses available materials to solve smaller problems. There's the lamp he made by covering a cardboard toilet-paper roll with an inside-out potato chip bag as a reflector; the solar water distiller he made from glass, poly film, conduit pipe, and a plastic water bottle; and the battery-charging windmills he fashioned from discarded plastic sewer pipes. Otten also hitched LEDs (donated by stateside friends) to flashlight batteries (ditto) as a cheap and safe alternative to burning diesel fuel or candles in a grass-roofed hut.

"Many people in Zambia are living on a dollar a day or less, so cost is a big thing for them," says Otten. He's now working on a battery recharger made out of refrigerator magnets and coils from telephone dialing equipment. "You wave the magnets around the wire coil and get the electrons moving enough to light LEDs," he says.

He also mentors local high school students; last summer he initiated a Seattle University Engineers Without Borders student group trip to the Zambezi River to build a water pump run by river current.

For his efforts in appropriate technology, the University of Missouri College of Engineering, in Columbia—one of Otten's many alma maters—gave him a Missouri Honor Award for Distinguished Service in Engineering.

As a child in St. Louis he relished making crystal and one-tube radios in his basement. At the same time, he adds, "there was always something about my personal relationship with God. I kind of felt called to the priesthood but didn't want to do it. Then, in high school, I got to know the Jesuits, their lifestyle, intellectual discipline, and mission work in the developing world. I realized I could do things like that in relation to technology, and it all tied together."

Otten started at the Jesuit-run St. Louis University as an electrical engineering major in 1950 but entered the Jesuit seminary three years later. He ended up with, among other degrees, a licentiate in sacred theology from the University of Saint Mary, in Kansas, and a Ph.D. in EE from the University of Missouri.

Otten first visited Zambia in 1994. Now 77, he lives in a Jesuit compound containing a high school, hospital, radio station, and staff living quarters within walking distance of the village of Chikuni. Africa seemed a natural choice for retirement. "Most of my contacts were in Zambia, and the needs in Africa were so much greater," he says. "One fellow, orphaned at 10, worked his way through teachers college making and selling LED lights," says Otten. "Now his life is different."

5 Engineering Selves
Hiring in to a Contested Field of Education

Gary Lee Downey and Juan C. Lucena

One day in an engineering thermodynamics class, the professor pulled out an Ann Landers column, titled "A Different Breed," from that morning's newspaper. He asked the seventy or so third-year students, "How many of you think this is reasonably accurate?" The column contained excerpts from letters, mostly from frustrated spouses, that portrayed engineers as technical, inflexible, and socially inept. The first provided a summary statement:

> Dear Ann: This letter, my first ever to a columnist, was sparked by your column about the engineer's wife who asked, "Are engineers really different?" The answer is ABSOLUTELY! My father was an engineer. My three brothers and four uncles are engineers. Engineers ARE a different breed. They are precise, logical, and great at problem solving, but they know very little about human interaction. My engineer husband makes a fine living, but when it comes to expressing emotions, on a scale of 10, he's about a 4.
>
> **A wife in Houston**

Three other letters added credibility to this characterization by attributing it even to the best engineers, describing the difficulties of living with such a person, and suggesting that things could be otherwise:

FROM TUCSON, ARIZ: Engineers ARE different. My engineer husband (graduate of MIT) tells me when my skirt is ⅛ inch shorter in the back.
If the floor in the bathroom looks uneven, he gets out a tape measure for "proof." A crooked window must be adjusted at once. If, however, I am crawling around the house with a killer migraine, he doesn't notice.
SANTA BARBARA: My husband the engineer has no tolerance for the gray areas of life. He sees everything in absolutes. It's black or white, right or wrong, yes or no. Never a maybe. He feels no joy, but he is never depressed either. Everything is in perfect order, or there is hell to pay. It is not easy to live with such a man.
NO CITY, PLEASE: Thirty years ago, I married an engineer. Our marriage was an emotional wasteland. He would have been a better father if our children had been robots he could program. Engineers can figure out everything except how to be human and caring.

The sole letter from a man removed gender as the determining feature:

CHICAGO: My wife is an engineer. She is precise, analytical, and definite in her views, and she always thinks before she speaks. She's as cold as ice and so sure of herself, she makes me sick. My next wife will be an empty-headed, bubbly moron, and it will be a relief.

Finally, letters from an optician and an interior designer established that the problems with engineers go beyond family relations, for engineers "are murder to work with" as well.

The only two letters expressing disagreement contested not that engineers were somehow a different breed but the judgment that this assessment is negative:

CARBONDALE, ILL: You're damned right engineers are different. I am still happily married to mine after 35 years. They tend to look before they leap and have stable marriages. By nature, they are problem solvers, sensitive, and caring. The woman who wrote to complain ended up with the wrong man, not the wrong profession.

INDIANAPOLIS: My engineer husband doesn't send me flowers. In fact, some days we don't even have a decent conversation, but I'll take this nerdy looking guy with his assortment of pens and eyeglass cases in his shirt pocket, his dull tie and wrinkled trousers over any of the men I've ever known. He is loyal, decent, dependable, and real. He'll never cheat or lie to me. That's worth a lot these days.

Prodded by the instructor to offer their own judgments, roughly a third of the students raised their hands to say they agreed that engineers were a different breed, nearly half said they disagreed, and the remainder did not respond. In sharp contrast with the usual practice when asking about today's homework or next week's test, not a single hand rose above shoulder level. The quick, low wave was the rule, as if students felt vulnerable and hoped to avoid calling attention to themselves. After all, why risk being wrong about something that would not be on the test? Despite the fact that the professor was a gentle man who was not afraid to discuss matters of the heart, he made no move to offer his own assessment. Did it even cross his mind to do so? To hear how he felt about things would have been interesting but unusual, and no one dared ask. Class began.

The two of us are interested in the making of engineers, the mechanisms of self-fashioning that take place in undergraduate engineering education. The professor and students in this class were grappling with what they understood to be the stereotype that engineers regularly criticize as a preconceived and oversimplified idea of the characteristics of the typical engineer trained in the United States. Questioning its accuracy involves asking whether or not people called engineers actually conform to the stereotype. During the course of our fieldwork among engineering students and teachers at Virginia Tech, a land-grant university with roughly four thousand undergraduate engineering students, we heard many discussions about the stereotypical engineer. In every case, someone disputed the image by claiming that it did not characterize accurately many engineers and hence was not true, only partly true, or at least too narrow. Yet the image persists, and students seemed to take for granted its existence and its power.

We were initially drawn to study engineering education by our own experiences as engineering students. Both of us completed undergraduate degrees but felt somehow that the sort of people we were being asked to become did not fit with the sort of people that we already were or wanted to be. Above all, we felt constrained. Over the years we have observed that people who left engineering, including ourselves, seemed to feel a need to explain why. It was not that they couldn't "hack it"—the engineering student's term for not having what it takes—but that somehow there was a lack of fit. Understanding learning solely as the transmission of knowledge from the heads of faculty to the heads of students did not begin to account for the bodily experiences of constraint so many of us experienced. How might we make these and similar feelings more visible? And might doing so suggest ways to shift engineering toward a place where people like us would want to be?

As a cultural anthropologist teaching in a graduate program in science and technology studies (STS) and a Ph.D. student in STS interested in cultural and anthropological studies, we organized a research project that would follow engineering students through their curricula to explore the changing demands they experienced as persons.[1] Do stereotypes count? How was the knowledge content of engineering related to the social dimensions of engineering personhood? Long-term participant-observation and extensive interviewing and document collection became strategies to explore students' experiences.

We quickly learned, however, that this work involved more than just studying students. As we elaborate below, we had organized our project in the midst of great debate among engineers over the contents of engineering education. Our study was even funded by the National Science Foundation, which had emerged as one of the leaders in this debate. Unless our written work appeared entirely irrelevant or uninteresting to engineers, it would likely be captured and positioned by this debate (see Rapp on abortion and Hess on capturing, this volume). Not only did running away or hiding seem pointless; we also wanted very much to participate in the process of retheorizing engineering education. We sorted out a pathway we call "hiring in."

HIRING IN

As a metaphor of employment, hiring in indicates a willingness on the part of social researchers to allow their work to be assessed and evaluated in the theoretical terms current in the field of analysis and intervention. It means becoming employees in a sense, whether paid or unpaid. Although one need not accept at face value what people say about themselves and what they consider desirable or worthwhile, hiring in involves, at minimum, acknowledging that established modes of theorizing constitute established power relations and that contributing new theorizing captures one within those relations. Maximally, hiring in involves following all the pathways to critical participation that one can identify and attending to all the details and doing all the work necessary to position, assess, and, if warranted, try to achieve a theoretical shift.[2]

Accepting the responsibility of hiring in, as many researchers studying science and technology, including anthropologists, have already done, may provide an opportunity to contribute directly and genuinely to the theorizing that takes place in a contested field. It offers the possibility of convincing people to shift their modes of theorizing from here to there by acknowledging and emphasizing that one can only start where one is. As a practice for researchers combining cultural perspectives and ethnographic fieldwork, hiring in can involve making visible modes of theorizing that are otherwise hidden, thus possibly legitimizing alternate perspectives that are rooted in the field itself.

However, complementary risks of cooptation and social engineering are substantial, each leading in its own way to marginalized ineffectiveness and self-delusion. The cooptation of a project involves its transformation into something indistinguishable from that which it studies. More than gaining participation or otherwise becoming located as part of the field, cooptation dissolves the identity of the researcher(s) entirely into the field. A coopted project not only goes native; it is nothing else. Social engineering involves presuming that one's expertise warrants the authority to legislate change through a research project. The arrogance of social engineering keeps a project permanently outside the door, preventing it from participating critically in that which it studies. A social engineer in the field never leaves the hotel.

We locate hiring in as one approach to a more general academic practice that one of us has elsewhere called "partner theorizing" (Downey and Rogers 1995), limiting it to those situations in which one is not already located in the field of intervention but seeks to gain entry. Hiring in contrasts with debates, for example, whose interlocutors are presumably located within the field of intervention. The main goal of partner theorizing is to encourage the growth of collaborative relations in academic work and relocate the agonistic politics of rebuttal from a necessity to an option in the everyday practices of academic researchers. It asserts that the practice of theorizing neither can nor should be a proprietary feature of academic work, for much theorizing, in fact the major proportion of theorizing, takes place outside the institutionalized Western academy. Alongside the anthropological perspectives represented in this volume and elsewhere, partner theorizing thus reconceptualizes relationships within and between academic disciplines, as well as between modes of academic and popular theorizing, as flows of metaphors in all directions rather than the necessary diffusion of truthful knowledge and power from the inside out.

Our project on engineering education illustrates three different moments of partner theorizing. First, it envisions all acts of theorizing as undertaken in partner relations with their interlocutors in collective, but temporary, negotiations of knowledge production. We are thinking of partner theorizing not as a market activity, a business partnership, but as a variety of activities of exchange among committed cohabitants, married or otherwise. One's work always intervenes in the context of other theoretical agendas. Competing theorists, both academic and popular, live together.[3]

The dominant theoretical agenda with which our project has to contend is the doctrine of "competitiveness." Since the early 1980s official United States dogma has redefined international struggle from a political and military to an economic idiom, transforming understandings of the nation from a site within which individual interests compete into a single economic actor maximizing a collective interest. The power of patriotic commitment to this economic call to arms became concentrated in the slogan of competitiveness and its logic of productivity, which locates humans alongside technology and capital as resources for the production of consumer goods. Popular theorizing about competitiveness seems to reach into everyday lives and selves much more than the military logic of the Cold War did, because it turns every action into an economic defense of the nation. Something is good if it enhances competitiveness and bad if it does not. Engineering education has gained particular salience in these developments because engineers figure as key participants in virtually every image of increased national productivity (see Business-Higher Education Forum 1983; National Academy of Engineering 1986; President's Commission on Industrial Competitiveness 1985).

National visibility is new for engineering education (Lucena 1996). In the years after World War II and before Sputnik (October 1957), engineering education stood alongside other forms of technical and scientific education as an integral component of a military struggle against communism. "Our schools are strong points in our national defense," said President Eisenhower early in 1957, "more important than our Nike batteries, more necessary than our radar warning nets, and more powerful even than the energy of the atom" (US Congress 1957). Sputnik shifted concerns somewhat because it was read as a shocking accomplishment of science rather than of engineering, and nationalist interest in education during the 1960s narrowed to an exclusive focus on science and the production of scientists for basic and applied research. The National Science Foundation took Edward Teller's advice that emphasizing engineering education would miss the point:

> It is my belief that it [engineering] should not be considered a weak link in our scientific and technological effort [and therefore has sufficient funding]. We should put the greatest possible emphasis on higher education in applied science. (National Academy of Sciences 1965:259)

The 1970s maintained an emphasis on basic and applied science but expanded the range of legitimate problems to include energy, transportation, pollution control, and other nonmilitary arenas, through such programs as NSF's Research Applied to National Needs.

The nationalist reinterpretation of economic competition as national struggle and strategic risk in the 1980s was sudden and dramatic, embodied and epitomized in the Reagan election. Engineers gained the opportunity to become leaders on the battlefield, as when President Reagan called upon the National Academy of Engineering in 1985 to

> marshal the nation's technical engineering-based expertise in a campaign that will ensure America's scientific, technological and engineering leadership into the 21st century.… These efforts … are essential to the goal of helping American businesses and workers to modernize and compete. (National Academy of Engineering 1986:3)

The NAE began by mapping engineering "education and utilization" visually in a computer-printed model that looked like a dense piping diagram. Tracing flows in from secondary school on the left to flows out through "death," "disability," "emigration," and so forth on the right, the model linked education and employment together in infrastructural movement with lots of connections and feedback loops. Education became a flow of bodies through engineering schools whose primary problems were, accordingly, "recruitment" and "retention."

The piping image stuck. Tracking flows of engineers made it possible to reimagine education and utilization in the economic terms of supply and demand and establish the goal of enabling supply to respond more flexibly to demand. The National Science Board, which sets NSF policy, said in 1988:

> If compelled to single out one determinant of US competitiveness in the era of the global, technology-based economy, we would have to choose education, for in the end people are the ultimate asset in global competition.... Economic performance and competitiveness will be particularly affected by undergraduate engineering education. (National Science Board 1988)

By 1988, NSF had not only explicitly adopted the pipeline image but also transformed it into a linear image that defined continuity of flow as the goal and leaks as the problem.

Within American industry, becoming more competitive meant becoming more flexible. The flexible accumulation of capital in a struggling nation needs flexible people, lots of them (see Martin 1994). NSF is now at work reengineering engineers, supported by the National Research Council:

> We have to be thinking now what we want to see 10 and 15 years from now in terms of what is coming out of the pipeline with respect to science and engineering ... human resources that will be flexible enough in terms of their training so that if they don't quite match what is at that time the need for their skills, they can be retooled very quickly. (Task Force on Science Policy 1985:43, 64–65)

NSF has pumped more than $200 million into research and innovations in engineering education to build more flexibility into engineering curricula and produce more graduates, mainly under its flagship program, the Engineering Coalitions.

Hiring in to engineering education in the 1990s thus involves recognizing that engineering educators in the United States can fairly easily construe what they do as being centrally in the national interest. Engineering education is widely understood not only as a place where good students prepare themselves for career tracks that promise financial stability and upward mobility but also as a test site for the refiguring of patriotism. Research that seeks participation in the fashioning of engineering selves risks simply contributing to this nationalistic fervor by improving students' abilities to pursue the goals of competitiveness without critically examining its contents. At the same time, it risks giving the impression that one seeks to prescribe change, to fix people who are presumably broken. One could easily come across as antiknowledge, antiengineering, and anti-American. Images count.

A second moment in partner theorizing involves viewing all theorizing as totalizing in content but not necessarily totalitarian in effect, in the sense that theorizing depends for its insights on a metanarrative, or background story, that builds a world within which its interpretations have meaning and power. From this perspective, the metaphor "temporary" may serve better than "partial" to describe the limited claims of totalizing theories participating in exchange relations. That is, the value of any form of theorizing is temporary, connected to changing circumstances.[4] If all forms of theorizing are temporary, then hiring in involves a historically and culturally specific encounter between distinct modes of theorizing.

The point of contact between our project and the ongoing retheorizing of engineering education lies in images of personhood. With the background story of competitiveness, we have to deal in particular with theorizing about "underrepresentation" and "flexibility." Because the pipeline model is an aggregate mathematical image, it counts people as individuals grouped according to distinct statistical categories. Sorting out persons biologically by sex and race, for example, the pipeline called attention to categories that were statistically underrepresented in engineering, namely women and minorities.

"If we want to supply our industries and government and our universities with the human power that we need in the future," NSF director Erich Bloch told Congress in 1987, "we need to concentrate on the groups which are underrepresented today in the scientific engineering areas—women and minorities" (US Congress 1987:9). Echoed the National Science Board, "From the perspective of economic competitiveness (as well as other perspectives), NSF programs and management efforts designed to help bring women, minorities, and the economically, socially, and educationally disadvantaged into the mainstream of science and engineering deserve continued focus" (National Science Board 1988).

By the late 1980s the dominant argument was that the country needed more engineers, but the pool of college-age people was declining, compounded by a declining interest in engineering among first-year college students (roughly 8% of college degrees). Since approximately three-quarters of the engineering bachelor's degrees are granted to white males, "greater participation on the part of women and underrepresented minorities in engineering studies would be one way of addressing the supply-side problem" (Bowen 1988:734). While constituting 51 percent of the population, women make up roughly 15 percent of first-year engineering students. Blacks make up 12 percent of the population but

> only 6 percent of first-year students, Hispanics 10 percent and 4 percent, and American Indians 0.7 percent and 0.5 percent, accordingly. In addition, while retention rates for white males hovered around 70 percent, retention rates for women were roughly 40 to 50 percent, 30 percent for blacks, 48 percent for Hispanics, and 33 percent for American Indians. Students of Asian descent were not considered a problem in this demographic profile, for with 3 percent of the population, they accounted for 6 percent of first-year students. Also, their retention rate exceeded 100 percent as more students subsequently transferred into engineering programs than departed. (National Science Foundation 1990)

The nagging problem for engineers in theorizing underrepresentation concerns how to explain it in the first place. If students enroll in engineering programs because of innate capabilities and dispositions, then are nonwhites and females less capable or otherwise naturally predisposed away from this career track? From our perspective, the problem of underrepresentation is a citadel effect, an effect of theorizing learning entirely in terms of a diffusion model of knowledge. Colleges of engineering have been able to respond only by establishing new recruiting strategies and support systems for minorities and women students to increase enrollments and retention rates. These must struggle to maintain legitimacy since providing support programs for students does not fit the dominant model of challenging students to prove they belong.

Statistical calculations turn biological groupings into socially significant objects, making each woman stand for all women and each African American stand for all blacks. An anthropologist might be inclined to challenge this essentialist view of the person because it seems to reduce people to biology by showing these biological categories themselves to be historically and culturally specific constructions. Maybe the problem of underrepresentation would simply dissolve away as a misleading construct. Although this pathway to critique and opposition is fairly straightforward conceptually, it is likely incomplete as an approach to intervening, at least in the near term. One might be able to alter the importance engineers attribute to biological categories, but the categories themselves are probably here to stay.

Furthermore, the allocation of resources through the pipeline image, which draws on biological categories, actually provided new access to the corridors of power for interest groups that explicitly defined themselves as representing women and minorities. One newly funded female professor gleefully told us, "I can't believe I'm so deeply involved in this. I'm making connections all over the place. We're building an old girls' network." Similarly, the National Association of Minority Engineering Programs Administrators (NAMEPA) proudly announced the theme, "Partners in the Pipeline," for its 1994 meeting in Washington, which included scheduled lobbying trips to Capitol Hill. Congress itself authorized formation of the Task Force on Women, Minorities, and Disabled in Science and Engineering. In other words, classifying students by race and sex made some people visible who were otherwise hidden (Lucena 1996:180–81). As Donna Haraway (1989, 1991a) has shown us, biology can be useful.

In order for our work to hire in to the problem of underrepresentation, or have any role at all, must we force our data into artificial, predefined groups of women, minorities, whites, nonwhites, etc.? For us, these are interesting as cultural categories of persons that people apply to themselves rather than as distinct types or categories of humans that should be taken as real because based in biology. What sorts of categories might following students' experiences produce, and would these help to account for underrepresentation as an outcome without dividing up the world by race and sex at the outset? Applied in this case, partner theorizing thus involves going beyond showing that students' experiences cannot easily be parceled up by race and sex and, hence, making the problem of underrepresentation seem illegitimate. We must also try to account for how students themselves identify people by race and sex and ask if this process has any implications for statistical underrepresentation.

The problem of "flexibility" raises different issues. At present, engineering policy makers are at a loss to establish what flexibility means in curricular terms, although many different groups, from NSF officials to Boeing engineers, are vying to control its definition. Generally, flexibility seems to mean "malleable," as in making the bodies of engineers sufficiently malleable to fit changing job definitions. Because no single mode of theorizing flexibility has become established, we have the opportunity to contest what flexibility could mean rather than having to limit ourselves to accounting for the effects of a given model.

A third moment in partner theorizing is that it focuses attention on the power relations between alternate modes of theorizing by accepting that knowledge is never simply knowledge *of something* but is also knowledge *for someone*. Accordingly, the practice of partner theorizing encourages one to look for ways of factoring into one's own thinking the views of others in the field of intervention without necessarily seeking the consensus that is often unrealizable. Rather, by looking for reasons to accept the legitimacy of others, even if one finds limitations in their perspectives, partner theorizing shifts the goal from simply advancing one position in a debate to advancing or replacing the debate as a whole. One theorizes in terms of both one's siblings and ancestors—the traditional mode of defining a theoretical stance—and one's interlocutors or the positions one is trying to engage. Together these locate one's theoretical position at the start of an analysis. Advancing a new mode of theorizing in a contested field through partner theorizing thus involves greater entanglement in existing power relations than either mastery through truth or opposition through resistance. Furthermore, when the goal is to hire in from an outside position, paying attention to the power dimensions of theorizing is crucial to make sure that one even gains the legitimacy to participate.

Our first step in hiring in to engineering education is to make visible the experiences of students as they move through their curricula, thus confining our intervention to what happens within engineering education. Were our own experiences idiosyncratic? Do the students themselves offer alternative ways of thinking about learning? The next section briefly summarizes a handful of these experiences, drawing material especially from three students labeled minority in order to engage theorizing about underrepresentation and flexibility.

We then search for pathways for intervention that take account of the current structure of engineering education and do not demand the resources that would be necessary to redesign curricula from scratch.

WEED-OUT

Jen Lopez is a Hispanic woman; Glenn Phillips and Rick Williams are African-American men. Although these labels identifying race and sex[5] were non-negotiable for students, the significance they played in students' lives is more problematic. Despite the expectations associated with her status as both minority and woman, fen appeared in many respects to be the prototypical engineering student who sought the appropriate goals and adjusted herself properly to fit the curricular demands. Although Glenn and Rick had difficulty finding ways of fitting themselves to engineering, in neither case was race the problem or issue.

The core knowledge content of engineering curricula is what engineers call "engineering problem solving." Learning how to draw a boundary around a problem, abstract out the mathematical contents and solve it in mathematical terms, and then plug the solution back into the original problem is central to the fashioning of engineers and a major challenge to the bodies and minds of students. Students regularly asserted that the goal of certain courses was to "weed out" students from engineering curricula.[6] For students who stayed, these and other courses also appeared to weed out a part of themselves as persons.

Establishing disciplined habits and attention to detail are typically a student's first adjustments to the practice of engineering problem solving.[7] Rick found the content of this discipline different from his experience in the Navy:

> Oh, it's a different type of discipline. When I was in the Navy, they would say you have to stand this guard pose, and you gotta do this, and you gotta do that. They would lay it out for you. Here, it takes more self-discipline because you gotta figure it out for yourself.

An associate dean of engineering articulated a key feature of this self-discipline when he told incoming would-be engineers and their parents at freshman orientation that "engineers have to learn how to have fun … efficiently." We later repeated this to a friend who was completing his Ph.D. in mechanical engineering. After laughing heartily for a minute or so, the student stopped suddenly and said, "You know, he's right." The discipline in engineering problem solving is a total body experience that extends across all engineering "disciplines," a term that seems particularly appropriate in engineering (see Foucault 1979). As a dean told graduating seniors in *engineering*:

> What you've all learned in engineering is how to attack and solve problems. It doesn't matter what discipline you go into. You all learn the same thing. Solving problems is what engineering is all about.

The first two engineering courses that students take at Virginia Tech, appropriately called Engineering Fundamentals, explicitly describe disciplining the body and the mind as essential prerequisites to later success. One instructor in the first course described his course as "probably as much training as it is education." On the first day of class, for example, this professor announced, "The wooden pencil is dead for you." Holding one up, he said, "This is gone. You don't use it anymore. You are an engineer in training. You are on your way to the top. This looks crummy." Then, holding up a mechanical pencil: "This looks great.… Zero point five millimeter HD lead." In the second class he frightened students with a pop quiz, had them grade it themselves on the honor system, and then informed them that it would not count, to their great relief: "I just wanted to get across to you [that] you are always responsible for knowledge contained in a previous class." In the third class he outlined the connections between good habits, including a regular bedtime, and success in engineering:

> One of the biggest mistakes students make is they have an irregular bedtime because tomorrow their classes don't start 'til eleven, the next day they start at eight … and your body, instead of developing a habit, you just change that every day. Folks … it won't work. I haven't met any student that's been successful that way.… The ones that oscillate back and forth, in the long run do not succeed.

New engineering students find out quickly that engineering problem solving revolves around homework exercises and that engineering courses have more daily homework than any other curriculum. To get a good grade, every correct homework assignment must include the student's name, course number, and date lettered properly at the top of a sheet of engineering paper, which is a cross between unlined paper and graph paper. One writes on the unlined side, guided by faint lines that show through from the other side.

It is significant that every problem undergraduates encounter begins with the word "Given." The boundary has already been drawn. Engineering problem solving confines itself to the ideal world of mathematics. All the nonmathematical features of a problem, such as its politics, its connections to other sorts of problems, its power implications for those who solve it, and so forth, are taken as given. This contrasts sharply with physics problem solving, in which the main challenge is to learn to "think like a physicist" (White 1996) so that one can bring that unique genius to bear in a process of discovery. In engineering, students learn they must keep reactions, intuitions, or any feelings they might have about the problem out of the process of drawing a boundary around it and solving it. These are irrelevant and can only get in the way.

Good problem solving follows a strict five-step sequence: Given, Find, Equations, Diagram, Solution. Students start by abstracting from a narrative description of the problem mathematical forms for both given data and what they must find in order to solve the problem. Then, invoking established equations and drawing an idealized visual diagram of the various forces or other mechanisms theoretically at work in the problem, they systematically calculate the solution in mathematical terms. They must write down each mathematical translation or risk losing credit. Also, if they write a numerical solution without including units of measurement, for example, as feet, meters per second, or pounds per square inch, the answer has no meaning and frequently receives no credit. The students we followed still learned the computer language FORTRAN, which translates engineering problem solving into computer code, even though far simpler programming strategies were available. It seemed to us that instructors continued to insist on FORTRAN because it penalizes small mistakes, thereby forcing attention to detail. During the undergraduate years, students are expected to solve thousands of problems either on paper or in programs.

All engineers learn something about "engineering ethics," but not to enable them to critique the ethical dimensions of the problems they solve. Rather, in order to be good problem solvers, engineers need to behave ethically. Being ethical is about controlling one's passions and impulses, making it more akin to a habit than a commitment. An administrative memo distributed to new students, for example, introduced the subject of ethics in the context of good problem solving. "Engineers in professional practice," it stated, "are required to perform their work in a neat, orderly, timely, and efficient manner.... Since they design and construct facilities upon which the safety and health of the public depends, professional engineers must conform to a strict code of ethics in their work." An introductory reader collated by faculty abstracted different types of ethical principles from the writings of philosophers. Students learned that "utilitarian ethics" alone could be dangerous because self-interest can lead to self-destruction, "duty ethics" alone could lead one to neglect the needs of individuals, and "rights ethics" alone could lead one to overlook the general welfare. "Value ethics," attributed to Aristotle, were perfect for engineers because these encouraged "tendencies, acquired through habit formation, to reach a proper balance between extremes in conduct, emotion, desire, and attitude."

As students become transformed into engineering problem solvers, what gets weeded out is everything else. That is, engineering students experience a compelling demand to separate the work part of their lives from the nonwork parts. Work is about rigorously applying the engineering method to gain control over technology and is simply not about any other stuff. Budding engineers can have the other stuff in their lives, but not in their practices as engineers.

We found that the constraints of engineering problem solving fit well in bodies where "other stuff" meant forces of identity or challenges to personhood whose meanings did not make competing demands of personhood. Jen, for example, was convinced of a deep connection with engineering because she had been interested as a kid in fixing cars and other things and had attended a science and technology high school. For her, the habits and attention to detail in engineering problem solving were simply an advanced form of tinkering, and high school had added the appropriate mathematics. Moving from tinkering to engineering problem solving was simply a shift from a private activity to a collective one. Jen thus found the demands she encountered entirely reasonable and appropriate. "It's just to make everything easier to read," she said, "like for instance if you need to present something to your boss, you're not going to just slap the answer on a piece of paper. You have to say, 'This is what I was given, this is what you asked me to do, and this is what I found out.'" Jen had made engineering a personal goal at a fairly early age. "I always knew I wanted to do engineering," she told us.

Accepting the rigors of engineering problem solving also fit very well Jen's desires for upward class mobility. Raised in a working-class family, she clearly felt class identity as a challenge and source of force: "I think eventually I will leave engineering because I want to work with management. I think I'm ambitious." She even arrived with some understanding of how being an engineer could be positioned as a job in a corporate environment. In high school she had worked ten hours a week as an intern at an engineering firm: "I got to see what it was like, and I liked it. I liked the atmosphere, the job atmosphere." Granting authority outside the self was no problem because Jen expected to work in a job with a boss. It was also important that, in Jen's image of class mobility, success in life was the outcome of sacrifice and individual determination. At freshman orientation, she recalled, "They were saying that, if you look around the room, half the people won't be graduating." Asked if that intimidated her, she replied, "A little bit, but I think if you really want something you can get it, if you try hard enough."

Finally, the expectation in engineering problem solving that one will control one's emotions fits the stereotypic definition of a mature man, one who is strong and in control and who exercises considered judgment. For Glenn, who had trouble keeping his emotions out of his problem solving, this part of the challenge in engineering courses posed a significant problem. He told us, for example, that he had a tendency to say "I feel" in expressing his views. "I don't think it was accepted too much," he said. "Maybe 'I think' was accepted more." As Rick explained in another interview: "I've never seen someone say, 'Well, sir, the answer is because I feel that.' The professor will just cut him to pieces." Referring to his feelings got Glenn in trouble in job interviews, where he appeared to be wishy-washy and lacking in self-confidence. His strategy for getting through engineering courses was to link himself to people whom he believed were in a similar situation, namely women. That is, rather than contest the challenge from engineering or the stereotypic man, Glenn made use of the stereotypic woman by seeking out emotional support from a network of female engineering students and avoiding men, who he believed would not talk about such things.

Because the stereotypic woman is emotional by nature, she is not the ideal engineering problem solver. For Jen, the continued currency of the stereotypic woman gave her a chance to stand out as an individual: "[Being a woman in engineering] feels kinda neat, because not many women are engineers. I kinda like that I stand out in a way because I don't want to be one of the nameless engineers. Girls aren't supposed to be [engineers], so I'm glad that I am one." Jen had indeed encountered overt discrimination from at least one male professor who drew on the stereotype in treating women, but she discounted this experience as exceptional. "You've always got those chauvinistic types," she said somewhat casually, such as the one who put on a worksheet a question "something like 'an engineer goes home to the housewife da da da.'" When she went in to ask for help, this professor "treated me like I was, you know, like a baby or something." But she insisted he was unlike the others: "Most professors are really nice."

Standing out as a woman did call for special strategies if Jen was to make sure that faculty were blind to the stereotype in judging her merit. One strategy she used to demonstrate her capabilities and commitment to male professors was to go and talk with each one, making sure they would have to treat her as a full person. "Most of the time professors like me a lot," Jen said. "I'll go talk to a professor just so they know who I am, not necessarily so they'll know my name or anything, but just so they know my face and they recognize me." She also had to compensate for a lack of study partners. Because she did not identify herself as a "woman engineer," she did not seek out other female students: "Not like real good friends or something." Also, there were few engineering students in her dorm:

Jen: It's especially hard. I live in an all-girls dorm. How many girls are engineers? Most of the guys,
you know, had at least two or three on their hall. So you just walk down the hall and
there's a guy doing his engineering homework too. But there weren't any other girls on
my hall that were engineers. So it's kind of like, "Well, what do I do?"
So, what do *you do?*
Jen: Well, I kinda figure it out by myself.

The dominant image of the graduate engineer is one who controls technology, who creates by
translating internal knowledge into object form; in short, one who designs. One department bro-
chure summarized its curriculum simply with the words: "Engineering teaches students to design."
The College of Engineering routinely advertised itself with photographs of solar cars designed
and built each year by students. First-year students meeting in small group interviews regularly
described how they long "to design something." Said one, "I want to be the person that draws it …
that kind of designs it in a way, and then hands it to somebody else and they go do it."

However, the images of design that incoming students carry with them often differ greatly from
those that discipline their work. Most students we encountered started out viewing engineering
design along the lines of the stereotypic architect, whose designs are a deep, personal expression
of some distinctive perspective, subjective orientation, or emotional reaction. Indeed, the romantic
fantasies that drew students toward engineering in the first place, such as helping society through
new designs, helping one's people improve economically, or advancing civilization through space,
generally included a heavy measure of agency or even autonomy for the individual engineer. But it
was not easy to hang on to these visions.

In engineering problem solving, design is the timely, disciplined application of the engineering
method to real-life problems. In this image, the genius of design shifts from the person to the method
itself, and authority shifts to the curriculum. Because seniors, were most likely to have mastered the
method, they were usually the ones who got to practice design. The solar-powered car, for example,
was designed and built each year by a group of senior students. But for newer students, becoming
successful engineers meant giving up on the fantasies that made them the geniuses behind great
designs. Instead, they had to accept the constraints imposed by the curriculum and learn to solve
problems properly. As we followed students through these challenges, images of creative invention
tended to dissolve away, and engineering design lost its romance.

Rick's main fantasy was to solve problems "relevant to ordinary life," but the commitment to mathematical problem solving in engineering course work soon left him with a strong sense of personal loss:

Drawing a cylinder with a hole in the middle and at two different angles had to be perfect. You had to
have it just right. That's what I hated the most. I really did. Took me at least fifteen hours a week sitting
at the computer to get it right. I hated it.

In Rick's view, the heart had to be connected somehow with the job:

My opinion is that anyone can do engineering but not everyone can love engineering. I didn't love it, so
I couldn't do it. My ex-girlfriend loves it. Every other word out of her mouth is in touch with systems
engineering. She loved it.

Rick united his work with his heart by switching to biochemistry, whose activities felt more
relevant to ordinary life:

I had gone from an engineering major that morning to a biochemistry major by dinner, and I already felt
the difference. I could really feel the difference. Now, my biochemistry class, I love it. I mean, I sit down
to do my biochemistry problems, and I can't believe the Navy is paying me to do these things. Just yesterday we were analyzing protein structures and how they interact with each other due to spatial orientation
and how they function. I was sitting there, you know, "Wow, I can't believe I'm getting paid for this."

Class identity played a significant role in Rick's career path, as it did with fen. A naturalized US citizen born in Ethiopia, Rick was graduated twenty-second from his high school class in upstate New York. He had enlisted in the Navy because his mother could not afford to send him to college. Moving to biochemistry rather than say, anthropology, preserved his goal of upward mobility, because the Navy would pay for his education and he could anticipate a well-paying job later.

In contrast with both Rick and Jen, Glenn's main fantasy had to do with race. Raised in urban Washington, D.C., he wanted to stand out as a black man in a white world, thereby challenging the racial stereotype of black people as mentally slow, physically lazy or undisciplined, and, hence, inappropriate:

> My high school was all black. I chose to come here over North Carolina A&T [a predominantly African-American school] 'cause I thought I needed to experience the world. It's not just going to school with white people; it's living with them.

Glenn elaborated on how he wanted to become a role model for other African-American students: "I want to inspire others to get into engineering. I guess me becoming [an engineer] encourages others to do that also. That's a contributing factor to why I stuck with it." A sense of loss would come later.

As we have already seen, Jen's fantasy embraced the identity of an engineer as a pathway to a job and income, but all this was alongside another passion: ballet. Jen had studied ballet since childhood and remained deeply committed to it, hoping also to learn choreography. However, doing both engineering and dance brought great stress:

> Every single night we had practices from after dinner 'til at least 12:00. My classes ended at 3:00 so there was no time to do homework. Sophomore year there's like three problems to do in every class and it takes you *a* long time. The people in charge of the dance were all like, "Go talk to your professor to get an extension on your homework." I was like, "You've got to be kidding me. No engineering professor is gonna do that."

Jen talked to an engineering professor after receiving a poor grade on an exam. "Worst choice," she said. "He's like, 'Why did you come to Virginia Tech? You've gotta decide for yourself, you've got to set your priorities straight. Did you come here to dance or to become an engineer?' He told me that directly to my face." Jen was angry about the image that "if you're an engineer at Virginia Tech, that's all you can do." She wanted to contest the view that mathematical problem solving was everything for a prospective engineer, "because you have to be well-rounded.... If you just sit at your desk or work at your computer all day long, you can't survive in society," she insisted, "because work is not just what you can do on your computer, it's how you associate with people." She reacted to the challenge with even greater determination: "It pissed me off and I was like, 'Well, no, I'm gonna do them both.'" Yet ultimately she gave in, adjusted, and reduced her involvement in dance:

> I just did the best I could. I did less dancing. I went to another dance company but I didn't do as much. I didn't choreograph because that takes a lot of time. I knew I wanted to be an engineer and I was gonna do it. I was only in one dance in that group whereas the year before I was in four. It worked out fine. I did well with my grades.

Students have no room or opportunity to challenge the priority given to mathematical problem solving in engineering lives and work. The curriculum is just there, demanding that they make all the adjustments and informing them through grades of the extent to which they were succeeding. Homework after homework, test after test, and course after course rank each student on a linear scale. Glenn struggled for a long time just to achieve the C average he needed as a minimum demonstration of membership; in other words, to graduate. When asked if he ever felt engineering was not for him, he said:

Probably every other day. Freshman year, I kind of just said, "Well, I have to adjust and do better next semester." The grades were the biggest thing that kind of told me. They made me tell myself to struggle to stay here and stay in engineering. I always believed that I could do better but I didn't know how. I messed up.

Fortunately for him, Glenn was able to locate one essential, that is, natural, connection to engineering: "I am more technically minded than most people." So he persevered. He had come to Tech with three black friends from D.C. Before the end of second year, all three of Glenn's friends had left school.

By the time students reached their junior year, the vast majority appeared to have found strategies for accommodating their bodies and minds to engineering problem solving. As courses became specialized within majors, the mathematical challenges in engineering problem solving became more complex, and strategies accumulated for commanding greater control of the world in mathematical terms. Back in engineering statics, for example, which many called the "first real engineering course," students had learned how to apply a single mathematical equation to a range of different circumstances. However, in the engineering thermodynamics course we attended, students found that any problem could have several pathways to a solution. They faced a whole menu of equations that may be appropriate in any given case and had to decide which assumptions to make in choosing which particular configuration of pathways to follow. "It's more intense," Jen said. "It's harder stuff, so it's [all about] how much you can handle … how much work you can handle."

One cost was a sense that the rain never stops. The experience of isolated struggle in the early years of engineering education had been replaced by a more shared struggle just to get through whatever came next. When we asked upper-division students how they were doing, we often heard the simple mantra, "Eat–sleep–study." By this point Rick had already left. Glenn was clinging to his female groupmates, still struggling to survive. Having failed thermodynamics, he said, "When I got into my major, it was like starting over for me. It really did feel like starting over freshman year."

Like many students we encountered, Jen coped not only by disciplining herself for engineering work but also by making sure her life had other things in it as well. The key lay both in maintaining a sharp boundary between the work part and the other parts and in making sure that their meanings for one's identity as a person did not conflict. Consider Jen's images of bounded, efficient play in a group interview of advanced students who had just outlined the pitfalls of dating nonengineers:

Deepak: I think it's engineers. They don't want you to have a social life. The ME [mechanical engineering] department, or any department.
Jen: You *can* have a social life.
You can? Do you guys all say that?
Thuy: We get as much as we have a chance to get.
Dan: Budgeted.
Deepak: Budgeted, yeah.
Did you hear what they said in freshman orientation, that engineering students are not like the other people? You have to learn how to have fun efficiently.
Jen: That's true.
Dan: That makes sense, I think.
Thuy: Very true, yes.
Dan: Some people's idea of having fun is just sitting around and just yakking. Just really nonproductive.
Jen: I think that's why I don't watch TV. I'd rather be out with my friends having a good time instead of sitting in front of the TV.
Dan: Do a good quality two hours of fun time.
Jen: Yeah.

Establishing an identity as an engineer can mean allowing engineering values to diffuse into other areas of one's life, even while holding these separate.

As the classroom encounter with Ann Landers illustrated, students by this point had come to treat instructors narrowly as functionaries who simply transmitted the knowledge students needed to pass tests rather than as independent sources of reflection and interpretation. Although professors might bring human characteristics to their work (one "is a lot of fun" while another "will slam you"), such factors were relevant only around the margins of standard pedagogy—presenting and testing mathematical knowledge. Students knew that the curriculum had been established by some past authorities and that the truth or validity of its contents was not subject to question:

Jen: I took a class in family and child development. It was like the biggest trip. The teacher's up there talking, people aren't paying attention. People are fighting over a multiple-choice test, saying the teacher didn't have the right answer. If they walked into an engineering class, everyone would be like, "Who the hell is she saying that professor is wrong in their answer?"

The value of engineering knowledge in the world persists over time. We were amused but not surprised when the thermodynamics class used the same textbook Gary Downey's own class had used in 1971.

Having survived the solitary struggles of the first two years, students had adopted a range of strategies for getting through their courses efficiently. Said one: "I now know that the homeworks and tests are what's important, so I'm a lot more efficient. I hate textbooks and never read them. I just listen to the lectures and work the problems." Also, a student no longer stood alone as an engineer but had become part of a larger group, an engineering major. "I see a lot more familiar faces in my classes now," said one student in a group interview. We heard all sorts of strategies for doing group work, including what qualities make good study partners and when group work helps the most or gets in the way. We followed one organized trio of students who divided up their three toughest courses in order to conquer them together. Each did the homework for one class and prepared the others for the tests.

In sharp contrast with the entering student, the engineering graduate who emerges from the curriculum is understood to be a disciplined, knowledgeable, and powerful person, at least in terms of engineering problem solving. Knowledgeable students have gained control over technology in a way that is unavailable to other persons, whether human or corporate. Through the logical, precise actions of identifying and solving problems in mathematical terms, each student has, by definition, succeeded in extending the realm of human control to include technology, transforming technology into a tool for human use. The stereotypic engineer is this and nothing else.

However, unilinear ranking also insures that students do not all attain the same level of control or receive the same credit. As students reached their last year of school and began looking for work, they focused on their grade-point averages to an extent they could only have imagined earlier. In a profession that does not make graduate school a prerequisite of employment, one's value as a potential employee depends in the first instance on the grade-point average. GPA is the key line in every resume because it is read as the main indicator of accomplishment. We once listened to an African-American recruiter for a corporation advise African-American students to keep their resumes from looking "too black," or employers might become suspicious that a student was putting racial identity before engineering identity. Glenn was one of several students we encountered who tried to resist this system of evaluation by leaving his GPA of 1.95 (C–) off his resume at the annual job fair:

At the first booth I went to, he pretty much told me, "Keep on walking." I remember one lady gave me a lecture. She was like, "Isn't your GPA any better?" She went on telling me all this stuff and she said, "Well, you need to work on your problem." I was like, "I don't have a problem." I felt my potential is much greater than what my grades say about me.

Glenn had been very active in student organizations and had accumulated considerable work experience, but he was unsuccessful in championing these as indicators of talent and motivation apart from his grades. Six months after graduation, Glenn had not yet found a job.

Interviewing for jobs also brings a new set of challenges to personhood as students emerge from the cocoon of mathematical problem solving and begin to face the vagaries of economic and social life. Jen's B average surpassed the minimum standard of acceptance, which students consider to be a B− average, yet she received far more interviews than students with much higher grade-point averages. She attributed this primarily to her status as a minority woman. In contrast with her experiences as a student, this time she felt the challenge deeply:

> I've never had any type of conflict [within engineering from being] Hispanic at all. I mean, me personally, I don't like the word "minority." That bugs me because the definition of minority means "less." So I really don't associate myself as being a minority. I'm Hispanic, not a minority.

Of course it was easier for Jen to avoid the label minority or woman as a student taking tests than as a potential employee trying to sell her labor in the marketplace. How could she maintain a sense of accomplishment for what she had achieved?:

So does it bother you, for example, policies like affirmative action?
Jen: Well, it helps me, so I don't mind, but I wouldn't like people saying "She only got the job because she's a girl," or "She only got the job because she's Hispanic." Because I don't think just because of that they should pick me. They should pick me on my merit, not from where my parents are from or what sex I am. It is an advantage because I get a lot more opportunities because I'm a woman and a minority.

This was a problem that mathematics could not solve. Jen's work life suddenly threatened to become Hispanic and/or female in content, undercutting her long-standing efforts to separate her work from her race and sex. She wanted a job, the best possible job with the most money and highest potential for advancement into management. Affirmative action had explicitly mixed race and sex with merit in a way that made these inseparable, yet without affirmative action would the currency of stereotypes by race or sex in corporations have prevented Jen's many opportunities from appearing in the first place? For her, this sort of problem could only be resolved through long-term job performance, demonstrating her worth as a person rather than as a Hispanic woman. She accepted a position and got started.

In sum, while the challenges to personhood that engineering students experienced overlapped challenges from the stereotypic man and a desire for upward class mobility, these conflicted with challenges from the stereotypic woman, the stereotypic black person, the stereotypic Hispanic, and a desire to link work with nonwork in some sort of organic whole or total self. In none of the three cases discussed here did stereotypic expectations drive students away. Wanting to challenge the stereotypic black person had actually motivated Glenn both to enter an engineering program and to stay, despite poor grades. The curricular insistence on keeping emotions out of problem solving added to his sense of being marginalized, however, for in seeking out female friends in engineering Glenn kept the distinction between work and nonwork permanently blurred. For Jen, being Hispanic was largely irrelevant, while being a woman provided great motivation. Her bodily adjustments involved trying to keep instructors or prospective colleagues from using stereotypes to classify or judge her. Finally, Rick left not because of a racial stereotype but because he could not live with a work self that not only had to be kept separate from everything else but that lived entirely in an ideal, esoteric world of mathematical problem solving.

PATHWAYS

But what about other students? How might this account hire in to theorizing and accounting for the problems of underrepresentation in engineering education and the flexibility of engineers? Because it will take far more than three cases to map students' experiences sufficiently to make the investigation plausible, we expect a book manuscript to be a necessary vehicle. Still, as we elaborate below, an argument or hypothesis that emerges from this account of students' experiences is that statistical underrepresentation is a special type of citadel effect, an effect of conflicting challenges to personhood. That is, the challenge from engineering problem solving to bifurcate the person between work and self, where work is dedicated wholly to solving bounded mathematical problems, might be driving away people who experience this as a denial of self rather than simply a novel encounter with discipline. Also, helping students understand and grapple with this bifurcation may have the effect of shifting the meaning of flexibility in engineering education from malleability to sophistication in critical reflection.

Stereotypes do count. For people who already feel challenged by stereotypes that somehow make them or the people they care about invisible or subordinate, might separating a raceless, sexless work self from a nonwork self feel like a double whammy, yet another demand for invisibility? Might one feel a need to leave or avoid engineering to demonstrate one's wholeness as a person? We have collected a great many stories that appear to fit such an account.

Wholeness in personhood is a Western problematic, built on a stereotypic image of society as a collectivity of autonomous individuals. This powerful image values coherence in personhood and challenges people to pursue coherent selves as a means to fulfilling lives. Indeed, any self in an adult body that is less than a coherent whole is considered downright pathological. Rick clearly sought such coherence in his person and life, as have the two of us. For Rick, the challenge from engineering education to live half his life in mathematical problem solving separated sharply from everything else he did or wanted was plainly intolerable. We propose there are many more who, like Rick, felt they had to weed out their persons in order to become engineers. One does not, for example, become an engineer-woman, engineer-African American, or engineer-world-helper, but only an engineer who happens to be a woman, an African American, or a person interested in helping the world. Unlike professional training in law and medicine, which understand themselves as adding expertise to an already educated, mature, and complete person, engineering education seeks out the high school graduate or, as one professor described it, the "blank slate." The disciplining in engineering education must be all or nothing.

Although our proposed generalizations are still tentative, Jen and Glenn may illustrate the strategies of many women and minorities who remain in engineering, resisting stereotypes by standing out as individuals. Knowing that "girls aren't supposed to be engineers" motivated Jen to prove that the stereotype did not apply to her. Similarly, Glenn wanted not only to prove to himself that he could "live with" whites but also to encourage other African Americans to follow him. Yet for the student who wants to "be around people like me," as several told us, standing out as an individual among white male engineers could feel like a lonely sacrifice of selfhood.

In sum, statistically significant differences of race or sex between students who stay and students who leave may depend upon a larger, more pervasive source of difference—the challenge made to diverse students to parcel off a mathematical work self from a nonwork self. In emphasizing such a separation, could engineering problem solving be driving away many passionate, motivated people who might be most likely to challenge the boundaries of engineering and/or want to spend their lives improving or changing the world? Based on following the experiences of students, we think that is the case. But must it be so? Must engineering be structured to fit best those who accept established authority and believe the world is pretty good right now? Might these same student experiences make visible pathways for linking engineering work to other dimensions of selfhood?

Attracting and retaining more members of underrepresented groups through support programs appears to be one such pathway. That is, making visible the experiences of engineering students may help affirm the value of recently developed programs for women and minorities that try to increase retention rates by reducing the extent to which such people might feel isolated or alone. One can argue that even white males need some nurturing and support, for the early disciplining in engineering education forces everyone to survive solitary struggles, and later courses keep the burden on students to prove they belong. However, because women and minorities also have to deal with stereotypes that label them invisible or inappropriate as they grapple with the challenge from engineering problem solving to ignore everything else about themselves, offering such groups extra opportunities to identify with other students makes good sense. For example, being able to frequent a "safe space" that is populated with other stereotypically "inappropriate" people may be one way to reduce the extent to which the stereotypes feel relevant or significant. Choosing to leave engineering should be acceptable as a legitimate decision by informed adults, but not if leaving is solely the product of having to cope with the stress of conflicting challenges to personhood.

A second possible pathway to hiring in is to reimagine the role of mathematical problem solving in engineering education and, accordingly, relocate the sharp separation between work and self. What if, for example, disciplined attention to mathematical detail and the avoidance of extreme emotion and desire were no longer celebrated as more important than anything else in an engineering problem solver? What if engineers located mathematical problem solving as simply one valuable resource among many they might use? Perhaps the stereotypic images of women and minorities would become irrelevant for prospective engineering students rather than congruent with images of incompetence, and the statistical problem of underrepresentation could potentially dissolve away.

This pathway could be elaborated in several ways, of which we have so far tried only one. We designed and teach a course called Engineering Cultures, which tries to demonstrate that placing the highest value on mathematical problem solving in the lives and decision making of engineers is both historically and culturally specific. That is, things could have been otherwise. Although this is only a single course offered as a humanities elective, it does stimulate students to reflect critically on the curricula that shape their lives.

The course begins in the present by illustrating how mathematical problem solving is never the totality, and frequently constitutes only a small part, of the activities of practicing engineers. In other words, the work of practicing engineers extends well beyond the limits of mathematical problem solving. Then, working backward in time to excavate a series of genealogical layers, the course helps students understand how the great emphasis on mathematical problem solving in engineering curricula was molded in response to the perceived threat of Sputnik to American science. Exploring a long-term tension between "design" and "manufacturing" in industry illustrates how competing perspectives with equal value can exist simultaneously and that the very act of drawing a boundary around a problem establishes a claim of authority over it. Tracing several disciplines through mechanical engineering and onto the factory floor establishes a novel connection between the identities of engineering management and labor.

The course then pushes hard on the boundary around engineering knowledge by briefly visiting a number of different systems through readings and guest lecturers trained in different countries and demonstrating that a range of ways of locating engineers is already available. Exploring connections through colonial traditions and the contemporary organization of multinational capitalism offers images of global relations that contrast with the doctrine of competitiveness, which pictures a population of autonomous nations competing with one another on a level playing field. The course concludes by encouraging students to hold on to their romantic images of life as engineers by treating engineering problem solving as but one resource among many in careers that could make a difference.

Other steps we might take to reimagine the role of problem solving in engineering education involve using our narrative of students' experiences as a test bed for assessing existing proposals to modify engineering education and for formulating some new ones. These proposals range from the structure of engineering curricula to the daily organization of classroom pedagogy. Proposals to increase the flexibility of engineering curricula, for example, often involve introducing design activities prior to the senior year while still conceiving design entirely as application of the engineering method to real-world problems. Perhaps modifying such activities to examine how engineering problems are connected to other sorts of problems and how the boundary around a given design problem can be drawn differently from different perspectives could help engineering students both to develop greater tolerance for different perspectives and to better assess how and when to apply the engineering method to the problems they encounter. Engineers trained to reflect critically on their own practices are not likely to be malleable but may indeed be more likely to figure out ways of making their disciplines and workplaces adapt to their dreams and fantasies of helping society or otherwise making a difference through engineering work.

Furthermore, must engineering education require a full body transformation? Might it be possible, for example, to formulate and structure a graduate degree in engineering for students with other undergraduate backgrounds, helping them to understand how engineering problem solving works and enabling them to map differences in engineering problem solving across fields and disciplines without necessarily having to master esoteric fields of engineering science? Couldn't one be just a little bit engineer, and might such people be capable of linking engineering problem solving to other social and personal agendas?

Lastly, what if engineering pedagogy admitted nurturing as necessary to its success? Could engineering instructors help individual students learn how to link engineering problem solving to their long-term fantasies and desires by offering their own personal stories and experiences? What if the thermodynamics professor offered his own reactions to the Ann Landers column or outlined what he was trying to achieve through research and teaching in thermodynamics? Could establishing legitimacy for welcoming and nurturing students improve their abilities to imagine new and different ways for using engineering problem solving as a resource *in* their later lives and work?

In sum, a theoretical approach that draws on the experiences of engineering students to intervene in engineering education need not demand a total transformation of engineering curricula to do so in potentially significant ways. Fully accepting the responsibility of hiring in to this contested field of education will require us to publish in places engineers respect, visit educators in their own spaces, and use reactions and critiques as opportunities to advance the discussion as a whole. This might involve as little as making presentations to engineering audiences and publishing specific recommendations in engineering publications, or as much as joining key organizations, committees, and ongoing policy debates. In either case, hiring in through ethnographic field work and applying a cultural perspective on the fashioning of selves might encourage engineers to foreground practices that constantly address and critically rethink the question: What is engineering for?

ENDNOTES

1. For lengthy reviews of research on engineers and engineering, see Downey, Donovan, and Elliott (1989) and Downey and Lucena (1994). Some other relevant work in the anthropology of education in science and engineering includes Chaiklin and Lave (1993); Eisenhart (1994); Eisenhart and Marion (1996); Lave (1990); Lave and Wenger (1991); Nespor (1994); Seymour and Hewitt (1994); and Tonso (1996a, 1996b).

2. This mode of critical participation joins with and draws from feminist critiques of science and literary cultural studies, both of which are challenging citadel effects of the academy from positions within it. It is lived most intensely by those people and perspectives that work to make a difference while accepting the risks of life in the so-called private, public, and/or nonprofit sectors, away from the guaranteed paychecks of tenured professorships.

3. Because the image of committed cohabitation is a white, middle-class ideal that can work to hide inequalities and abuses of power, we see the danger of narrowness. But even in everyday language partners can be more than two, be of various sexual orientations, develop a variety of styles of commitment, extend partnering to some areas of life but not to others, etc. Because the key assumption in partnering is that each position presumes the

legitimacy (but not necessarily the value) of others in principle, we mean the practice to be multidimensional, constantly involving work, and having varying power dimensions at the capillary level. We worry that the image of commitment may be limited by its individualistic overtones, but it does emphasize the mutual dependence of alternate modes of theorizing.

4. In anthropology, for example, the linear evolutionism that thrived during the nineteenth century could not account for a twentieth-century world of nation-states. Likewise, the structural-functionalism of the 1940s and '50s lost its relevance in a '60s world of rapid change, and a '70s symbolic anthropology could not maintain its holistic image of cultures in a contemporary world where the production of hybrids appears more the rule than the exception. Today, theories of postmodernism vie with theories of late capitalism for control over how to interpret the contemporary scene; eventually both will fall silent in the face of changing circumstances. Arguably, each step has been a historically specific improvement, but to label the whole as progress is to suggest that later forms of theorizing would have been appropriate, indeed desirable, in earlier periods. In fact it seems unlikely that, say, structural-functionalism would have played well in the nineteenth century, or postmodernism in the '40s.

5. We use the word "sex" here rather than "gender" to call attention to the fact that, in popular theorizing, the words "woman" and "man" are uttered and heard as labels rooted in biology. The word "gender" loses this dimension by displacing the label into academic theorizing that treats "woman" and "man" as cultural terms. We must maintain a focus on how our everyday theorizing uses biology to naturalize our categories.

6. Whether or not faculty actually tried or wanted to weed out students does not matter, for weed-out was a student category.

7. In this account we do not address the problem of variation in curricula among the three hundred or so schools that offer engineering degrees in the United States.

Sometimes these differences are indeed significant. Often they are not, owing to the fact that almost all schools seek accreditation from ABET, the Accreditation Board for Engineering and Technology, which offers strict guidelines for engineering curricula. One instructor told us, "I tell parents ... that a student who goes to Clemson, Ohio State, Michigan, Penn State, or Georgia Tech will come out looking pretty much like one who comes through here."

REFERENCES

Business-Higher Education Forum. (1983). American Competitive Challenge: The need for a national response. Washington, D.C.: The Business-Higher Education Forum, 1983.

Federal Actions for Improving Engineering Research and Education A Report to the President of the National Academy of Engineering, 1986.

Report of the President's Commission on Industrial Competitiveness: Hearing Before the Subcommittee on Economic Stabilization of the Committee on Banking, Finance, and Urban Affairs, House of Representatives, Ninety-ninth Congress, First Session, March 5, 1985.

J.C. Lucena, Making Policy for Making Selves in Science and Engineering: From Sputnik to Global Competition. Dissertation submitted to the Faculty of the Virginia Polytechnic Institute and State University, 1996.

Public Papers of the Presidents of the United States, Dwight D. Eisenhower, 1957.

National Academy of Sciences. 1965. Basic Research and National Goals. A Report to the Committee on Science and Astronautics U.S. House of Representatives, 1965.

6 Race, Sex, and Nerds
From Black Geeks to Asian American Hipsters

Ron Eglash

CONTENTS

The development of technological expertise requires not only financial resources but also cultural capital. Nerd identity has been a critical gateway to this technocultural access, mediating personal identities in ways that both maintain normative boundaries of power and offer sites for intervention.[1] This essay examines the figure of the nerd in relation to race and gender identity and explores the ways in which attempts to circumvent its normative gatekeeping function can both succeed and fail.[2]

NERD IDENTITY AS A GATEKEEPER IN SCIENCE AND TECHNOLOGY PARTICIPATION

Turkle [1] vividly describes nerd self-identity in her ethnographic study of undergraduate men at MIT. In one social event "they flaunt their pimples, their pasty complexions, their knobby knees, their thin, underdeveloped bodies" (196); in interviews they describe themselves as losers and loners who have given up bodily pleasure in general and sexual relations in particular. But Turkle notes that this physical self-loathing is compensated for by technological mastery; hackers, for example, see themselves as "holders of an esoteric knowledge, defenders of the purity of computation seen not as a means to an end but as an artist's material whose internal aesthetic must be protected" (207).

While MIT computer science students might be an extreme case, other researchers have noted similar phenomena throughout science and technology subcultures. Noble [2] suggests that contemporary cultures of science still bear a strong influence from the clerical aesthetic culture of the Middle Ages Latin Church, which rejected both women and bodily or sensual pleasures. He points out that the modern view of science as an opposite of religion is quite recent, and that even in the midst of twentieth-century atheist narratives, science (and "applied" technological pursuits such as creating artificial life or minds) continues to carry transcendent undertones. Noble's historical argument easily combines with Turkle's social psychology of nerd self-image.

Normative gender associations are not the only restrictions that nerd identity places on technoscience access. In an essay whose title contains the provocative phrase "Could Bill Gates Have Succeeded If He Were Black," Amsden and Clark [3] note that the lack of software entrepreneurship among African Americans cannot simply be attributed to lack of education or start-up funds, since both are surprisingly low requirements in the software industry. Rather, much of the ability of white software entrepreneurs appears to derive from their opportunities to form collaborations through a sort of nerd network—either teaming with fellow geeks (Bill Gates and Paul Allen at Microsoft) or pairing up between "suits and hackers" (Steve Jobs and Steve Wozniak at Apple).

But if nerd identity is truly the gatekeeper for technoscience as an elite and exclusionary practice, it is doing a very inadequate job of it. First, while significant gaps are still present, there has been a dramatic increase in science and technology scholastic performance and career participation by women and underrepresented minorities since the 1960s [4]; yet during that time period nerd identity has become a more and not less prominent feature of the social landscape. Second, this change has been far stronger in closing the gender gap than in closing the race gap. For example, in the 1990s the gender gap in scholastic science performance for seventeen-year-olds was significantly lower, while the gap between black and white seventeen-year-olds remained the same. Yet Noble and Turkle portray gender/sexuality, not race, as the overriding feature of nerd identity (Turkle does not, for example, offer any reflections about the possibility of racial identity in her comments about "pasty complexions"). Finally, we might note that in comparison to, say, Hitler's Aryan *Übermensch*, the geek image is hardly a portrait of white male superiority.

Indeed, the more we examine it, the more nerd identity seems less a threatening gatekeeper than a potential paradox that might allow greater amounts of gender and race diversity into the potent locations of technoscience, if only we could better understand it. Of course, to the extent that geekdom fails to create such barriers—to the extent that it allows women and underrepresented minorities to fully participate in technoscience without being nerds—one can simply ignore it. But what happens when we fuse the ostensibly white male subculture of nerds with its race and sex opposites? To what extent might nerd identity become one of the *fracta* that can help open the gates?[3]

THE NERD IN HISTORICAL PERSPECTIVE

A good history of the American nerd has yet to be written, but its starting point might be in the radio amateurs of the early twentieth century, starting with teenage "wireless clubs" in the 1920s.[4] In an interview with Mark Dery [5, p. 192] science fiction (SF) writer Samuel Delaney notes this connection: "The period from the twenties through the sixties that supplies most of those SF images was a time when there was always a bright sixteen- or seventeen-year old around who could fix your broken radio. … He'd been building his own crystal radios and winding his own coils since he was nine. … And, yes, he was about eighty-five percent white."

These (predominantly) young white males were, however, distinctly lower in class status than the figure of the intellectual or "egghead" of the same period. A good illustration of the distinction can be seen in the historical drama *Quiz Show*. In this film about a television game show in the 1950s, upper-class WASP Charles Van Doren beats geeky, working-class Jew Herbie Stempel, to the great relief of the quiz show staff: "At least now we got ourselves a real egghead, and not a freak." The implication is that Stempel's nerd challenge threatens both race and class boundaries for intellectual status.

After World War II the broad category of "electronic hobbyist" fused ham radio operators with dimestore science fiction, model trains, stereophonic sound, and mail-order kits. The Cold War era emphasis on science education (as well as veterans' education funding) drove these hobbyists and their more scholarly counterparts closer together. While the wholesome image of a Boy Scout merit badge in chemistry underscores the normative side of these postwar nerds, there was always the danger of their attachments to categories of the artificial or unnatural. In the 1955 film *Rebel without a Cause*, Sal Mineo's character, John Crawford, gives us one of the first screen appearances of the nerd. Nicknamed "Plato" for his bookish habits, he rides a scooter rather than a motorcycle and is seen at one point primping his hair before a photo of screen star Alan Ladd. A loner who lacks the tough demeanor exhibited by his male classmates, he appears to have a crush on the film's protagonist, James Dean. Plato's implied homosexuality is a warning for future generations of would-be geeks. Nerd identity will come at a price, threatening the masculinity of its male participants.

In the cultural logic of late-twentieth-century America, masculinity bears a particular relation to technology. Being a "real man" is to claim one's physiology in muscle and testosterone; male-associated technologies tend to involve physical labor (lawnmowers and power drills), subduing nature through force (trucks and tractors), and physical violence (tanks and guns). More masculine technologies tend to be seen as concrete, massive, and having direct physical effects. The more abstract artifice of science does not seem nearly so testosterone-drenched; it is easy to see how the artificial spaces of mathematics and computing can be framed in opposition to manly identity. Thus the opposition between the more abstract technologies and normative masculinity keep nerd identity in its niche of diminished sexual presence.[5] How does this normative gender dichotomy compare to similar contrasts in racial identity?

In his analysis on the history of race in biology, Gould [6, pp. 401–12] notes that although the racial categories proposed by Linnaeus in 1758 were based only on geographic distinctions, in 1776 German naturalist J. F. Blumenbach extended the Linnaean categories to form an evolutionary framework: two lines of "degeneration" from an original "Caucasian" (a term he coined for the supposed origin near Mt. Caucasus) to Asians and Africans. Ironically, Blumenbach was motivated by his conviction in the unity of human beings—he opposed the claim for separate origins of humans on different continents—but that did not stop succeeding generations of racist scholars from using his work for their claims of an evolutionary hierarchy (and thus a hierarchy of genetically determined intelligence). Blumenbach's categories were quickly collapsed into a single ladder of evolutionary "advancement," with Africans at the bottom, Asians in the middle, and whites on top. In the postmodern era we have seen a return to Blumenbach's dichotomy; the best publicized have been *The Bell Curve* by Murray and Herrnstein [7] and the pseudoscience of Phillip Rushton [8]. Much like Emily Martin's [9] analysis of flexibility in postmodern representations of the immune system, these examples of postmodern racism are also marked by a flexible designation of particular characteristics: orientalism and primitivism.

Primitivist racism operates by making a group of people too concrete and thus "closer to nature"—not really a culture at all but rather beings of uncontrolled emotion and direct bodily sensation, rooted in the soil of sensuality. Orientalist racism operates by making a group of people too abstract and thus "arabesque"—not really a natural human but one who is devoid of emotion, caring only for money or an inscrutable spiritual transcendence.[6] Thus exists the racist stereotype of Africans as oversexual and Asians as undersexual, with "whiteness" portrayed as the perfect balance between these two extremes. Given these associations, it is no coincidence that many Americans have a stereotype of Asians as nerds and of African Americans as anti-nerd hipsters. Pop musician Brian Eno, for example, starkly states this race/geek alignment in a *Wired* magazine interview: "Do you know what a nerd is? A nerd is a human being without enough Africa in him" [10, p. 149]. But what does it mean to use nerd identity as the grounds for contesting these links between race, sex, and technology? The following four examples of black nerds illuminate some of the possibilities for dislocating (or at least broadening) these narrow normative roles in the ecology of race and technoculture.

> Thus exists the racist stereotype of Africans as oversexual and Asians as under-sexual, with "whiteness" portrayed as the perfect balance between these two extremes. Given these associations, it is no coincidence that many Americans have a stereotype of Asians as nerds and of African Americans as anti-nerd hipsters.

AFRICAN AMERICAN EXEMPLARS

Let's begin with the personal style invoked by Malcolm X. At first nothing seems more incongruous than associating a founding father of black nationalism with pimple-faced computer geeks.

But Malcolm's horned-rimmed glasses and insistent intellectualism recall the earlier figure of the egghead—not quite a nerd, but only because he needed to challenge the class restrictions as much as the mental stereotypes (in other words, challenging Herbie Stempel would not be nearly as powerful as taking on Charles Van Doren). In the section of his autobiography covering his dramatic self-education in prison, Malcolm repeatedly attributes all credit to Allah, his messenger Elijah Muhammad, and his struggle for black identity. Yet the most overtly eggheaded example in his autobiography is his passion for the debate over the identity of Shakespeare: "No color involved here; I just got intrigued over the Shakespearean dilemma" (1992, 213) [11].

While the Shakespeare example proves Malcolm's cultural intellectualism, his persistent references to mathematics provide a kind of underlying nerd power: "I've often reflected upon such black veteran numbers men as West Indian Archie. If they had lived in another kind of society, their exceptional mathematical talents might have been better used" (135). "When [Jackie Robinson] played ... no game ended without my refiguring his average up through his last turn at bat" (179). "Allah taught me mathematics" (quoting Elijah Muhammad, 237). "[The University of Islam] had adult classes which taught, among other things, mathematics" (240). And in a television interview, his explanation for the new surname: "X stands for the unknown, as in mathematics." By invoking the abstract rationality of math, Malcolm stood in shocking contrast to primitivist expectations of white America.

Taking Malcolm's oppositional equation to a logical extreme, in January 1996 African American computer wiz Anita Brown launched the Web site Black Geeks Online.[7] Dedicated to "bridging the widening gap between technology haves and have-nots," she explains the aims of this community service organization in the following introductory passage:

> Why? Our experience indicates that from South Central to South Jersey computing is a hard sell in "the hood." Unlike baggy pants, hip-hop music and drugs, Information Technology (IT) is rarely marketed to African Americans. Black "geeks" rarely appear in media ads; there are few (if any) hardware and software ads in *Emerge, Essence, Vibe, The Source, Black Enterprise*; and the "nerd" and "geek" images associated with computer professionals are still considered "uncool."

Even in outer space futures and alien landscapes, white access to technocultural identity remains supreme.

Brown's "uncool" assertion is certainly supported by what is probably the best-known public figure of the black nerd, Jaleel White's Steve Urkel from the television sitcom *Family Matters*.[8] Urkel was originally written into the show merely as a guest for one episode, but he quickly became the most popular character in the show. The winning combination of Urkel's uncool persona and black racial identity was partly due to White's own comedic genius, but his appeal also derives from a combination of popular American fascinations: on the one hand, opposing the myth of biological determinism, on the other, continuing the myth of Horatio Alger, who in this case must pull himself up not the financial ladder but the social status rungs of youth subculture.

While Urkle's geek persona is a signature, other technology-associated black television figures remain less nerd-identified. Consider, for example, the black characters on various iterations of the *Star Trek* series, such as communications officer Lieutenant Uhura (of the original series), chief engineer Geordi La Forge, chief of security Lieutenant Worf, and Whoopi Goldberg's Guinan (on *Star Trek: The Next Generation*), Captain Benjamin Sisko (of *Star Trek: Deep Space Nine*), Vulcan security officer Officer Tuvok (of *Star Trek: Voyager*), and Travis Mayweather (of *Star Trek: Enterprise*). Out of a total of seven, only two—LeVar Burton's Geordi LaForge and Tim Russ's Tuvok—really qualify as nerds, and neither of them compares with the extraordinary geekiness of the teenaged

Wesley Crusher from *Star Trek: The Next Generation*.[9] Such limitations for black nerds can be illuminated through a comparison of the first series' Vulcan, Mr. Spock, with *Voyager*'s black Vulcan, Tuvok. Leonard Nimoy's Jewish identity readily orientalized Spock,[10] and as a result, Tuvok comes off as a kind of alien Tiger Woods: less nerdish than Spock since he is a security officer rather than a science officer (thus implying that black Vulcans are more physical or athletic). Even in outer space futures and alien landscapes, white access to technocultural identity remains supreme.

The career of African American actor Samuel L. Jackson also illuminates the figure of the black nerd in popular media. During the 1980s Jackson played a series of drug dealers and junkies,[11] but his increasing popularity allowed him greater control over his roles. As a result, he quickly switched to playing black nerds, including a computer hacker in *Jurassic Park*, a Pulitzer Prize–winning writer in *Amos and Andrew*, and a mathematical prodigy in *Sphere*. His role in *Sphere* is particularly illuminating in light of work by ethnographers of scientific culture such as Sharon Traweek.

Traweek [12] describes an event in which a graduate student of physics repeatedly stuffed bread into his mouth at a restaurant. Rather than discouraging these poor manners, his professors were delighted, calling the waiter to bring more bread. This and similar scenarios brought Traweek to the realization that the ability to "ignore the social" (and thus express one's dedication to the asocial, universal realm of physics) is considered to be a sign of a good physicist.

Similarly, Jackson's mathematics nerd in *Sphere* is so socially unaware that he unwittingly causes the vessel to run aground (while he is immersed in his favorite science fiction, Jules Verne's *2000 Leagues under the Sea*) with blissful ignorance. Jackson's own real-life dedication to the sci-fi genre is not trivial: after confessing his geek love for the *Star Wars* films to producer George Lucas, he achieved the ultimate nerd fantasy of playing a Jedi knight—Mace Windu—a role that originally called for a white actor.

PROMISES AND PROBLEMS IN STRATEGIES OF REVERSAL

What can we conclude about the oppositional possibilities for the figure of the black nerd? Even if it was only in the world of fantasy, Jackson's agency in changing the racial composition of the Council of Jedi Knights was a hard-won victory. As Anita Brown of Black Geeks Online maintains, the contradiction between the cool of African American identity and the uncool of nerds is no coincidence; it is precisely this racialized intersection of technology and personal identity that functions as a selective gateway to technosocial power. There are, of course, limits to this strategy of technocultural identity reversal. We might, for example, focus on the ways in which hegemonic whiteness allows itself to be defined as an unmarked signifier and thus can affirm its own identity through asocial or antisocial behavior, while blackness depends on an explicitly social identity (for example, if Traweek's geek grad student had followed proper decorum, or if Jackson's mathematician in *Sphere* had been obsessively reading Malcolm X, neither would properly perform as nerds). But such limits are best understood not as specific to African Americans but as a general problem in resistance to hegemonic norms. In order to understand the more general problematic, let's see how such reversals operate for other racial groups, such as Asian Americans, and other social categories, such as gender.

The compulsory cool of black culture is mirrored by a compulsory nerdiness for orientalized others such as Middle Eastern groups, groups from India, and Asian Americans. Just as the black nerd fuses the desexualized geek with a racial identity stereotyped as hypersexual, Asian American hip hop allows racial groups stereotyped as desexualized nerds to fuse with the hypersexual funk of rap music. Oliver Wang's [13] superb analysis of Asian American hip hop (1999) points to the oppositional power of these Korean American Seoul Brothers and Chinese American homies; he notes that their work helps to expose some of the realities of struggling Asian immigrants in America. But Wang's analysis runs the danger of turning Asian American hip hop into a narrative of sameness; his argument could be read as saying that Asian American youth and black youth perform hip hop because both encounter similar challenges. Drawing such a conclusion would miss some of the ways that the local contexts of these two varieties of hip hop work in opposite directions. While African American hip hop affirms a kind of unapologetically stereotyped identity (which, as Rose [14] points out, works as a mode of resistance when the refusal to apologize for "keepin' it real" is linked to demands for broader structural change), Asian American hip hop seeks to challenge comparable stereotypes of Asian American identity. Asian American hip hop is useful not because it embraces previously disparaged attributes, but because it questions what were previously the cherished attributes of America's "model minority"—not affirming negritude, but negating nerditude.

Similarly, female exclusion from the male domain of technology is mediated by the opposition between nerd sexual formations, which focus desire into male antisocial forms, and female youth gender formations, which emphasize strong sociality. Wakeford [15] makes this point in her analysis of gender in Web site constructions. Focusing on sites such as GeekGirls and NerdGrrrls, Wakeford critiques the easy assumption that sexism is rampant throughout the Web yet makes clear the motivations for creating these hybrid technogender identities. She suggests that "the words themselves are codes to explicitly subvert the easy appropriation of women, and to resist stereotypes" (60). These stereotypes are both external—mainstream sexist portraits of women as unable or unwilling to engage with computer technology at the level of personal identity—and internal—stereotypes from what GeekGirl creator RosieX calls "an older style feminist rhetoric which tended to homogenize all women" (60). Similarly, the triple "r" in NerdGrrrls signifies an alliance to the punk-feminist bands (such as Riot Grrrls) that produce a break with humanist or romantic strands of feminism while calling for new forms of gender identity and affinity. Just as Black Geeks Online was battling against both external racism and the internal affirmation of essentialist concepts (essentialism that forced an opposition between black identity and technological prowess), these grrrl geeks vow dual oppositional use of their technocultural identity.

The problem with this line of resistance is that, in the words of Donna Haraway, it is never enough to "simply reverse the semiotic values."[12] Despite their identity violations, these figures of technological and cultural hybridity often reproduce the very boundaries they attempt to overcome: not surprising since they are focused on attaching the "wrong" race to the "right" identity. While the figure of the black nerd contradicts the normative opposition between African American identity and technology, it does so only by affirming the uncool attributes of technological expertise. The consequences can be tragic for the many African American students and teachers whose interest and identification with science and technology lead to accusations that they are "acting white." This phenomenon is sometimes referred to as "peer proofing" by education researchers [16]. But the public reaction to such reports is often problematic, implying that the need for change is purely internal to the black community,[13] rather than seeing a need to challenge the ways in which nerd identity itself is constituted or to loosen the geek grip on technoscience access.

Afrofuturists blur the distinctions between the alien mothership and Mother Africa, the middle passage of the black Atlantic and the musical passages of the black electronic, the mojo hand and the mouse.

THE AFROFUTURIST ALTERNATIVE

It is for this reason that we see the turn to Afrofuturism. Rather than merely reverse the stereotypes, the Afrofuturists have attempted to forge a new identity that puts black cultural origins in categories of the artificial as much as in those of the natural. Afrofuturists blur the distinctions between the alien mothership and Mother Africa, the middle passage of the black Atlantic and the musical passages of the black electronic, the mojo hand and the mouse. Categories like "black nerd" lean too heavily on the crutch of universalism; they assume that nerd identity is only racially aligned by a kind of shallow, arbitrary association and is otherwise universally available. Afrofuturism, in contrast, challenges both the implicit whiteness of nerds and the explicit technological absence of both realist and romantic black essentialisms.

That is not to say there is an absence of oppositional power in the reversal strategies; those who pioneered categories like "black nerd" or "Geek Girl"—Anita Brown, RosieX, the Seoul Brothers, and their fellow travelers—are my heroes. Nor should we have utopian illusions about Afrofuturism; it is fraught with problems stemming from its derivative relations to the original futurist movement, the elitism of academic influence, and, most problematically, its preference for artistic and literary approaches over science and technology, economics, politics, and other disciplines. But its ability to disrupt and redefine the boundaries of technocultural identity—the putative opposition between blackness and technology—rather than merely relocate the figures that inhabit them is important and controversial. Take, for example, the following discussion from the AfroFuturist listserv concerning DJ Spooky:

> I've heard more about who Spooky is than people playing his music ... I never hear about how great the music is ... just that he's a nice guy. ... Spooky has always seemed to me to be an over-intellectual nerd draped in hip drag ... sort of like the "Junior" ("My Mama Used to Say") of electronica without the preppy clothes. ... I'll be through to see if somethin' new is goin' down with Spooky next week on 18 January at Joe's Pub ... maybe the cat'll put a foot in my grill with his power. ... I hope so.

Even in the context of Afrofuturism, the figure of the nerd continues to haunt us.

CONCLUSION

Primitivist racism and orientalist racism maintain their power through mutually reinforcing constructions of masculinity, femininity, and technological prowess; yet mere reversal is never sufficient as an oppositional strategy. *Nerd* is still used in the pejorative sense; its routes to science and technology access are still guarded by the unmarked signifiers of whiteness and male gender. Groups such as the Afrofuturists seek alternative routes to circumvent the technocultural gateways of the geek. Black nerds, Asian hipsters, and geek grrrls both succeed and fail in challenging these boundaries, showing the limits of social transgression and the promise of reconfigured technocultural identity.

ENDNOTES

1. A discussion on the origin of *nerd* ran on the Humanist listserv in May 1990. Although the Oxford English Dictionary cites *If I Ran the Zoo* by Dr. Seuss [17] as the earliest written occurrence ("And then, just to show them, I'll sail to Ka-Troo and Bring Back an It-Kutch, a Preep and a Proo, a Nerkle, a Nerd, and a Seersucker, too!"), the earliest use in its contemporary sense was cited from student-produced burlesque at Swarthmore College in 1960. The term was not in common usage until the 1970s, when it became a stock phrase on the television show *Happy Days*.
2. I will be using the terms *geek* and *nerd* interchangeably here only for the sake of reducing repetition. The amount of writing devoted to making this distinction is surprising (see Ref. [18]). Coupland's [19] *Microserfs* offers several comparisons; perhaps the most illuminating is that "a geek is a nerd who knows that he is one."
3. Derrida [20] (278) introduces the concept of rupture or disruption as an unacknowledged contradiction in what appear to be seamless structures of modernity. Lyotard [21] (60), referring to these as "fracta" (from Mandelbrot's fractal geometry) more explicitly links such epistemological fissures to beneficial social change and recommends their study through an interdisciplinary "paralogy." While Lyotard's account comes dangerously close to implying that fracta *automatically* lead to a more democractic society, I would

agree with Ernesto Laclau and Chantal Mouffe's [22] assessment that such "dispersions" or "unfixity" only represent opportunities, not guarantees, toward the praxis of radical democratic politics (in the case of this essay, toward a more democratic technoscience).

4. Bass [23], for example, notes the obsession with home-brewed radio among two generations of physics students, and Stone 1995 [24] cites the crystal radio as an epiphany in geek self-construction. The popular electronics company Radio Shack still bears this legacy. Smith and Clancey [25] provides several essays on these "hobbyist worlds."

5. This association is illuminated by the tension it creates in the face of the rising economic value of information technologies; how can corporations mass-market products that are culturally associated with wimps and geeks? The film industry's answer is often to adopt an elaborate apparatus that replaces the keyboard and mouse with impressive physical agility carried out in a virtual reality interface: Michael Douglas in *Disclosure*, Keanu Reeves in *Johnny Mnemonic*, and Matt Frewer in *Lawnmower Man*. Another strategy is providing contexts that try to link information technology with sexual undercurrents; thus the recent spate of television commercials in which potential lovers are in a physically proximate space (a loud rock concert, a sudden cloudburst) but have relations mediated by a gadget they just bought. See Cockburn and Ormrod [26] for more general discussion.

6. The foundational use of *orientalism* comes from Said [27], but his definition is more concerned with a Western dichotomy of self/other than the contrast to primitivism used here. For other such contrasting examples, see Gilman [28] on the orientalist/primitivist contrasts in conceptions of the body and Campbell [29], 60 on differences in the "primitivizing" and "orientalizing" rhetoric of various narcotics discourse (for example, marijuana versus opium). See Chinn [30] for a more general discussion in relation to technology.

7. Brown's many achievements range from fashion entrepreneurship to national Web awards. See www.blackgeeks.net for more information.

8. The show ran from 1989 to 1998. A top-ratings performer as part of ABC's Friday family night, the series moved to CBS in its last season.

9. Admittedly, Wil Wheaton's character would be hard to beat; in a recent interview the actor himself admitted: "I consider myself to be really nerdy. I like things that are traditionally nerdy, like role playing games. ... I consider myself a geeky person and I revel in it. Geek pride and all those things" (see www.aint-it-cool-news.com/display.cgi?id=6627). But the racial roles for *Star Trek* characters have been disappointingly limited; consider, for example, the ways in which Uhura's duties were suspiciously close to those of a secretary. See Bernardi [31] for a detailed survey.

10. For example, the Vulcan four-fingered "live long and prosper" salute was an impromptu adoption from Nimoy's childhood experience watching the *kohanim* give the hand gesture for *Shin* (first letter of *Shaddai*) at synagogue services.

11. For example, he played "Gang Member No. 2" in *Ragtime* [32], "Hold-Up Man" in *Coming to America* [33], and a crack addict in *Jungle Fever* [34]. Jackson recently commented: "We've been given a lot of stock roles over the years. The pimp is one of them, the drug addict another. Criminals, bank robbers, rapists. .... When you get those roles, people will ask, 'Why did you take a role like that?' Well, number one, I needed the job" (see www .moviemaker.com/issues/21/jackson/21_jackson.htm).

12. This quotation [35, p. 162] refers to a postcard that reversed the King Kong/Fay Wray relationship: it shows a gigantic blonde woman reaching in through a skyscraper and snatching a terrified gorilla from its bed. Haraway remarks that such reversal always fails; later in her text she notes the same failure for feminist evolutionary theories that attempt to establish a primeval matriarchy in human origins. Her broader point is that hegemony is too much a world-making enterprise to be undone by a simple act of reversal; such acts can become part of, but never fully constitute, the path toward more just and sustainable futures.

13. Similarly, the Asian American community gets blamed for generating the need for stereotype contradictions. A *Time* magazine article titled "Kicking the Nerd Syndrome" concludes: "The fact that the best and brightest among Asian Americans are veering away from programmed patterns of success may be, in fact, another sign that the over-achievers are settling into the mainstream" [36, p. 66].

REFERENCES

1. Turkle, Sherry. 1984. *The second self: Computers and the human spirit.* New York: Simon and Schuster.
2. Noble, David F. 1992. *A world without women: The Christian clerical culture of Western science.* New York: Knopf.
3. Amsden, A., and J. Clark. 1999. Software entrepreneurship among the urban poor: Could Bill Gates have succeeded if he were black? ... or impoverished? In *High technology and low-income communities: Prospects for the positive use of advanced information technology,* edited by J. Mazzeo. 1999. *Trends in academic progress: Three decades of student performance.* Washington, D.C.: NAEP.
4. Campbell, Hombo, and Mazzeo 1999
5. Dery, Mark. 1994. Black to the future: Interviews with Samuel R. Delany, Greg Tate, and Tricia Rose. In *Flame wars: The discourse of cyberculture,* edited by Mark Dery. Durham, N.C.: Duke University Press.
6. Gould, Stephen Jay. 1996. *The mismeasure of man.* New York: Norton.
7. Herrnstein, Richard J., and Charles Murray. 1991. *The bell curve: Intelligence and class structure in American life.* New York: Free Press.
8. Rushton, J. Philippe. 1995. *Race, evolution, and behavior.* New Brunswick, N.J.: Transaction.
9. Martin, Emily. 1994. *Flexible bodies: Tracking immunity in American culture from the days of polio to the age of AIDS.* Boston: Beacon.
10. Kelly, Kevin. 1995. Eno: Gossip is philosophy. *Wired* (May): 145–58.
11. Malcolm X and Alex Haley. 1992. *The autobiography of Malcolm X.* Reprint, New York: Ballentine.
12. Traweek, S. 1982. Uptime, downtime, space-time, and power. Ph.D. diss., University of California at Santa Cruz.
13. Wang, Oliver. 1998. Asian Americans and hip-hop. *Asian Week,* 12–28 November. www.asianweek.com/111298/coverstory.html.
14. Rose, Tricia. 1994. *Black noise: Rap music and black culture in contemporary America.* Hanover, N.H.: Wesleyan University Press.
15. Wakeford, Nina. 1997. Networking women and grrrls with information/communication technology: Surfing tales of the World Wide Web. In *Processed lives: Gender and technology in everyday life,* edited by Jennifer Terry and Melodie Calvert. New York: Routledge.
16. Fordham, Signithia. 1991. Peer-proofing academic competition among black adolescents: "Acting white" black American style. In *Multicultural education and empowerment,* edited by Christine Sleeter. Albany: SUNY Press.
17. Seuss, Dr. 1950. *If I ran the zoo.* New York: Random House.
18. Katz, Jon. 1997. Geek backtalk, part II: Geek? Nerd? Huh? *HotWired,* 1 August. hotwired.lycos.com/synapse/katz/97/30/katz4a_text.html.
19. Coupland, Douglas. 1996. *Microserfs.* New York: HarperCollins.
20. Derrida, Jacques. 1978. *Writing and difference.* Translated by Alan Bass. Chicago: University of Chicago Press.
21. Lyotard, Jean-François. 1984. *The postmodern condition: A report on knowledge.* Minneapolis: University of Minnesota Press.
22. Laclau, Ernesto, and Chantal Mouffe. 1985. *Hegemony and socialist strategy: Towards a radical democratic politics.* London: Verso.
23. Bass, 1985
24. Stone, Allucquère R. 1995. Sex and death among the disembodied: VR, cyberspace, and the nature of academic discourse. In *Cultures of computing,* edited by Susan Leigh Star. Chicago: University of Chicago Press.
25. Smith, Merritt Roe, and Gregory Clancey. 1998. *Major problems in the history of American technology: Documents and essays.* Boston: Houghton Mifflin.
26. Cockburn, C., and S. Ormrod. 1993. *Gender and technology in the making.* Thousand Oaks, Calif.: Sage.

27. Said, Edward. 1979. *Orientalism*. New York: Vintage.
28. Gilman, S. L. 1999. *Making the body beautiful: A cultural history of aesthetic surgery*. Princeton, N.J.: Princeton University Press.
29. Campbell, Nancy D. 2000. Using women: Gender, drug policy, and social justice. New York: Routledge.
30. Chinn, S. E. 2000. Technology and the logic of American racism: A cultural history of the body as evidence. New York: Continuum.
31. Bernardi, 1998
32. Ragtime (1981)
33. Coming to America (1988)
34. Jungle Fever (1991)
35. Haraway. 1989. Primate visions: Gender, race, and nature in the world of science. New York: Routledge.
36. Allis, Sam. 1991. Kicking the nerd syndrome. *Time*, 25 March, 65–66.

7 Can Real Women Be Imaginary Engineers?

Rania Milleron

Negligible increases have been observed in the number of women undergraduates from engineering programs in twenty first century United States. For comparison, physical or biological science programs graduate at least double the percentage of undergraduate women in engineering. Those with a birds-eye look at engineering, very often advocate for increasing the percentage of women engineers as a way of heightening the competitive edge. Data from the Programme for International Student Assessment (PISA) has begun to document that girls and boys have similar cognitive abilities in mathematics and science assessments within their respective countries.[1]

The question of how to increase the number of women in the engineering workforce has been discussed for decades. One solution is often to encourage girls to become engineers. For example, the feature film "Dream Big: Engineering Our World,"[2] sponsored by the American Civil Engineering Society, profiles several women who are making impressive contributions in their fields of engineering. Like many other previous efforts, the movie encourages girls to choose engineering and the featured women engineers serve as role models. A ubiquitous part of the discussion has been the belief that what it takes to train women engineers is to find ways to stimulate their interest in engineering. On the surface this common sense idea seems tenable. However, in the twentieth century, programs that encourage girls to choose engineering did not produce the desired results. While there have been successes on the individual level, the overall numbers point to the need for multiple solutions that will broaden the engineering identity.

One of the most thought provoking research articles explaining women's continued underrepresentation in engineering is Cindy Foor and Susan Walden's "Imaginary Engineering"[3] (2009). If we want to produce more female engineers, then we need to take a more integrative approach. "Even after decades of attempts at opening the borders of engineering to women, young engineers still perceive it as a hostile and transgressive place for women to occupy."[4] Foor and Walden conducted a study at the University of Oklahoma (OU) to explain why more women are attracted to the field of Industrial Engineering (IE) as compared to other engineering disciplines at OU. The author's title, "Imaginary Engineering" is the moniker given to the field of IE, perhaps, due to the larger proportion of women students. An integrative approach requires not only encouraging girls to become engineers but also changing the current culture within engineering. "Imaginary Engineering" explains why the "leaky pipeline" metaphor does not adequately account for the low numbers of women engineers in the US. This innovative piece of research demonstrates how the boundaries created by cultural norms within the fields of engineering limit participation by women.

Foor and Walden skillfully investigate how boundaries not only affect women but also impact their male counterparts. Women who serve as role models help draw more women into the field when "female students understand that they can pursue engineering without the threat of committing gender inauthentication."[5] These women professors are role models who are on the one hand actively engaged as professionals and on the other hand have an active personal role with their family while avoiding being labeled as "manly," "butch," or "weird." However women today remain "faced with negotiating both an educational and life experience within two competing discourses: 'Engineering is Men's Work' but 'Women can (and must) do Engineering.'"[6] Cultural boundaries also impact male engineers but they have been able to break through boundaries by developing new constructions of masculinity in contrast to dominant norms such as the "geek."[7] The male engineer sees the field of industrial engineering as a place where he can be a "cool guy" who can hone his social skills and be a leader because "all engineers are gonna end up working for industrial engineers."[8] In this way the male engineer reclaims engineering status and distinguishes himself from the feminization of IE.

The case study of IE shows how "the discourse of "natural," heteronormative gendered prescriptions for women and men will continue to serve as gatekeepers to female inclusion to other engineering disciplines."[9] Underrepresentation of women in engineering is explained, in large part, by the persistence of gender stereotypes.[10] Therefore, professional engineer's awareness of "these mechanisms of constructing and negotiating identities is imperative for designing programs to enable and encourage all potential participants to maneuver within the competing discourses of engineering."[11] All engineers would benefit from reading and thinking about a "Re-imagined Engineering."

ENDNOTES

1. PISA tests the knowledge of 0.5 million 15 year-old students representing 28 million in 72 countries every three years. The results of the 2015 assessment were published in December 2016.
2. MacGillivray, S (Producer), & MacGillivray, G. (Director). (2017). *Dream Big: Engineering Our World* [Motion Picture]. U.S.A.: MacGillivray Freeman Films.
3. Cindy E. Foor and Susan E. Walden. "'Imaginary Engineering' or 'Re-imagined Engineering': Negotiating Gendered Identities in the Borderland of a College of Engineering." NWSA Journal, Vol 21 No. 2 (Summer) (2009): 41–64.
4. Ibid. p. 52.
5. Ibid. p. 53.
6. Ibid. p. 42.
7. Ibid. p. 55.
8. Ibid. p. 56.
9. Ibid. p. 57.
10. Cracking the Code: Girls and Women's Education in Science, Technology, Engineering, and Mathematics. UNESCO. 2017.
11. Cindy E. and Susan E. Walden. "'Imaginary Engineering' or 'Re-imagined Engineering': Negotiating Gendered Identities in the Borderland of a College of Engineering." NWSA Journal, Vol 21 No. 2 (Summer) (2009): 41–64.

8 Discipline and Play
The Art of Engineering

David P. Billington

NEH Chairman Bruce Cole spoke recently with David Billington about the beauty, economy, and innovation of structural art. Billington is professor of civil engineering and operations research at Princeton University and has written several books, including *Thin-Shelled Concrete Structures* and *Robert Maillart: The Art of Reinforced Concrete*.

Bruce Cole: You teach a very popular engineering course at Princeton for nonengineers. Why is it important, say for humanities students, to study engineering, and vice versa?

David Billington: I teach, actually, three courses like this: one on structures, one on all engineering, and one on rivers. The first two are heavily enrolled as courses that satisfy a lab science for liberal arts students. I think part of the reason is because through engineering one can make science more accessible and more relevant.

The second idea is that we do live in what is often called a technological society, and it is important, therefore, that all students understand how our society got built and how it functions today. That's, I think, the basic idea of why humanities students should study it.

Why engineers should study the humanities is the corollary to that, because there is a deep connection between engineering and the humanities. The way we teach engineering in these courses is from three perspectives: what we call a scientific, a social, and a symbolic perspective. You can't really understand engineering if you don't understand the scientific perspective. That needs no explanation. But social perspective is the fact that you can't build the kinds of things we build unless you have a society that is organized in a way that permits and encourages it. The third thing is that engineering has a significant influence on the art and literature of its time. So we want to make engineers aware of that.

For example, in the art history department—since we make students in one of our engineering programs take at least two courses in that department—the art history people tell us that our engineering students perform better than many liberal arts students in their courses. That's partly because our students have already had our course that makes a connection between art and engineering and have had experience analyzing structures visually.

Cole: I think about how technology now impinges on almost every facet of our life and how engineering not only allows us to communicate better and live better but also shapes what we do. There's that old adage about how men make buildings and then buildings make men.

Billington: Well—we have to watch out for the gender statements there, but still—

Cole: That's right. Well let's put it this way—people make buildings. So engineering is very much a vital part of a liberal arts education, I think.

Billington: I believe it, yes.

Cole: Let me ask you a little bit about your background. What led you to engineering, and then what led you to what I think is an unusual approach, to the teaching of engineering and to the bridging of engineering and the liberal arts and humanities and sciences?

Billington: As an undergraduate at Princeton, I took a very strange engineering course which was sort of barely engineering. It was called Basic Engineering. I took it because it allowed me to take more liberal arts courses than any other engineering program and it didn't specialize in anything. So I took courses in art history, in music, and literature. Those were very important courses to me. Now I didn't make any connection between those and engineering at the time, but I did take them.

Then I won a Fulbright Scholarship to study in Europe, and studied there for two years. That was in a way, my engineering education, because I learned structural engineering there. I came back and practiced for eight years and was completely unaware of any kind of tradition that structural engineering at its best would be an art form.

I began to teach in 1960, and because I had worked with architects in practice, I was invited to give a course to architects—graduate students of architecture—even though I was a full-time civil engineering professor.

The architects hated what I was teaching because it was all stick diagrams and formulas and very dry and dull and they despised it. They would bring me pictures of what they considered to be beautiful things—bridges and buildings—and they'd say, "Why can't we study these?" It seemed to me like a reasonable question, but I found nothing in the engineering literature about these things.

There was nothing in the teaching literature of engineering that had any aesthetic appeal at all. Therefore, I began—and I began with the National Endowment of Humanities because Princeton's president Robert Goheen and professor Whitney Oates told me about it. They said, "The Endowment should be interested in this." So with a colleague, Robert Mark, who was also confronted with the same questions, I went down there and got an early grant that got us started.

The idea was to study structures that were beautiful and see if they were good engineering. That was the challenge for us. We agreed to split history up into two parts. Mark studied things before the Industrial Revolution and I studied things afterwards. The net result was that I came to the conclusion that there was this tradition of structural art that began in the Industrial Revolution in Britain. The tradition we call structural art continues and flourishes today but was unknown—almost literally unknown—in educational circles in this country.

So that impelled me to introduce a course in 1974—again, with help from the National Endowment, and now getting the National Science Foundation a little bit interested, too. That course proved immediately to be popular because it was so visually oriented.

Cole: I'm very proud that the NEH had a role.
Billington: Yes, it did. I hope it has a continuing role now.
Cole: We started talking about the Industrial Revolution.
Billington: Yes.
Cole: What changes in engineering, major changes in engineering, came out of the Industrial Revolution?
Billington: The simple answer to that is everything.
Cole: Everything?
Billington: Everything. Everything we study now, everything we do in engineering, none of it existed before the Industrial Revolution. Nothing. It's all new. That's why the Industrial Revolution is the greatest event in modern history.
For example, the two great early innovations were industrialized iron and efficient steam power. Those innovations did have antecedents. There always was iron, and there always was steam power. But they were so inefficient to produce you couldn't do anything with them until you had the great engineers—they were British—of the late eighteenth century, who changed things radically. So those things brought in a kind of a whole new world.
Cole: Is this true for concrete, as well? I'm thinking of the Roman Pantheon.

Billington: Sure, the Romans had concrete, but they didn't have reinforced concrete—

Cole: Right.

Billington: —and they used concrete in very—what we would today call—very primitive ways, very clever ways often, but very primitive. As you said, the Pantheon was concrete, but it was very, very heavy, and it cracked substantially. They learned from that, of course. But they were not essentially what we would call modern concrete designers at all. In fact, there was such a break that concrete had to be essentially reinvented, and the Pantheon had no influence on that.

Cole: Interesting. So this is what you're talking about, these antecedents that then really are transformed by the Industrial Revolution—

Billington: Transformed, yes. But then, of course, there are completely brand new things that come in—reinforced concrete and structural steel, cars and planes, electrical power and networks that send power out or send sound out, and the refining of steel and oil in completely new ways—that made it possible to use things in ways they never had been used for before.

Cole: You've got a new book coming out on engineering in the late nineteenth and early twentieth century.

Billington: Right. Right.

Cole: What's the title?

Billington: The book is cowritten with my son David Jr., who is an historian. We call it *Innovators II* because I had written a previous book called *The Innovators,* which dealt with the period up to the late nineteenth century. So the tentative title is *Innovators II,* but Princeton University Press doesn't like that much, so we're probably going to discuss that in the next couple of months.

Cole: The sequel.

Billington: It's definitely a sequel. It picks up where the other one left off.

Cole: So who are some of the great engineering/architecture heroes of the late nineteenth and early twentieth century?

Billington: The first one we deal with is Thomas Edison and then George Westinghouse. The next chapter deals with Alexander Graham Bell. These are pretty obvious choices. The third chapter deals with the oil industry, which involves, of course, Rockefeller as the entrepreneur, but the two engineers are William Burton and Eugene Houdry, names you probably never heard of. But they were fundamental people who changed the way oil was refined and made it a modern industry.

Then the next chapter deals with the automobile, and that's where we deal with Henry Ford, obviously, and Alfred P. Sloan Jr. Then the airplane, where we discuss the Wright brothers and Samuel Pierpoint Langley.

The following chapter is on the radio, where we begin with Marconi and then move to all the battles that ensued between Armstrong and Sarnoff and Deforest. After that comes steel, where we deal primarily with Othmar H. Ammann and the big steel bridges around New York. Then concrete, where we deal with two engineers who developed major innovations in concrete: Anton Tedesko and John Eastwood. Then, finally, we end up with a chapter on the 1930s, where we are preparing for the World's Fair. The bookends for the volume are the Centennial Fair of 1876 and the World's Fair of 1939. The last chapter deals with the so-called streamlined era, the art deco era of the 1930s, and we focus on Donald Douglas and the DC-3, and Walter Chrysler and the airflow car. So those are the figures we deal with.

Cole: It just struck me that almost everyone you mentioned came from the United States.

Billington: We are only focusing on the United States. That doesn't mean things didn't happen elsewhere.

Cole: Right. But these are really monumental inventions—I've forgotten who said that all around
        the world, almost every minute of every day, the vast majority of people's lives are
        made better by American innovation.

Billington: That's right. But the reason we had this focus is because we want to connect this to the
        history of the United States, and to politics and art in the United States.

Cole: Of course, that interests us here because, as you know, we have a new initiative called *We the*
        People, which is designed to improve the teaching and study of American history.

Billington: Yes. Right.

Cole: I know this is a very complicated answer, and I'm asking you to give it in sort of a short form,
        but why did this all happen in the United States? I mean what were the conditions that
        created this incredible flowering of such innovation.

Billington: The first thing, I think, is the fact that there was in this country, very quickly, a techno-
        logical infrastructure. Now we got it all from Britain because that's why Britain was the
        leader in the Industrial Revolution. It was because they had a technological infrastructure.

What I mean by that is that in Britain, all over that country, there were little machine shops. There
were people who were learning how to work. They first, of course, were working with wood, but they
began working with metal in the middle of the eighteenth century. By the end of the eighteenth century
there were just all kinds of machine shops. That meant that somebody such as James Watt, for example,
who was a great inventor and innovator, could connect with John Wilkinson, who was really, essentially, a machinist, who knew how to bore cannon and knew how to make, therefore, precise cylinders.

Now we got that from Britain. Most of our people in the late eighteenth century were British anyway, and we came with that tradition. We were suppressed by the English, by the British monarchy,
from actually having our own industries until the Revolution. But that didn't stop us from having
illicit ones and having a lot of technically sophisticated people. When independence came, there
was a flowering of this technological infrastructure. People were really interested in making things.
So that is, I think, the first prerequisite.

You remember the statement by de Tocqueville? When he came over here, he said he couldn't
understand it: there was nobody really interested in science, but they had created something that
shook the world. He was talking, of course, about the steamboat, America's first great innovation.
But the point was, what he saw was there wasn't much reflective study. It was all active study.

Along with the infrastructure, of course, comes an ethos of progress, you know all those clich's,
progress and a better living standard, of bringing up the lower middle classes and so forth, all those
things.

I don't pretend to be an expert on all that, but from an engineer's point of view, that's what I see.

Cole: That's very interesting. Is there a difference between invention and innovation?

Billington: Oh, sure, a huge difference. I mean there are millions of inventions in the patent office
        in America. Very, very few of them ever see the light of day.

I believe the proper definition for an invention is just an idea, an idea that you can show to the
patent office as somehow new, but an innovation is something which is—it often comes from an
invention, but not always—a new idea that becomes, in effect, practical.

Edison's electric light system is an example. It's one of the foremost examples of an innovation.
Or Bell's system of telephone is an innovation.

Those happen to be inventions, and—

Cole: It's an invention that became an innovation.

Billington: An innovation would be the George Washington Bridge. That was a huge innovation,
        but it wasn't an invention. In other words, there was no patent associated with that, but
        there was no bridge even faintly close to that in span.

Cole: When you're talking about that kind of practical, sort of seat-of-the-pants idea of innovation and invention and progress, isn't Edison a good example of that?

Billington: Yes, but I would never call it "seat-of-the-pants." This idea that Edison was a tinkerer or the Wright brothers were tinkerers is absolutely wrong. They were first-rate engineers. They operated the way the best engineers always operate—even today.

Cole: Then how did that idea of Edison arise?

Billington: It arises partly because he fabricated it himself. To have a kind of an image of this sort of, you know, rough-cut genius—the Edison method—a lot of industries talk about that, "Well, we just try one thing after another until we get an answer." Edison knew what he was doing always. Sure, he had to try a lot of things.

Nobody had ever found a solution before for the high-resistance filament, for example. What are you going to use for the filament so it doesn't burn itself up right away? He did have to try and err on a lot of things, but that was all done in a very short period of time. He came to the idea that he was working on in 1878, and he produced the working lamp in mid- or late 1879. It was less than a year. That's not a very long time, when he was doing a lot of other things, too.

And he had right next to him one of the best trained mathematical physicists in the country, Francis Upton, who was guiding him in the theoretical sense. He had some of the world's greatest machinists—John Kruesi and some of the others—to build things. So he was a smart guy. He ran a research and development operation that was first rate.

Cole: But this creation of the myth that he was this tinkerer was pervasive.

Billington: Yes. It's a myth that unfortunately persists today. People say, "Well, he wasn't really an engineer." Well, that's just wrong. I mean Fulton was a first-rate engineer, too. They operated like engineers. That is to say they based their work on calculations, but they also tried things out. They made experiments to make sure that it was going to work. And they read the literature and they studied what other people had done. They did everything that good engineers do.

Cole: You may not want to speculate on future innovations, but what do you think is next?

Billington: I'm not a futurist in any remote sense. The present is about as good as you can do. Edison never said, "Ah, this is what's going to happen in ten years." He was not interested in that at all. And had he been interested in it, I think he would have failed.

None of these people were interested in that. They were interested in making money and they were interested in establishing a business. Edison, in a particular example, once he started a company, he became completely uninterested. He left it and went back to his lab.

Tackling the problems that you see around you now and trying to solve them are both very important. And, of course, predicting what is going to find a market is also very important.

Cole: Are we in a technological revolution with the computer?

Billington: I don't think so. Throughout the period since the late eighteenth century people have claimed industrial revolutions. They say the second industrial revolution was the late nineteenth century.

Then, following World War II was another industrial revolution. Now they say there is another industrial revolution. There's change, but there is nothing at all like the change then. Just imagine the differences between 1780 and 1840.

You had no railroad. You had no telegraph. You even had no steamboat. The difference is monumental. We don't have that kind of change today.

When I was a young man, life was a certain way, with cars and planes and radio and even television. Those are still, today, the major things.

Look at the Fortune 500. Today's Fortune 500 are some of the same companies that began in the late nineteenth century. They haven't changed. They have changed some of the products they produce, but they are basically the same kinds of things. But they simply didn't exist before 1876. So those were when big changes happened.

Cole: Let me ask you a little bit about beauty and engineering. I think when people think about the study of engineering they don't really have beauty in mind. Could you digress on that a bit?

Billington: Well, based on the stimulus from the architecture graduate students, I began to study and I went to Europe. I went to Switzerland and met Robert Maillart's family, his daughter in particular. She was very welcoming and interested. So that's when I began that study.

That required French and German and I knew those two languages more or less. Photographs of Maillart's works struck me, and I went to see them then. They were all in Switzerland, and they were so striking that I decided to work on Maillart. So for the next twenty years I spent my time doing the research that was necessary, I thought, to create a written life and works of Maillart and a series of scholarly papers and books.

Now how do you determine beauty? Almost all engineers say, "Oh, well, beauty is in the eye of the beholder," "That's a relative thing," "It's not interesting to us because you can never define it, "Everybody has their own opinion," and all that stuff. The only answer that I could find was two-fold. One is the sense in which art, real art, becomes classic. People who spend their life with music don't argue about Mozart and Beethoven and Brahms. There is no longer any argument about them.

The same thing is true with painting, maybe not the painting that is done right now, but painting that has sort of matured in the art history world. So I used that classical argument by saying certain works of structure have become classical, such as the Eiffel Tower and Brooklyn Bridge. That was the first part of my argument.

The second part was to say, all right, I'll find out if these things are really beautiful or not by confronting our art museum with them and saying, "Is this worthy of exhibition in an art museum?" In 1972, we held the first exhibition of Maillart in the Princeton University Art Museum. That led to a whole series of exhibitions in the 1970s, culminating in two of them on contemporary Swiss engineers.

So that was a way of convincing other people that if it could be exposed in an art museum—you couldn't bring the big works in there, but if you could bring pictures and models—then the works had beauty.

In 2003, we put on a major exhibition in the art museum "The Art of Structural Design: A Swiss Legacy." Last fall it spent three months at MIT—that's the bastion of engineering, after all—trying to convince them, which I think we did, that aesthetics or art was a really important aspect of engineering. It is in Kansas City now; it will be in Zurich, and then it will go to Toronto, Grinnell College in Iowa, and Smith College in Massachusetts.

By doing that, I think, it was able to overcome the normal engineering objection to whether engineering itself could be art.

Cole: I see. I like your example of the Eiffel Tower, which to me is, in a way, pure engineering, but also one of the most beautiful structures.

I think of other things like the Pantheon, for instance, in Rome, and others. But do you think now, let's say in terms of architecture, that the idea of engineering—I'm thinking about skyscrapers—

Billington: Yes.

Cole: I'm thinking particularly about Mies van der Rohe—the idea of engineering, how a building works and how that is exposed by modern architecture. I mean that we are more conscious of that now and that is part of the aesthetic of modern architecture.

Billington: Let me make a very sharp distinction between structural art and architecture. Mies van der Rohe was stimulated, as all good architects are, by the culture that surrounded him. The Crown Building, you know, that building in Chicago—

Cole: Yes.

Billington: —It is, I would say, an example of a perfectly horrendous structure. It makes no sense whatsoever to a structural engineer.

But that's a different story. Illinois Institute of Technology has worked closely with SOM— Skidmore, Owings & Merrill—and there, that's the only really major example we have in this country where structural art has appeared in works that would normally be considered architecture.

But the Crown Building is not that. If you give me fifteen minutes I could explain why not. I have the original drawings of that building, and it's just all wrong. He's showing structure, for sure, but he's not expressing real structure.

Cole: That's fascinating.

Billington: That's a very big difference. Whereas the Hancock Tower, in Chicago, which gets the same opprobrium as the Eiffel Tower got, is a work of true structural art. That's because an engineer made the form—Fazlur Khan, one of the great engineers of our story. Bruce Graham, the architect, was willing to let Fazlur Khan make the form, and when Graham tried to change it, Khan got so angry that Graham stepped down and let him keep it.

So there are a few skyscrapers like that. Not many. Not many at all, but a few. They are mostly by Fazlur Khan. So he's a big, important figure in our story.

Cole: Interesting.

Billington: Don't misunderstand me. I'm not criticizing the architects now. I'm just saying it's very different.

Cole: Right.

Billington: And I'm not pretending to be an architectural critic, so—

Cole: I understand. Is this just sort of the triumph of the architect's aesthetic ideas over good engineering structure?

Billington: Yes. I'd say it's the feeling that many architects have that engineering structure is simply secondary. They make the form and then they call in the engineer and say, "Make it stand up." The trouble is that unless the engineer is really good—and that's very difficult to get a really good engineer to work with an architect like that—then the engineering is likely to be second-rate.

Cole: Well, what about the training of architects? Do they have a lot of engineering?

Billington: There are two types of training that architects get. In some schools they get a very solid training in engineering, while in others very little.

Cole: Let me ask you a couple of not too personal questions. Your brother is the Librarian of Congress. Were you at Princeton together?

Billington: Yes. Yes.

Cole: Did you come from an academic background?

Billington: Well, no. You can say our father should have been a college professor, but his father died young, and he had to go to work when he was a teenager to support his family. So he could never go to college, which he always regretted. He became a successful insurance broker but not an academic, as he should have been. My mother should have been an engineer—her father was an engineer—but in those days you didn't let women be engineers. It was definitely taboo in almost all circles.

Cole: I'd like to talk more about your work on the Swiss engineer, Maillart. You don't think that bridges are architecture? Because, of course, in the history of art, bridges are included.

Billington: No, sir. They are definitely not architecture. Maillart was not trained as an architect. He
          didn't practice in architecture. He didn't think like an architect. Those works are works
          of pure structural art. So it's fine for them to be included in the history of art. That's fine.
Cole: So tell me why they are not architecture.
Billington: Architects are educated to think in terms of space. Architects want to control spaces.
          Engineers are taught to control forces. Architects are taught to deal with constructions
          that are going to be intimately used by people.

Engineers are not taught that. Engineers' spaces or engineers' buildings or bridges are used
by industry or machines, but not intimately by people. Therefore, a bridge, except for very small
pedestrian bridges, are not essentially built for intimate use by people. The prototypical work of an
architect is a private house, and the prototypical work of an engineer is a public bridge.

As you move in between them, you get to industrial buildings where—and I worked on these
when I was in practice—we had architects, but all they were doing was shuffling around room
spaces. They had nothing to do with the structural form.

So bridges and certain kinds of buildings are the work of engineers, whereas schools and
churches and office buildings are the work of architects. They use engineers, but the engineers are
often secondary. Then you get to a middle group where you have very large buildings with long
spans or very high buildings. There's an opportunity—and that's what SOM did when Fazlur Khan
was alive—there's an opportunity for structural art.

But the structural art must be done collaboratively with architects. When you have a long span
roof, like covered stadiums, then there's an opportunity for the engineer to be the designer, although
there usually is an architect involved.

I don't know if you know the work of Pier Luigi Nervi in Italy, or Felix Candela in Mexico. Those
are works, buildings, some of them are even churches, where the engineer actually makes the form.
They become works of structural art.

Candela built hundreds of things. Some are thin shell concrete structures. And these are spectacular structures. Now he was trained as an architect, but he practiced as a builder and a structural
engineer. He says that Maillart greatly influenced him. If you ever go to Italy and you look at
the Little Sports Palace in Rome, designed by Nervi, you'll see what I mean. I think it's the most
spectacular interior space of the twentieth century. And it is designed entirely by an engineer. It's
inspired by being in Rome—you walk around Rome, you go to the Forum, you go to the Vatican
Museum, and you'll see ribbing patterns that Nervi used, but he's done it in reinforced concrete, and
he's done it in a unique way. He was also a contractor, a builder.

Cole: It's a good example of engineering as art, as great art.
Billington: That's right. And that's why we can put these things in our art museum. Christian
          Menn, who is practicing today, is the world's greatest bridge designer. He never works
          as an architect, and he would boggle if you called him an architect. Othmar Ammann,
          who designed all the New York bridges, was a great engineering artist. Actually, his
          bridges are so big that he did hire an architect to help him with some of the overpass
          details and approaches. But the bridges themselves, the basic forms, were done by him.
Cole: The Brooklyn Bridge has a romantic quality. Why do you think that is?
Billington: It's because of the way it was designed.

The most impressive feature of it is its central elevated walkway, which Roebling purposely put
there, and you can therefore walk through the structure and you can see the city through the structure, just like you can with the Eiffel Tower. That makes it a spectacular thing for artists.

Cole: Isn't it, though, that these bridge builders use form and function in the way they calibrate their
          structures? Those are the basic elements of art.

Billington: Yes, of art, but not architecture, even though they are art. In other words, everything in that bridge by Roebling is there for an extreme structural purpose. Now he wanted to show it off. That's why the elevated walkway walks through it.

But what he's walking through is a structure, all parts of which are essential.

Cole: On some of these bridges where they have hired artists or sculptors, it's a kind of embellishment—

Billington: Well, it diminishes the quality of it and makes it often very difficult to maintain. It usually makes them very expensive. They just finished a bridge out in Redding, California, which is very famous, many people praising it. It cost eight times the budget. Eight times the budget. That's not structural art at all because the principles of structural art are first, efficient form, that is to say, waste no materials. The second is economical form, waste no money. That's the discipline of structural art. Within that discipline, there is plenty of opportunity for the designer to play with the form and express his own or her own aesthetic vision. That's the way these engineers work: discipline and play.

When people say, "Well, but you have to have lots of money to do the play," then I can show them the works of Maillart. Maillart was chosen for his bridges almost always because they were less expensive. That was the reason he got his commissions. The high art world thought they were ugly.

Cole: What drew you to him?

Billington: The beauty of his work. That's what drew me to him. These architecture students showed me the pictures, and they were very stunning, and I thought, I've never seen anything like that. Here I was already a professor and had practiced for eight years and I had never even heard of Maillart. Few engineers had heard of Maillart. The architects, even though I'm not talking about them as bridge designers, some of them are very good critics of structure. And it was architectural historians or architects such as Siegfried Gideon and Max Bill, who wrote about Maillart and brought him to the attention of a lot of people.

That was what drew me to him. It was purely aesthetic to begin with. My challenge was to find out whether it was good engineering.

Cole: Okay.

Billington: I found out that he was the best engineer of his time, the best one purely technically. That's what led me to understand that all of these artists were also, from a technical, innovative point of view, the best. It was not a question of designers making pure pictures or sculptural objects. These structural artists were combining the very best, most innovative, most advanced engineering ideas and making them also beautiful.

Beauty gets you first. That's why all the lectures are visual, and the emphasis, initially, is always visual. That's the way you interest people nowadays. We live in a very visual world now, and the students are oriented this way. So you get them by the visual thing and then you say, "Now we're going to make the calculations, which are very simple, and show you how efficient these are."

Cole: In Switzerland, of course, you can see not only these wonderful bridges, but how they fit in to their environment. It's a spectacular aesthetic experience.

Billington: That's correct. That's exactly right.

Cole: Well, this has been terrific. Thank you very much.

Part II

Daily Practice

It is not enough to teach a man a specialty. Through it he may become a kind of useful machine, but not a harmoniously developed personality. It is essential that the student acquire an understanding of and a lively feeling for values. He must acquire a vivid sense of the beautiful and of the morally good. Otherwise he—with his specialized knowledge—more closely resembles a trained dog than a harmonious developed person. He must learn to understand the motives of human beings, their illusions, and their sufferings in order to acquire a proper relationship to individual fellow men and to the community.

Albert Einstein (1952)

Part II is a critique of the daily practices of engineering with a focus on the engineering workers, the profession, and designs. The articles are a sampling of engineers' solutions to the travails of daily practice, an indication of what engineers can teach each other. Practice is perhaps used ironically since the articles in this part are not generally what engineers think about in daily practice. However, engineers can and do learn from the mistakes and critiques of other engineers for, as the readings illustrate, mistakes and negative critiques can lead to well-thought-out positive projects.

The historian of engineering Edwin Layton pointed out that American engineers' late nineteenth-century framing of their work as a professional activity that creates social good was an *ideological claim* prompted by public confusion regarding the negative impacts of technology (Layton, 1971). As a result, during the early phases of professional engineering development, the *conceptualization* of engineering's role in society moved from an implicit loyalty (to the corporation) to the now popular idea of engineering as "knowledge and skill for the enhancement of human welfare" (Mitcham, 2009). In fact, the consideration that engineering means knowledge "for the enhancement of human welfare" is most common among the disciplinary rules included in contemporary engineering codes of ethics.[1] In reality, though, relationships with clients and the conduct of engineering as a *business* have been central to the agendas of national and international professional societies. Long-term social and political implications of engineering projects have only occasionally been in the mainstream of American engineering.

According to Edwin Layton, the engineer's everyday dilemma "has centered on a conflict between professional independence and bureaucratic loyalty, rather than between workmanlike and predatory instincts" (Layton, 1971, p. 1). Whether this conflict is driven by bureaucracy, as Layton believed, or by "specialists in technological knowledge" and "the captains of industry"—as has been maintained by Thorstein Veblen and more recently by historian of technology David Noble—remains the subject of scholarly debate.[2] If Layton is correct, however, any attempt to revolutionize engineering seems unlikely to bear fruit, for it "would violate the elitist premises of *professionalism* and because revolution would not eliminate the underlying source of difficulty" (Layton, 1971, p. 1; emphasis ours).

Several pieces in "Daily Practice" touch on the issue of engineering ethics: that "conversation without end" (Florman, 2002). Layton's essay serves as an important reminder that codes of ethics suggest *historical* and *cultural* products. Although the 1930s signaled the beginning of an era in which the dominant narrative in American engineering ethics was that of "social responsibility," Layton argues that the early codes of ethics actually served as "barriers to engineers who wished to serve the public."[3]

Layton would not be surprised that, although the exact percentage is not known, the majority of engineering graduates work in profit-seeking bureaucratic settings. Most importantly, of the 43 licensing boards that responded to a survey designed by the National Council of Examiners for Engineering and Surveying, 29 have exemption provisions for those engineers who occupy positions in industry (Musselman, 2009, 2010). Although the National Society of Professional Engineers (NSPE) has repeatedly lobbied for phasing out the so-called industrial exemption—meaning that "licensing laws only applied to 'engineers responsible for engineering design of products, machines, buildings, structures, processes, or systems used by the public'"—these standards remain a part of the engineering status quo in the United States (Norman, 2005).[4] Given that engineering work effects people daily, what is the social cost of maintaining these exemptions? "How free is an engineer?" to paraphrase Ralph Nader. As a response to a thread conversation posted on the web page of NSPE, a professional engineer recounted the following story:

> After completing my Masters in engineering at Stevens Institute of Technology in 1979 I became an MBA student in NYU's graduate school of business. At about the half way point in the program I took a business ethics course which posited the FDA banning of a product by a fictitious American pharmaceutical manufacturer. It was the study group's responsibility to decide how best to protect the company and shareholders from catastrophic financial loss. Sadly, it did not take too long for the group's majority to decide to sell the product in a third world country, free from the control of the FDA.
>
> Some 30 years later those folks in my MBA class have now risen to be the titans of industry and of Wall Street. Need [we] say more? (Musselman, 2009)

Certainly, engineers do good works, and the articles in Part II discuss those qualities that truly make an engineer a professional free to pursue socially conscious convictions, working in the context of the public good. Being socially minded, learning how to handle the dynamics of the workplace, averting disasters, setting standards, and ethical whistleblowing require thought and early preparation. Nothing can prepare an engineer for dealing with such challenges more than can personal experience, or near-personal experience. This process requires trial and error, testing each aspect again and again. However, certain issues faced in daily practice, such as political contributions require a different strategy. Trial and error is too time consuming when "the market rarely allows [the engineer] the time, freedom and funds to explore the social implications of his work."[5]

Ethics, Politics, and Whistleblowing in Engineering can help engineers to create an environment where socially appropriate technological practice is freely debated. Biomedical engineering—with a projected employment growth of 72% in the next decade—is still engineering, but may also take the issues of ethics to a different level (e.g., bioethics) (US Bureau of Labor Statistics, 2010). To address this limitation, the authors propose three canons for inclusion in the NSPE's code.

Teaching ethics is difficult when using abstract examples and can be frustrating, boring, and irrelevant for both students and teachers, especially if the latter do not view their work as a process of continuous self-reflection. Leaving aside the critical questions of how, where, and by whom engineering ethics is taught, a review of engineering codes of ethics suggests that "the[ir] similarities exist in what has *not* been included in the ethical codes" (Catalano, 2006, p. 20; emphasis ours)[6–9] Let us envision codes of ethics that would promote peace, poverty reduction and sustainability. To further the conversation on ethics, Appendix I contains a number of the most recent versions of codes of ethics approved by various professional societies of engineers.

The discussion on ethics ultimately seeks to empower engineers through litigation and whistle-blowing. No doubt, this is easier said than done. Engineers who blow the whistle are truly coura-geous people who are taking professional and personal risks. Even if an engineer decides to appeal his or her penalty, the chances for success are discouragingly low. In an interview, Tom Devine, who currently serves as legal director of the Government Accountability Project, was quoted to have said that "since 1994, only 3 out of 210 cases of whistleblowers appealing their punishments have been successful because whistleblowers can appeal only in a special whistleblower court, which inter-prets the law in a way that all but guarantees their appeals will fail" (Gladstone, 2010). Does the absence, then, of a systematic way to protect whistleblowers and other engineers who act morally constitute the question of the "invisible engineer," held up to a different light? What reasons make some engineers consent to their own state of invisibility?

We advocate that visibility, accountability, and dissent must be organic parts of engineering! Whistleblowers need support from engineers who have a sense of injustice, have experienced the making of ethical decisions, and who understand the legal ramifications. The selections in "Whistleblowing" discuss how engineers can protect themselves while speaking up, and include various examples of engineers who have done so.[10]

In "Daily Practice," disasters,[11] standards, and the human ramifications of engineering design are discussed. Although many engineering disasters are attributed to "human failure," bureau-cratic compartmentalization and poor technology management may have equally negative impli-cations.[12,13] The technology we have developed for human beings has perhaps overwhelmed us. Initially, we wanted to make our daily lives simpler, but as many engineers and others have com-mented, many of our technologies have effects that are often unpredictable and uncontrollable, making our lives even more complex. Who could have predicted that the washing machine would create more washes instead of less because the expectation of how often clothes should be washed has increased (Cowan, 1983)? Technology changes our cultural practices. In this part, we bring together many decades of critique in engineering and give a sense of the cultural impact of engi-neering practice.

Learning from disaster is important, not only in changing future designs, but also in setting stan-dards. Nowadays, our lives are determined by "systems of classification limned by standard formats, prescriptions and objects"; hence, the politics of regulation is key to defining the infrastructure of our technological society (Bowker and Star, 1999, p. 1).[14] The selections in "Setting Standards" give information on trade product standard systems and product liability laws as well as the importance of tort litigation in protecting consumers from harm.[15] With these tools, engineers are beginning to be more realistic about assessing risks, which results in a change in our perceptions of the impor-tance of safety and designing engineering as if people mattered. In 1973, *Standardization News* published the following:

> It makes little sense to deny someone a product because he cannot afford to purchase the completely
> safe version of it.... If the only justification for setting high minimum standards is to save consumers
> from having to think about their purchases and read the accompanying labels, it is hard to imagine it is
> worthwhile baring poor consumers from the market. (Lare, 1973)

Today this viewpoint would almost certainly be considered on the margin. Increasingly, safety is not considered a privilege but a necessity and a right that engineers owe to the community and their general public. For example, once safety began to sell cars, discussions of cars becoming "too safe," such as in the 1970s, were no longer commonplace (Myers, 1971).

Humanizing the engineering profession is not a token gesture to quell disgruntled consumers and citizens. Humanizing the engineering profession means keeping sight of the people upon whom technological projects and designs have impact. A simple inquiry on the web gives evidence of how engineering colleges advertise themselves as educational centers for the "people-serving profession." Moreover, commentators have stressed that people who do engineering work have historically served a particular cause other than itself (Downey and Lucena, 2010).[16] If, however, engineers deny that conditions vary and prevent communities from employing their own specialized knowledge, then both "engineering" and "people-serving" must be challenged.

The future of the engineering profession has stimulated questions for contemplation by both engineers and philosophers of technology (Mitcham, 1994). But it is only in the last decades or so that STS practitioners have suggested that technological design per se may be shaped by particular social and political considerations (Bijker and Law, 1997). Arguments of that sort are important, for they transcend the commonplace conceptualization of engineering as pertaining solely to the technical sphere and emphasize that the black box of technology must be opened and subjected to democratic scrutiny.[17] But "deconstructing" technologies, from a point of view of the human ramifications of design, may be a limited practice, unless it is accompanied by critical reconstruction. The goal is to engineer alternative technologies with assumptions of inclusiveness, social justice, democratic governance, and environmental sustainability embedded into them. Shall engineering design, then, become less imperial, to include "the other ninety percent"?[18] Also, could it be that a transformation of engineering morality be part of the solution?[19]

We began the introduction to Part II of *Ethics, Politics, and Whistleblowing in Engineering* by quoting Edwin Layton, who maintained that revolutionizing the engineering world in America would necessarily require the violation of "the elitist premises of professionalism." Engineers have increased autonomy because engineering is a profession. Nevertheless, could a claim to "engineering professionalism" prove functional within contemporary engineering? Those who want to analyze take the pessimistic view that, in fact, the possibilities of establishing "an autonomous engineering profession oriented toward ideals of broad social responsibility ..." seem very low (Williams, 2002, p. 80). "The social contradictions are too great, and the epistemological core and the institutional support are absent, whereas the educational program does not seem to be of great help" (Williams, 2002, p. 80). Furthermore, as Ruth Oldenziel remarked, taking Layton's concept of professionalism for granted fails to consider that for engineers, "professionalization" was part of a wider master plan of "making technology masculine," that is, to make it difficult for society to conceptualize women as important agents in the processes of designing, producing, and manufacturing technological artifacts (Oldenziel, 1999).

One cannot discuss engineering professionalism without paying special attention to the demarcation between engineering and management. We present some of the nuances of engineers' practice by going beyond that demarcation. In any case, as the literature suggests, "[i]t is *not* possible to sustain the view that all engineers share similar work values; nor is it possible to argue that they are divided in two opposing groups, one with professional values, the other with organizational values" (Watson and Meiskins, 1991).

The historical evidence suggests that American engineers have not been able to lead a revolution. The historian of technology Matthew Wisnioski argues that "[t]aken together, ... when we look across engineering cultures from the 1960s to the present, we see on the one hand the emergence of alternative visions of technology that struggle to gain traction but nonetheless continue *to draw an eclectic minority of reformers*" (Wisnioski, 2012). But does that mean that engineering is deemed to retain its hubris and hegemonic characteristics? We most definitely hope and believe not. Designing ingenious methods to address socio-technical problems has always been the engineer's forte. It is through reimagining technical work that a new future may be shaped.

ENDNOTES

1. Appearing in 11 out of the 15 major engineering professional societies' codes. For codes of ethics in engineering, Appendix I.
2. These terms are Veblen's (1921, p. 39). Noble (1979 [1977]), in his *America by Design: Science, Technology, and the Rise of Corporate Capitalism*, maintained that the engineers' revolution was structurally impossible, due to the domination of engineering by corporate interests.
3. For a really informative historico-philosophical account of the development of engineering codes of ethics in the United States, see Mitcham's (2009) piece.
4. Musselman (2009) notes that "[i]n the states that do not have industrial exemptions licensure requirements for engineers in industry aren't commonly enforced, in part for fear the legislature would enact an industrial exemption if enforcement were to begin."
5. See "The Engineer's Responsibility for Product Safety," by Herbert Egerer.
6. Haws (2001) classified the methods currently used to teach engineering ethics to students into six categories: professional codes, humanist readings, theoretical grounding, ethical heuristics, case studies, and service learning.
7. Should teaching of engineering ethics be part of the engineering curriculum, or should it not be? Should it be an integral part of capstone design classes or not?
8. How engineering ethics is taught, of course, depends on who teaches the subject. Engineering ethics instructors are individuals with expertise in some of and/or a combination of the following: engineering, political philosophy, moral law, or STS. For a discussion of the alleged distinction between American and European approaches in teaching engineering ethics, see the papers of Davis (2003) and van de Poel (2001).
9. The literature on engineering codes of ethics is extensive. For an evaluation of the usefulness of the American Society of Civil Engineers' (ASCE) *Code of Ethics* from the perspective of environmental sustainability, see Vesilind and Gunn's (1998) paper.
10. Tekla S. Perry's piece "Knowing How to Blow the Whistle" (Chapter 17) offers a helpful guide to the engineer whistle-blower. In addition, we recommend IEEE's (2014) webpage "Engineer's guide to influencing public Policy."
11. See "Disasters as Object Lessons in Ethics: Hurricane Katrina," by Heinz C. Luegenbiehl (Chapter 11).
12. See "The Human Factor in Failures," by George F. Sowers.
13. See "Trading Disasters," by Aron M. Bernstein and Mark Kuchment.
14. See "Standards in Engineering Society," by Albert Batik (Chapter 12).
15. See "Questions and Answers about Trade Product Standards: A Primer for Consumers," Subcommittee on Antitrust and Monopoly of the Committee on the Judiciary of the United States of America.
16. See also Alder, Ken. *Engineering the Revolution: Arms and Enlightenment in France, 1763–1815.* Princeton, NJ: Princeton University Press, 1997.
17. See "Nuclear Power: More than a Technological Issue," by Ralph Nader (Chapter 14).
18. See "Design for the Other Ninety Percent," by Paul Polak.
19. See "Engineering Design in a Morally Deep World," by George Catalano.

REFERENCES

Alder, Ken. *Engineering the Revolution: Arms and Enlightenment in France, 1763–1815.* Princeton, NJ: Princeton University Press, 1997.

Bijker, Wiebe and Law, John, eds. *Shaping Technology, Building Society, Studies in Sociotechnical Change.* Cambridge, MA: The MIT Press, 1997.

Bowker, Geoffrey C. and Star, Susan Leigh. *Sorting Things Out: Classification and Its Consequences.* Cambridge. MA: MIT Press, 1999.

Catalano, George D. *Engineering Ethics: Peace, Justice and the Earth.* Synthesis Lectures on Engineers, Technology, and Society. San Rafael, CA: Morgan & Claypool Publishers, 2006.

Cowan, Ruth Schwartz. *More Work for Mother: The Ironies of Household Technology from the Open Heart to the Microwave.* New York: Basic Books, 1983.

Davis, Michael. What's philosophically interesting about engineering ethics? *Science and Engineering Ethics* 9, no. 3 (2003): 353–361.

Downey, Gary Lee and Lucena, Juan C. The need for engineering studies: engineers in development. In *Engineering: Issues and Challenges for Development*, edited by Tony Marjoram, 167–171. Paris: United Nations Educational, Scientific, and Cultural Organization, 2010.

Einstein, Albert. Education for independent thought. *New York Times*, October 5, 1952.

Florman, Samuel. Engineering ethics: The conversation without end. *The Bridge* 32, no. 3 (2002): 19–23.

Haws, David R. Ethics instruction in engineering education: A (mini) meta-analysis. *Journal of Engineering Education* 90, no. 2 (2001): 223–231.

Gladstone, Brooke. WikiLeaks and protections for federal whistleblowers. On the Media, December 3, 2010. http://www.onthemedia.org/transcripts/2010/12/03/04. Accessed September 3, 2011.

IEEE (Institute of Electrical and Electronics Engineers). Engineer's guide to influencing public policy. IEEE, 2014. https://www.ieeeusa.org/policy/guide/index.html. Accessed August 30, 2014.

Lare, Lester B. *Standardization News*, 1973.

Layton, Edwin. *The Revolt of the Engineers*. Cleveland, OH: Case Western Reserve University Press, 1971.

Mitcham, Carl. *Thinking through Technology, The Path between Engineering and Philosophy*. Chicago, IL: University of Chicago Press, 1994.

Mitcham, Carl. A historico-ethical perspective on engineering education: From use and convenience to policy engagement. *Engineering Studies* 1, no. 1 (2009): 3553.

Musselman, Craig. The industrial exemption: What, if anything, should the profession do? National Society of Professional Engineers, October 1, 2009. http://community.nspe.org/blogs/licensing/archive/2009/10/01/the-industrial-exemption-what-if-anything-should-the-profession-do.aspx. Accessed December 20, 2011.

Musselman, Craig. The 80% myth in the engineering profession. National Society of Professional Engineers, September 13, 2010. http://community.nspe.org/blogs/licensing/archive/2010/09/13/the-80-myth-in-the-engineering-profession.aspx. Accessed December 20, 2011.

Myers, Phillip S. Technological morality and the automotive engineer. SAE Technical Paper 710003, 1971.

Noble, David F. *America by Design: Science, Technology, and the Rise of Corporate Capitalism*. Oxford: Oxford University Press, 1979 [1977].

Norman, Neil A. Where does NSPE stand on industry exemptions? National Society of Professional Engineers, November 8, 2005. http://www.nspe.org/resources/pdfs/blog/industry_exemptions-neil_norman.pdf. Accessed December 2011.

Oldenziel, Ruth. *Making Technology Masculine: Men, Women, and Modern Machines in America, 1870–1945*. Amsterdam: Amsterdam University Press, 1999.

US Bureau of Labor Statistics. *Occupational Outlook Handbook*. Washington, DC: US Bureau of Labor Statistics, 2010.

van de Poel, Ibo. Ethics and engineering courses at Delft University of Technology: Contents, educational setup and experiences. *Science and Engineering Ethics* 7, no. 2 (2001): 267–282.

Veblen, Thorstein. *The Engineers and the Price System*. New York: Huebsch, 1921.

Vesilind, Aarne P. and Gunn, Alastair S. Sustainable development and the ASCE Code of Ethics. *Journal of Professional Issues in Engineering Education and Practice* 124, no. 3 (1998): 72–74.

Watson, James M. and Meiskins, Peter F. What do engineers want? Work values, job rewards, and job satisfaction. *Science, Technology, and Human Values* 16, no. 2 (1991), p. 155.

Williams, Rosalind. *Retooling: A Historian Confronts Technological Change*. Cambridge, MA: The MIT Press, 2002, p. 80.

Wisnioski, Matthew. *Engineers for Change: Competing Visions of Technology in 1960s America*. Cambridge, MA: The MIT Press, 2012.

9 The Engineer as a Local, Not Global, Citizen

Jonathan D. J. VanderSteen and Usman Mushtaq

CONTENTS

EXECUTIVE SUMMARY

Twenty-first century engineering graduates are expected to be skilled in a variety of competencies beyond the profession's traditional technical skills. An engineering curriculum that fosters community engagement can help engineering students acquire many of these additional abilities, such as teamwork and cross-cultural communication, while at the same time helping the profession achieve its broad goals of improving the quality of life.

Although some experts suggest that focusing on global engineering is a useful way of gaining these attributes, we maintain instead that local community engagement provides many of the benefits gained from international experience with fewer practical, ethical and pedagogical risks. As well, focusing on local communities strengthens engineering's relationships with the people it is most intimately connected with and helps practitioners prepare for a future in which they are likely to be more locally engaged.

Engineering is a field in flux, and like any other modern institution, the engineering profession must evolve as the conditions in which we work change. In light of this, the 2009 Declaration at the Montreal National Engineering Summit commits Canada's engineering profession to improving ecological and societal health, as well as ensuring sustainable development and contributing quality of life. The declaration concluded: "As a profession, we are committed to helping provide the best possible quality of life for all Canadians." To realize the declaration's goals, engineering educators must increasingly engage in discussions on the role of engineers in society. Issues such as accountability, equity, professional and personal ethics, and the impact and unintended consequences of engineered systems need to be considered.

Indeed, engineering education is paying more attention to human development and ecological and social justice. Vesilind [1], for example, suggests that engineers historically have been employed as "hired guns, doing the bidding of both political rulers and wealthy corporations," but there is a new kind of engineer emerging, one "rooted in the greater ideas and aspirations of engineering as a service to all humanity."

In order to better position engineering as "service to humanity," some engineers have sought to better understand their responsibilities as global citizens. Engineering authors such as Johnston [2] and Chan and Fishbein [3] point out that being a global citizen often encompasses diametrically opposite ideals. Thinking globally involves an attempt to understand global connectedness and the source of worldwide injustices. But at the same time, global citizens must embrace that connected-ness and strive for global competitiveness [3,4].

Regardless of this schizophrenic identity (the global competition is not unrelated to the injus-tices), it is clear that the call for more global engineers has encouraged Canadian students to get involved in organizations, projects and courses that deal with engineering outside of their country's borders. In best-case scenarios, these experiences have led to increased cultural awareness, broad-ened perspectives, improved communication skills and practical design experience.

Much has been written about the benefits of global experience [see Ref. 5–7], but this paper proposes that engineers who think and act locally can gain many of the same skills while avoiding some of the pitfalls. For example, working within a local community eliminates the ethical issue of making the developing nations the lab of the Canadian engineer, whose learning experience is one more thing the Western world takes from low-income communities.

Most importantly, engineers gain much more than skills by working in their own backyard. Engineers who engage locally through imagining a better place to live and improved relations contrib-ute to richer, transformative and more resilient neighbourhoods that can withstand a wide variety of stresses. Focusing on faraway communities or on large global institutions or corporations has the risk of making us forget about the relationships that define us the most. International experts leave spouses and children at home for large portions of the year. Travel is both time-consuming and exhausting and can put strains on relationships with family, friends and neighbours. It is not a coincidence, for example, that many of the practitioners who engage in international development are single or young.

CAN WE REALLY BE GLOBAL CITIZENS?

We start this analysis by asking what it means to be a global citizen or a global engineer. Obviously, we cannot deny the fact that our actions often have a global impact. The products we use are fre-quently made from materials and manufacturing processes from other countries. The companies we work for collaborate with firms on many continents. Sometimes we buy goods that could have been made only through the extraction of materials that marginalize indigenous people. It is, no doubt, beneficial for us to attempt to understand these global interconnections as well as possible. One should not do development work in sub-Saharan Africa, for example, without understanding the politics behind the Structural Adjustment Programs sponsored by the World Bank and International Monetary Fund.

Even if we understand these programs, we can't intimately know the global world like the pre-industrial people understood their villages [8]. In the end, we contribute most meaningfully to sys-tems that we truly know well. At the same time, westerners interested in solving issues of disparity in low-income countries often, even subconsciously, promote a globalized, Western culture [8]. This globalized culture does not promote strong resilient systems, but rather it encourages communities to compete internationally with few regulations or rules to protect them from the instability of the global market. Global engineering, therefore, could unintentionally reinforce existing injustices, clearly not an encouraging path forward.

We are not saying that no good has come out of the work of some global institutions. We are saying that focusing on global thinking is often mixed up with the hubris of utopian thinking and fixing the world. (What would it look like once it was fixed?)

Much great wisdom throughout the ages suggests that change starts with looking inward. There certainly are enough problems crying out for engineering expertise at home. Instead of going abroad, let us begin to improve our own neighbourhoods.

STRONG LOCAL ENGAGEMENT

Global engineering work is said to make stronger engineers because it fosters the development of communication skills, multidisciplinary experiences, the development of social responsibility, opportunities to be entrepreneurial and an ability to deal with complexity and systems thinking [3]. Yet the same skills and experience can be obtained through domestic engineering work. For example, the bulk of the challenges our society faces are "interface" problems among disciplines. Thus, local work requires the same multidisciplinary and interdisciplinary teamwork required of global engineers.

Local engineering can even involve cross-cultural experiences. Working with the most marginalized in our communities often requires significant "border-crossing," whether the border to be crossed is that of race, class, gender or immigration status [9]. Finally, the technical challenges and complexities of global engineering work are also present in our own communities [10].

In a recent study, a Canadian civil engineer interested in community-engaged scholarship asked, "What kind of engineer are you if you can't find a technical problem [locally]? There's no end to it" ([11], forthcoming).

There are other benefits to focusing on strong local engagement. Domestic investment provides people with a long-term interest in the prosperity, health and beauty of their own neighbourhoods. Second, the better we know our communities, the deeper and richer our understanding of our role as engineers in Canadian society. If our goal is to work for justice as engineers, we must understand our own society deeply. Third, a deeper understanding of local nuances helps one to better understand the human and ecological cost of our actions and work. Finally, Canadian engineers can do much in their own communities by making them resilient against the fluctuations of the global market. After all, the people have no power and the land no voice without prosperous local economies [12].

It is important to note that this paper is not advocating that global inequality and injustice be ignored. We are not opposed to dedicated engineers who want to commit their lives to work for justice elsewhere. At the same time, while we should be deeply troubled by the poverty and despair that occurs in much of the world, we must also realize that much of this situation derives from a complicated past that does not allow for simple, even engineering, solutions.

HOW TO FOSTER A LOCAL ENGINEER

The way in which the curriculum engages the local community is an issue of utmost importance for every Canadian engineering school. Since engineering is a professional degree, the process of education must take seriously the obligations the profession has set forward, including the improvement of the quality of life.

Much positive transformational change is possible through a formal education, but changes to the engineering curriculum are never easy. There are many external forces that understandably lobby to influence engineering education. Since there are pressures from multiple sources, many professors present material according to their own priorities. For example, some professors are pressured from industry to suggest that universities should better serve business. Other professors feel that students need to be more computer literate and more capable with computational tools. Still others argue that such programs are at the expense of fundamental understandings.

Meanwhile, the Canadian Engineering Accreditation Board is demanding that engineers be competent in communication, ethical, equity and engineering impact assessment skills [13] without recognizing that these qualities can be achieved outside the classroom by educating the engineer as a local citizen. This realization could form the basis of reintegrating the curriculum.

There are a number of ways that so-called global skills can be learned, but it is important to remember that the appropriate lessons cannot be taught prescriptively. They can only be fostered. For example, open-ended and design problems can introduce ideas that encourage local thinking. This can occur within a core course, a design course or as part of a capstone project. Ultimately, however, local thinking should be integrated throughout the curriculum. A work placement or internship within the local community could be an effective way to introduce concepts of local engineering.

There are many local potential projects that engineering students can get involved with. Consider, for example:

- Grey water recycling for public buildings or local shelters.
- Food-waste composting and community gardens for large-scale community meal providers.
- Developing an engineering and technology certificate for youth and young adults with physical and mental disabilities.
- Technology, design, math and science tutoring in after-school programs.
- Bridging the information gap between residents and city engineers for needed infrastructure through learning alliances.
- Design of more energy-efficient food-storage methods for local restaurants in order to encourage more local foods during the winter season.
- Installation and policy work related to rainwater harvesting.
- Design and delivery of a technical literacy school curriculum that involves critical analysis and goes beyond the "gee-whiz" state [14] for local school boards.
- Channel improvements to avoid riverbed erosion and to control flooding in conjunction with local watershed authorities.
- Promotion of composting and urine-separation toilets.
- Design of an affordable bicycle trailer for a local bicycle advocacy group.
- Creating opportunities for local engineers to get involved with pro-bono engineering.
- Creating a physical accessibility audit of locations and buildings within the local community.
- Building educational toys for use by childcare agencies [15].

The benefits of service-learning projects, whereby academic instruction is combined with community service, are well documented [9,15]. Many engineering schools in North America have engaged in service learning with positive results, such as the work reported by Riley and Bloomgarden [10] where a group of master's students at Smith College teamed up with a community-based organization in Holyoke, Massachusetts, to address the problem of wood-fired oven emissions posed by a local bakery.

For our part, we suggest that local placements not be approached with charity in mind, but with a notion of justice [16]. The focus should be on community transformation and not on giving back or simply on enhancing the educational experience. The success of this sort of project will depend on the institution at which the student is placed and the student's approach, attitude and self-awareness. Thus, these sorts of placements should not be mandatory.

CONCLUSION

Ultimately, Canadian engineers can create positive, transformative change in society. This occurs best when engineering students work in their own communities, thus gaining the benefits of globalized work without many of its risks. An engineering curriculum that focuses on developing strong local communities can be a part of the trans-formative process within students, in which they learn to value co-operation above competition. To borrow Ursula Franklin's words, we believe that "the purpose of a university is not only to be a place where knowledge and understanding find a home, but also to provide a bridge for interaction with the larger community" (2006).

Jonathan VanderSteen is a post-doctoral fellow at the University of Guelph's School of Engineering. His research interests include adaptive engineering education and sustainable urban water. Previously, VanderSteen studied thermal/fluid engineering and alternative energy systems. VanderSteen did his undergraduate work in mechanical engineering at the University of Waterloo and received a PhD in civil engineering from Queen's University.

Usman Mushtaq is an M.Sc. candidate in the Department of Civil Engineering at Queen's University. Mushtaq's research interests include socially just design processes and the values embedded within technology. He is a graduate in computer engineering from the University of Pittsburgh.

REFERENCES

1. Vesilind, P. Aaren "Peace Engineering." *Journal of Professional Issues in Engineering Education and Practice*. Volume 132, Number 4, 2006.
2. Johnston, S.F. "Towards Culturally Inclusive Global Engineering?" *European Journal of Engineering Education*. Volume 26, Number 1, 2001.
3. Chan, Adrian D.C., and Jonathan Fishbein. "A Global Engineer for the Global Community." *The Journal of Policy Engagement*. Volume 1, Number 2, May 2009.
4. National Engineering Summit. Leading a Canadian Future: The New Engineer in Society. The Montreal Declaration. May 2009.
5. VanderSteen, J.D.J. et al. "International Humanitarian Engineering Placements: Who Benefits and Who Pays?" *IEEE Technology and Society Magazine*. Winter 2009.
6. Epprecht, Marc. "Work-Study Abroad Courses in International Development Studies: Some Ethical and Pedagogical Issues." *Canadian Journal of Development Studies*. Volume 25, Number 4, 2004.
7. Sichel, Benjamin. "I've Come to Help: Can Tourism and Altruism Mix?" *Briarpatch Magazine*. November 2009.
8. Esteva, Gustavo, and Madhu Suri Prakash. *Grassroots Post-Modernism: Remaking the Soils of Cultures*. London: Zed Books, 1998.
9. Riley, Donna, and Alan H. Bloomgarden. "Learning and Service in Engineering and Global Development." *International Journal for Service Learning in Engineering*. Volume 2, Number 1, 2006.
10. Coyle, Edward J., Leah H. Jamieson and William C. Oakes. "EPICS: Engineering Projects in Community Service." *International Journal of Engineering Education*. Volume 21, Number 1, 2005.
11. VanderSteen, J.D.J. et al. "Humanitarian Engineering Placements in Our Own Communities." *European Journal of Engineering Education*. 2010 (forthcoming).
12. Berry, Wendell. "The Idea of a Local Economy." Orion Magazine. Winter 2001.
13. Canadian Engineering Accreditation Board (CEAB). "Accreditation Criteria and Procedures 2008." September 2008, www.engineerscanada.ca/e/files/report_ceab_08_txt_only.pdf.
14. Franklin, Ursula M. *The Ursula Franklin Reader: Pacifism as a Map*. Toronto: Between the Lines, 2006.
15. Eyler, Janet. "Reflection: Linking Service and Learning—Linking Students and Communities." *Journal of Social Issues*. Volume 58, Number 3, 2002.
16. Marullo, Sam, and Bob Edwards. "From Charity to Justice." *American Behavioral Scientist*. Volume 43, Number 5, 2000.

10 Thinking Like an Engineer

Michael Davis

Employed engineers sometimes claim that their status as employees denies them the autonomy necessary to be "true professionals." Such claims also appear in important scholarly work. For example, in *The Revolt of the Engineers*, Edwin Layton observed: "Employers have been unwilling to grant autonomy to their employees, even in principle. They have assumed that the engineer, like any other employee, should take orders. . . . [But] the very essence of professionalism lies in not taking orders from an employer."[1] What are we to make of the claim that professionalism is inconsistent with being an employee (or, at least, with taking orders from an employer)? Is it a conceptual truth (a deduction from definitions) or an empirical one? How might it be proved—or disproved?

The purpose of this chapter is to answer these questions by developing a conception of *professional* autonomy. We have no such conception now (though some claim otherwise). What we have instead are conceptions of *personal* (or moral) autonomy applied to the workplace. These conceptions help us understand neither the specific contribution of professional constraints to personal autonomy nor the ways corporate organization and professional responsibility might be consistent.

The literature on autonomy may be divided into three categories: (1) a general philosophical literature on "personal" autonomy, (2) a philosophical literature explicitly concerned with "professional autonomy"; and (3) a sociological literature concerned with autonomy in the workplace. These three literatures, though related in principle, seem in practice to have grown up largely independent of one another. Let us consider them in order. Having seen what they have to offer, we should be in position to develop our own conception of professional autonomy and see what research, if any, it suggests.

Our first hypothesis was that the boundary between engineer and manager would be relatively clear in most companies because the staff-line mode of organization would force the distinction to be made clear. In fact, we found almost no trace of the staff-line distinction. What we found in its place was something much more like the distinction made in universities between faculty and administrators.

In most universities, senior administrators (president, vice presidents, and deans) hold faculty appointments. Many still do some teaching. Ordinary faculty, on the other hand, do considerable administrative work, whether as department chair or through various departmental, college, or university committees. Faculty differ from senior administrators only in degree (though "administrative staff" are more like what engineers call technicians). Some ordinary faculty may be paid more than any administrator, even the university president.

In most companies at which we interviewed, the distinction between engineers and managers was similarly one of degree. A bench engineer was an engineer who spent most of his time at his bench (like an "ordinary faculty member"). A pure manager was an engineer who no longer did any engineering himself. Especially in large companies, there might be several grades of engineer-manager. In general, the distinction between engineer and manager did not seem to determine pay, benefits, or weight in technical decisions.

The one company that seemed to make the distinction between engineer and manager as sharply as we originally expected did not seem to have any more of a staff-line organization than the other companies at which we interviewed. Yet, though only for accounting purposes, the sharpness of the distinction seemed to hurt relations between its engineers and managers, making engineers feel as if they and managers belonged to "separate camps." This bad feeling may have contributed to the poor communications we found there.

Our second hypothesis was that engineers would be primarily concerned with safety and quality while managers would be primarily concerned with costs and customer satisfaction. This hypothesis was generally confirmed but in a way suggesting that the concerns overlap more than commonly thought. Managers in most companies usually paid more attention to costs and customer satisfaction in their *initial* response to an engineer's recommendation than the engineers *initially* did. In all companies at which we interviewed, however, decision was generally by consensus, not by management fiat. Decision by consensus required managers to inform engineers about considerations of costs and customer satisfaction they may have overlooked. No doubt as a result of that, most engineers we interviewed had a much better appreciation of such business matters than we expected. Even allowing for the fact that most managers we interviewed were trained as engineers, decision by consensus seemed to have a corresponding effect on managers. They seemed to have a better appreciation of engineering considerations than we expected. *Decision by consensus itself appears to be an important means of maintaining good communications* between engineers and managers.

Our third hypothesis was that engineers would tend to defer to management judgment because management had ultimate responsibility for decisions. This hypothesis derived from our assumption that engineering would be treated as a staff function (with no responsibility for decision) while management would be a line function. Yet, the hypothesis was in fact independent of that assumption. It could have been confirmed even if, as it turned out, engineering was a line function. Engineers could still have routinely deferred to management.

Our findings here are therefore significant in their own right. Deference to management was *not* what was expected of engineers. Quite the contrary. Engineers were expected to "go to the mat" on any question of safety or quality they considered important. Even managers who expressly reserved the right to overrule an engineering recommendation emphasized the need for engineers to "hammer" at them anyway. Engineers themselves expressed no deference to management on questions of *safety*. There they expected their recommendation to be "final." Only on questions of quality, customer satisfaction, or cost were they willing to let management have the last word—and, even then, they were willing to give management an "earful" first. Here again the analogy with decision making in a university (where faculty "advise" but expect to have administrators take their advice) seems much closer than decision making in the military (where—we are told—officers "command" and "subordinates" are expected to "obey").

Our fourth hypothesis was that the more hierarchical organizations were more likely to suffer a communications breakdown than the less hierarchical. This hypothesis, like the previous one, was derived from the assumption that the companies at which we interviewed would have a traditional (quasi-military) hierarchy. Though their tables of organization made them look as hierarchical as we assumed they would be, none of the companies at which we interviewed was in fact organized in that way. The small companies were too personal for formal hierarchy to matter much. Even in the large companies, the use of consensus and bringing other people in meant that individual managers could not control information or access in the way they would have in a traditional hierarchical organization. (And, in addition, the managers generally did not want to.) Even the communications gap we found in Company B did not result from hierarchical organization but from a combination of other factors, including too narrow a definition of engineering considerations, too much interference from the top in the details of engineering, a failure to consult directly with those most likely to know, and the use of motivational techniques likely to discourage the reporting of bad news. The absence or presence of a code of ethics or formal appeal procedure seemed to have little part in technical communications between managers and engineers.

Our fifth hypothesis was that we could develop a procedure for identifying a communications gap between engineers and managers if one existed. We now have some support for this hypothesis. Our open-ended interview identified what seemed to be a serious communications gap at one company (Company B). The interviews also provided us much useful information about how engineers and managers generally work together.

Our sixth hypothesis was that we could add to the stock of procedures for preventing a communications gap or at least to procedures for helping to eliminate such a gap once it appears. We came across two, the informal bringing others in and the formal technical review.

RECOMMENDATIONS

I (and the research group) believe this research justifies the following recommendations:

1. *Companies should try to soften the distinction between engineer and manager as much as possible.* Too sharp a distinction (as in Company B) seems to create resentment that can interfere with communication. Providing for a promotional ladder for bench engineers parallel to management's may help reduce the feeling that managers are "above" engineers. Managers, especially, seem to welcome the possibility of bringing in a *senior* engineer (that is, a "technical person" with rank equivalent to "manager") when they disagree with an engineer's recommendation. Companies should also look for other ways to treat engineers and managers as professional employees, differing only in specific function and responsibilities (for example, by avoiding differences in benefits based on classification as "manager" or "engineer").

2. *Engineers should be encouraged to report bad news.* Communication is most likely to break down between engineers and managers when procedures or other aspects of the work environment discourage engineers from reporting bad news (for example, design problems). Top-down engineering may be justified at times, but it should be accompanied by on-site visits with the bench engineers doing the work ("management by walking around"). Senior management needs to remember how much bad news is likely to get filtered out by several layers of management. Senior managers should also be wary of motivational techniques that discourage bad news or otherwise inhibit the give-and-take that is a precondition of decision by consensus working well.

   While on-site visits, especially informal surprise visits with bench engineers, can undercut the authority of mid-level managers, that is not a necessary consequence. Undercutting can be avoided by *open* discussion of the rationale for the visit, emphasis on the helping (rather than the controlling) role of managers, and (when a problem *is* discovered) a focus on solving the problem rather than finding someone to blame.

3. *Companies should check now and then for signs of trouble in relations between managers and engineers.* Such trouble may not be obvious to managers inside the company even if it is obvious to the engineers there. How many subordinates tell a superior more bad news than he asks for? One way for senior management to discover trouble is to meet informally with small groups of bench engineers and ask. Another way is to have outsiders interview engineers and managers in the way we have.

4. *Companies should encourage both engineers and managers to settle technical disagreements by informally bringing other experts in.* Companies should also consider adopting an open-door policy, ombudsman, or other formal appeals procedure. Though such formal procedures are seldom used to settle technical disagreements between managers and engineers, they nonetheless seem to help establish an environment in which even technical information flows more freely.

5. *Companies should look for formal procedures that bring out bad news that might otherwise be missed.* The most effective procedure of this sort we came across was Amoco Chemical's HAZOP study. Though this procedure is probably too elaborate for most companies (that is, those with a less dangerous technology), it may provide a useful ideal against which any company can measure its own technical review procedures. Of particular value, is that: (1) the reviewing body consists entirely of engineers who, though having the appropriate experience, have had no part in developing the plans (or process) they evaluate

(and so, no built-in conflict of interest); (2) the plans have to stand on their own (the drafters not being there to defend them); and (3) all recommendations are put in writing, rejecting a recommendation requires a written justification, and both recommendation and rejection are kept on file (thus assuring later accountability). Such an independent review gives everyone directly involved in a project considerably more incentive than they would otherwise have *not* to play down bad news in the early stages of a project. At a minimum, however, we think companies should encourage engineers to put their doubts in writing and circulate them among all those concerned.

6. *Companies should not expect a general code of ethics to have much impact on engineering decisions.* Any company that wishes to make safety or quality more central in its engineering decisions probably has to do so through specific technical specifications. It may also find training engineers in their profession's code helpful, because these codes are generally more specific about problems engineers face than is a general business code. Such training may also confirm engineers in the belief that their employer wants them to be advocates for engineering standards.

7. *Companies should try to improve the way they use bad news.* Companies cannot learn from their mistakes if they do not remember them. In particular, companies should consider including information about how parts failed in technical manuals (or data bases) engineers use or, at least, bring engineers together from time to time to discuss failures they have learned from.

8. *Technical engineering courses should include more about the place of cost, manufacturability, and other business considerations in engineering.* One manager in fact told us that, except for the graduates of co-op programs, engineers fresh out of college were poorly prepared to think about the range of considerations routinely part of good engineering. There seemed to be general agreement that engineering education is now too narrow.

9. *Engineers should be trained to make a case for their recommendations.* Ability to present data clearly, orally or in writing, and the ability to make arguments from the data seem to be essential to participating effectively in decision by consensus. Right now, engineers seem to have to learn these skills on the job. They are, however, skills any school of engineering can teach.

ENDNOTE

1. Edwin Layton, *The Revolt of the Engineers* (Press of Case Western Research University: Cleveland, 1971), p. 5.

11 Disasters as Object Lessons in Ethics
Hurricane Katrina

Heinz C. Luegenbiehl

CONTENTS

When large-scale disasters occur, one response is to look for causes of the catastrophe, whether they lie in human error or the forces of nature, in order to assign praise or blame. Most often, of course, in events of great magnitude no single cause is determinable, making such assessments fraught with uncertainty and the tendency to scapegoat. A seemingly more positive response is to acknowledge that "there is enough blame to go around" and to argue that instead we should focus on the future so that similar events can be prevented. In the case of 2005's Hurricane Katrina both of these responses became part of the public discourse, often seemingly to the exclusion of each other. Yet it is apparent that the two are not incompatible and in fact should form part of a total assessment. Without understanding why events occurred, it is difficult to see how much can be learned from them. If there is suspicion about assigning responsibility it therefore likely revolves around the motives of those who are doing the blaming. Perhaps they want to divert attention from themselves, want to score points against an opponent, or exhibit a sense of moral superiority.

A second issue is that of complexity. Given the confluence of a large number of causal conditions, it will be difficult and perhaps unfair to single out individual actors or institutions for responsibility. As the Independent Levee Investigation Team report on the hurricane put it:

> In the end, it is concluded that many things went wrong with the New Orleans flood protection system during Hurricane Katrina, and that the resulting catastrophe had it[s] roots in three main causes: (1) a major natural disaster (the Hurricane itself), (2) the poor performance of the flood protection system due to localized engineering failures, questionable judgements, errors, etc. involved in the detailed design, construction, operation and itself), (2) the poor performance of the flood protection system due to localized engineering failures, questionable judgements, errors, etc. involved in the detailed design, construction, operation and maintenance of the system, and (3) more global "organizational" and institutional problems associated with the governmental and local organizations responsible for the design, construction, operation, maintenance and funding of the overall flood protection system [1, p. xviii].

While this list points to three major causes, it actually describes a myriad of individual sources of responsibility, many of which can only be properly assessed in conjunction with other underlying factors. The task of assigning responsibility will therefore at best be incomplete and at least somewhat subjective.

When the possibility of suspect motives is combined with the inherent difficulty of assigning responsibility to individuals or organizations in cases of large-scale disasters it might seem that a *prima facie* case against undertaking ethical assessments of such cases has been established. In learning lessons about ethics, perhaps it would be better to focus on other types of situations in which ethical questions arise. If the objective is to learn lessons for future use, perhaps simpler is better. Contrary to this possibility, I want to argue in this paper that indeed examples like Hurricane Katrina are a robust and appropriate source for learning about ethics, revealing many important lessons about ethics that usually will not be learned from other types of cases. Study of disasters also shows how dramatic consequences often result from the confluence of many relatively minor, seemingly insignificant, decisions. Difficult as they might be to deal with, examination of large-scale disasters are a necessary part of raising the ethical awareness of engineers.

TYPES OF CASES

It is possible to distinguish a number of different approaches to looking at engineering ethics cases. Among these are: using positive role models as opposed to engineering failures, focusing on everyday events versus large-scale rare occurrences, looking at individual actions or organizational responsibilities, and doing prospective analysis where a given outcome is unknown rather than examining an already completed scenario through a retrospective analysis. The first possibility in each of these pairs has a number of advantages associated with it. However, there are also problems associated with them that point to the utility of using examples such as Hurricane Katrina.

A focus on positive ethical experiences would certainly reflect the typical reaction to events by engineers. As well, in raising children or teaching them, we have learned that positive reinforcement can be more effective in influencing behavior than can punishment. Stories such as structural engineer's Bill LeMessurier's proactive response to becoming aware of possible problems with the design of New York's Citicorp tower are inspirational to other engineers, filling them with pride in their profession [2]. However, most such stories go unnoticed by the press, following the dictum that dog biting a person is not news, only the reverse is. Thus, there are actually very few examples of positive ethical experiences of major impact to be found for analysis. Furthermore, once a positive example has been set out, it becomes more difficult to think of alternative solutions, even other beneficial ones. The advantage of a negative example is that it forces one to use one's moral imagination to consider alternative courses of action, since the given solution is from the onset unsatisfactory. As Henry Petroski has put it:

Success may be grand, but disappointment can often teach us more. Thus the colossal disasters that do occur are ultimately failures of design, but the lessons learned from those disasters can do more to advance engineering knowledge that all the successful machines and structures in the world. Indeed, failures appear to be inevitable in the wake of prolonged success, which encourages lower margins of safety [3, pp. 9 and xii].

It has also been argued that an emphasis on disasters is not productive [4]. A focus on everyday events has the advantage of relating to the life of the typical engineer. Most engineers will seldom, if ever, be involved in an ethical crisis of major national impact. Concentrating on large-scale events may thus mislead them into the belief that they are not likely to have to deal with ethical issues at all during their career or leave them unprepared for dealing with the type of ethical issues they are likely to experience. Also, analysis of everyday issues has the advantage of easier formulation of hypotheses regarding a case and of restricting the number of questions that need to be considered. It is, however, the very limitation of scope that is also a major disadvantage of these types of cases. Ethical analysis is by its very nature a complicated and often messy enterprise and to limit the scope of analysis to the immediate consequences of everyday decisions has the potential of resulting in only a restricted picture. When thinking about ethical issues one needs to be able to take into consideration a large variety of factors. In cases drawn from everyday life such as cases of data manipulation or conflicts-of-interest, broader issues of concern to the public are easy to ignore. Large-scale events such as Hurricane Katrina, on the other hand, by their very nature require taking into account a variety of societal considerations.

The challenges associated with rehabilitation and improvement of the New Orleans Flood Defense System (NOFDS) need to be addressed in an integrated way combining public and social, organizational and institutional, natural and environmental, and commercial and industrial considerations. This is a "system problem" that has many parts which are interactive, interdependent, and highly adaptive [1, pp. 14–19].

Focusing analysis on the behavior of individuals has the advantage of engendering a sense of individual responsibility for actions. By concentrating on a main character, it makes it difficult to avoid questions of personal accountability, which should be a significant strand in any ethical analysis. Since we tend to take an individualistic perspective on people in Western society, this type of focus also makes it easier for students of ethics to relate to such examples. However, to a large extent this perspective fails to reflect the nature of contemporary engineering work, which is primarily team and organizationally based. It is important for case studies to reflect the role of individuals in this context if students of ethics are going to be adequately prepared for real-life situations. Otherwise, when actually functioning within larger groups, they might be unable to make the connection between their accountability as individuals and their role in a larger whole where a variety of pressures will be exerted on them.

While it is possible to do prospective analysis of large-scale events, such considerations will tend to point toward viable solutions and the avoidance of potential disasters. Prospective analyses reflect true decision-making and highlight the positive elements of ethical behavior, but they are more suited to a study of everyday events. Prospective study of large-scale events is a complicated affair which requires sophisticated analytical tools such as fault tree or event tree analysis [5, pp. 156–159] for which young engineers may not be prepared or may be unable to carry out in the time allotted for ethical consideration. While an important lesson in the study of ethics is to discover ways in which unethical actions could have been avoided through the exercise of moral imagination, one common response in prospective analyses is instead to point to a lack of information as a way of avoiding consideration of hard ethical questions. This seems to be the wrong lesson to learn from the study of ethics, since in real life decisions typically cannot be avoided. The avoidance syndrome thus more often justifies lack of ethical analysis, something which is forced if an event has already occurred and judgments have to be made.

DISASTER ANALYSIS

The preceding is not to argue against using positive, everyday, individual, or prospective cases to think about engineering ethics. It does, however, make the point that such cases by themselves are insufficient to capture the full scope of engineering ethics and thus should, at the very least, be supplemented by consideration of large-scale events. Given the advantages of studying such occurrences, it might even be argued that these are the best starting point for the study of the subject, unusual as such events might be. Following are some of the built-in features of disasters which make them suitable for study.

LARGE AMOUNT OF AVAILABLE DATA

Any disaster is going to be widely reported and analyzed. In the Katrina example, not only was there significant press coverage, but it was given from a variety of different perspectives. At least five major professional reports on the event were produced. There is thus a ready source of information available to be used in ethical analysis, information that will not be restricted to one point of view.

COMPLEXITY OF THE SITUATION

By their very nature, disasters involve a variety of different factors. Analysis thus requires being able to recognize the interaction of such factors in doing ethics. Study of everyday events often leads to neglect of factors which are actually relevant to an analysis, for example considerations of family responsibilities.

MULTIPLICITY OF ISSUES

An important aspect of the study of ethics is being able to recognize ethical issues in the first place. As many philosophers have put it, in a sense it is the questions that matter more than the answers. In large-scale disasters a whole host of ethical issues will arise. Discussion of such events with others will help learners discover such issues. It will also permit a variety of different considerations of the same case, allowing each analyst a unique perspective, rather than pointing to a sole "ethically correct" judgment about an event.

INTERTWINING OF ISSUES

Not only will a large number of ethical issues be present in the analysis of a disaster, these will not be isolated from each other. This will demonstrate that seeing an ethical issue as an isolated phenomenon can lead to superficial and even misleading analysis of a total situation. Political, historical, and cultural factors, among others, need to be taken into account in developing a judgment about complex ethical situations.

VARIETY OF INDIVIDUALS INVOLVED

In any large-scale disaster a variety of actors will have a role to play. It is important for engineers to recognize that their professional ethics perspective is not the only one that will be brought to bear on a situation and that they need to take these other perspectives into account in their own analysis.

SOCIETAL CONSEQUENCES

All actions have long-range consequences. It is disasters, however, that make this fact most apparent and require consideration of these outcomes. Some analyses tend to focus on the inherent nature of the act itself as a mode of ethical analysis in the Kantian strain, but seeing the results on the lives of many humans forces recognition that different ethical theories should typically be considered in evaluating actions.

RESPONSIBILITIES TO SOCIETY

Disasters make more obvious than any other possible scenarios that engineers cannot just be concerned with themselves and their own welfare, but that it is an essential aspect of professional duty to be concerned about the "safety, health, and welfare" of society, as is proposed in almost all engineering codes of ethics.

CENTRAL ENGINEERING ETHICS ISSUES

Responsibility for public safety, health, and welfare is not only an aspect of engineering codes of ethics, it is the core ethical responsibility of engineers. Study of disasters is inevitably going to implicate this aspect of engineering ethics while study of a few smaller ethical examples might not.

VARIETY OF INTERPRETATIONS

By their very nature disasters are subject to a variety of interpretations. There is not going to be a sole right or wrong solution, especially since so many different factors and issues are involved. This demonstrates that ethical discussions seldom have a sole correct answer.

"SHOCK AND AWE"

One of the aims of ethics education is to demonstrate the importance of ethics in the lives of engineers. There is no better way of demonstrating this than where things have gone seriously wrong. If some engineers will not become involved in a discussion of whether they should accept a $50 gift, their attention will surely be caught by death and destruction.

KATRINA AS AN EXAMPLE OF ENGINEERING ETHICS

When the above advantages of looking to large-scale disasters are considered, Hurricane Katrina unfortunately serves as an ideal example. As the report of the Independent Levee Investigation Team put it:

> This event [Katrina] resulted in the single most costly catastrophic failure of an engineered system in history. Current damage estimates at the time of this writing [May 22, 2006] are on the order of $100 to $150 billion in the greater New Orleans area, and the official death count in New Orleans and southern Louisiana at the time of this writing stand[s] at 1,293, with an additional 306 deaths in nearby southern Mississippi [1, p. xviii].

Before assessing the case as one of engineering ethics, however, the possibility that it falls outside the scope of the field must at least be briefly considered. It might be argued for example that it was an "act of God," due to the forces of nature rather than human action, and thus ultimately not preventable. Or it might be claimed that the origins of the situation lie so far in the past that assignment of responsibility is a fruitless enterprise. Or one might point to the limited authority and power of engineers in the flood prevention system. Or that the people of New Orleans were responsible for their own fate, having made the decision to reside there. All of these possibilities have been debated, but it is beyond the scope of this paper to consider them in detail. It is enough for our purposes to establish that there is an engineering ethics dimension to the disaster. To that point, the head of the Independent Levee Investigation Team, Civil Engineering Professor Raymond

Seed, asserted on the *PBS NewsHour*: "The levee system failed in large part because of embedded deficiencies and because safety and reliability were put at risk, they were traded for economic efficiencies" [6]. Among the specific engineering failures pointed to in reports regarding the hurricane were the following.

INADEQUATE MARGINS OF SAFETY

The factor of safety used for design of the levee system was inappropriately low for protecting an urban population. It was more suitable for agricultural farm land [1, pp. xxiv, 10–15].

AN INCOMPLETE SYSTEM

At the time of the hurricane, sections of the New Orleans levee system were not complete, leaving open gaps in the system [1, pp. 8–16]. Also, parts of it were left below their design height specifications [7, pp. 1-1–1-2].

A FRAGMENTED SYSTEM

A variety of local organizations had control over different parts of the levee system. This resulted in incompatible structures with inherent weaknesses at points of transition [8].

INADEQUATE DESIGN

The failure of some of the levees could have been prevented by relatively simple measures such as installation of concrete splash pads at the base of the levees to prevent erosion or by using T-shaped rather than I-shaped walls [1, pp. 7–8]. Floodgates could have been installed at the entrance to the New Orleans drainage canals, but this was prevented by the infighting of local agencies [1, p. 8–15].

FAILURE TO LEARN FROM OTHERS

Studies for improvement of the levee system were commissioned by the Corps of Engineers, but these were ignored in subsequent design decisions [1, pp. 8–13]. A flood control system is used in the Netherlands with the capability of protecting against a Category 5 hurricane, while Katrina was a Category 3 at landfall and even weaker by the time it hit New Orleans [1, pp. 8–13], [9, pp. 1–2].

LACK OF CENTRAL CONTROL

Part of the failure of the system is attributable to the fact that no one had overall oversight of the levee system. The system was divided into sectors under the control of different local agencies, with inadequate coordination among these [1, pp. 5–8].

The above is not a complete list of possible engineering issues, but it is sufficient to show that the case is suitable for analysis in terms of engineering ethics. It should also be mentioned that the engineering failures pointed to above reflect a general consensus among the different reports on Katrina, although there are differences among the reports in regards to the assessment of specific responsibility.

STUDYING KATRINA

Making the claim that Katrina is in part a case of engineering ethics is of course not the same as making a judgment that engineers are in part or whole to blame for the results of the hurricane, although that is the conclusion some would immediately arrive at based on a statement like that of Lt. General Carl Strock of the U.S. Army Corps of Engineers, who stated on the *PBS NewsHour*:

At the end of the day, we have accumulated a level of risk and I don't think we truly understood that. So I think what you'll see is that, before we defer to others on elements that involve engineering decision-making and our ethical responsibilities to ensure what we build is going to serve its purpose, I think you'll see a greater propensity on the part of the Corps of Engineers to stand up and say "No" [6].

Using a disaster like Hurricane Katrina as a learning tool instead points to the complexity embedded in sophisticated ethical analysis. Just because safety was "put at risk" or engineers failed to exercise engineering autonomy does not automatically imply that they are to be blamed for what happened. The factors pointed to as advantages of disaster analysis show that disasters are good learning tools just exactly because they are not susceptible to simplistic conclusions. Examples like Katrina should instead be used as a subject for robust consideration of a variety of possible interpretations and conclusions, with the understanding that no one conclusion is necessarily the only right one, or even the best one. The ultimate point of looking at disasters in engineering ethics is thus not to assign blame, but rather to engender discussion of issues in engineering ethics. Both the nature of the disaster and the numerous sources about the case make Katrina an example that will provide rich opportunities for such discussion in the years to come. In conclusion, then, I want to point to some possible questions relevant for consideration of the case without, however, arriving at decisions regarding answers. Assigning responsibility is not the purpose of this article. That should occur in the process of discussion or individual consideration of the Katrina disaster. The purpose of raising the questions is to stimulate discussion. In light of the previous points made, I would also urge the reader to reflect on the possibility of other possible questions that are relevant to reflection about the case.

- To what extent should engineers be able to exercise their professional autonomy? Was the Corps of Engineers sufficiently autonomous in its decision-making processes, given the quote from General Strock above?
- What is the responsibility of engineers to ensure that people have a safe exit if a disaster strikes an engineered work? Should the Corps of Engineers have been involved in evacuation planning rather than this being left to local governmental agencies?
- Should engineers think of themselves as social experimenters with a duty of gaining informed consent from affected populations for their actions? Was the local population sufficiently aware of the potential risks associated with the levee system?[1]
- To what extent do engineers have a responsibility to ensure the safety of the local population, despite whatever countervailing demands might be made? Should the engineers have refused to build the levees within the design constraints if they believed those to be inadequate?
- What is a level of acceptable risk in different kinds of situations? Was preparing for a Category Three hurricane adequate for a densely populated area prone to hurricanes?
- What is the responsibility of engineers to monitor the ongoing status of completed projects they have designed? Did the Corps of Engineers have a responsibility for continued follow-up even though control of the system had been turned over to a variety of local agencies?
- Are there specific professional ethical responsibilities that engineers have that go beyond ordinary ethical duties? Should the issue of whistle blowing have arisen for any of the engineers involved in the building of the levee system?
- What is the role of professional organizations and outside parties? Should some of those who warned about the state of the levee system prior to the hurricane have taken stronger action?
- To what extent can engineers be expected to be aware of the interactive nature of decisions? Are the levee designers responsible for factoring in the effects of dredging of the Mississippi and of the loss of wetlands?
- Should engineers consider differential consequences on different groups in their design decisions? Did the levee designers have a responsibility to consider that different social strata in New Orleans would be affected in different ways by major flooding?

- Do engineers have a positive duty to do good, beyond the simple duty of avoiding harm? At what juncture do the risks involved in levee construction outweigh the potential benefits to the local population?
- To what extent should engineers submit to financial restrictions that are imposed on their projects? Are there limits to which safety should be compromised because of limited governmental funding for the levee project?
- What are the limits of engineers' duty to be aware of, and to utilize prior knowledge in their designs? Did the engineers have a responsibility to utilize the best available model from the Netherlands in their design decisions? Is the "not invented here" phenomenon ever a justification?
- Is a military organizational structure suitable for constructing urban civil works which involve the safety of large numbers of people? Can engineers be expected to answer to several different "masters?"[2]

When Henry Petroski wrote *To Engineer Is Human* he was surely referencing the saying "to err is human, to forgive divine." Engineers, like all others, are far from perfect, yet like many, most strive to do their best. In order to do so, they must learn from the past. This in itself is an important lesson to be learned from ethical analysis. As Petroski put it: "Yet no disaster need be repeated, for by talking and writing about the mistakes that escape us we learn from them, and by learning from them we can obviate their recurrence" [3, p. 227].

ENDNOTES

1. For discussion of safe exits and social experimentation see [10].
2. Different levels of discussion of the ethical questions posed are possible. For most engineers discussion will be based on their engineering background, their use of reason, and the moral beliefs they have been taught. At the next level discussants might seek out relevant passages of the Code of Ethics of an engineering group to which they hold allegiance. Ultimately, the author hopes that discussion of Katrina-like cases will encourage further delving into the literature on engineering ethics and more formal classroom study or the reading of textbooks like the Schinzinger and Martin one cited above, which address methodologies to deal with many of the specific questions raised.

REFERENCES

1. R.B. Seed *et al.*, *Investigation of the Performance of the New Orleans Flood Protection Systems in Hurricane Katrina on August 29, 2005*. draft final rep., no. UCB/CCRM-06/01. May 22, 2006; http://www.ce.berkeley. edu/~new_orleans/report.
2. J. Morgenstern, "The fifty-nine story crisis," *The New Yorker*, pp. 45–53, May 29, 1995.
3. H. Petroski, *To Engineer Is Human*. New York, NY: St. Martin's Press, 1985.
4. M S. Pritchard, "Beyond disaster ethics," *Centennial Rev.*, vol. 34, no. 2, pp. 295–318, Spr. 1990.
5. C.E. Harris, Jr., M.S. Pritchard, and M.J. Rabins, *Engineering Ethics: Concepts and Cases*, 3rd ed. Belmont, CA: Thomson Wadsworth, 2005.
6. PBS, *NewsHour*, June 12, 2006; http://www.pbs.org/newshour/bb/science/july-dec05/levees_10-20.html.
7. U.S. Army Corps of Engineers, "Performance Evaluation of the New Orleans and Southeast Louisiana Hurricane Protection System," June 1, 2006, Executive summary; https://ipet.wes.army.mil/.
8. Select Bipartisan Committee to Investigate the Preparation for and Response to Hurricane Katrina, *A Failure of Initiative*, Feb. 15, 2006; http://katrina.house.gov/.
9. A. Graumann *et al.*, *Hurricane Katrina: A Climatological Perspective*. NOAA National Climatic Center, Oct. 2005; http://www.ncdc. noaa.gov/oa/reports/tech-report-200501z.pdf.
10. R. Schinzinger and M.W. Martin, *Introduction to Engineering Ethics*, New York, NY: McGraw-Hill, 2000, ch. 3.

12 Standards in Engineering Society[1]

Albert L. Batik

CONTENTS

Albert L. Batik is a standards consultant. He was formerly associated with Information Handling Services as director of standards marketing. From 1958 through 1977, he was a member of the ASTM staff and in his last five years was deputy managing director of publications and marketing.

Is there any need or any use for voluntary consensus standards? Like it or not, we live in a world dominated by science and technology. In fact, we have reached the point that, like Frankenstein, we have such an advanced science and technology society, that we can literally remove all life from this planet. But nuclear holocaust is preventable by statesmanship. A short-circuit in a toaster cannot be prevented by even the most enlightened general secretary of the Communist Party. Directly or indirectly, safety for human beings is the single most important driving force for standards.

ON THE RIGHT TRACK

The first standard that ASTM ever produced dealt with steel rails. At that time, some rails were cracking, sending locomotives and trains off the roadbed. Apart from the loss of trains and freight, humans were being badly hurt—even killed. The standard that evolved dealt with testing a steel rail to see if it had the properties to keep the train on the track. Indirectly, it saved lives.

It is hard to believe that in 1884, in excess of 10,000 boilers exploded or ruptured. Is it any wonder that a Boiler and Pressure Vessel Code evolved and that in the period 1974 to 1984 there was not one single boiler or pressure vessel failure in the United States.

Standards are the language of trade.

Early in this century, there was a fire in Baltimore, MD. Fire companies from at least 100 miles around rushed to the aid of the stricken city, only to find that the hose couplings did not fit the hydrants or other hoses. The city was a tower of Babel; screw threads were not standardized.

Thankfully, these problems no longer exist. Would anyone from government or industry employ an engineer that put safety last on his agenda when he designs, manufactures, processes, installs, or delivers a product or service? If safety is a basic requirement of engineering, then how can an engineer effectively and ethically pursue his or her profession if he is ignorant of such things as the National Fire Code, the Boiler and Pressure Vessel Code, or the National Electrical Code. These are standards.

OUR NATURAL RESOURCES

For years, as the world industrialized, mankind blindly assumed that the water, the air, and the land were free to use as seen fit. Hopefully, man has awakened in time to restore the environment. Were it not, however, for standardized tests, how could the past, present, and future road mankind must travel to an ecological safe world be recorded? The engineering profession is required, legally and morally, to build in environmental safeguards into the products and processes they develop. Without standards for facilities and production, for testing effluents and emissions, and for statistical analysis, there would be a fundamental problem. The problem is not the polemics of environmental zealots. Rather, the problem is where can man live if industrial garbage is not controlled? Engineers ill educated in the role of standards to protect this tiny planet are ill educated to serve mankind.

It seems strange that more than a century ago civil engineers recognized, in part, that sanitation was a significant area of their professional concern. The development of water collection and treatment plants; the disposal of sewage; and the development of sanitary food handling all came from a single premise. That premise was that if these cares were not tended to or were tended to poorly, disease would sweep the land, killing as it went. Standards were developed to ensure safe water, and some protection from sewage and contaminated food. Have those needs subsided? Of course not. In fact, they have intensified and that is one additional reason that engineers must be made aware of the tools at hand.

In the 1970s the United States learned that petroleum was not an unlimited resource. Unbeknownst to the general public, but well-known to strategic planners, is the fact that many materials are in short supply. There are those that say one material can be substituted for another—composite auto bumpers for chrome plated steel. This is quite true. Be that as it may, matter cannot be created nor destroyed. So for all intents and purposes, we work with a fixed reservoir of earthly goods. Resources must be wisely juggled and equivalents scientifically selected. There is the rub. To select equivalents, standard methods of comparison are required, and once developed, must be intelligently used.

Engineers working in an atmosphere of abundance may see little need now to consider alternatives. But the professional engineer has an obligation to his employer, the community, and to his own self-respect to select materials on the basis not only of past considerations, but on the basis of existing or potential economic conditions. An adequate understanding of standards is a most important tool.

STANDARDS AND INNOVATION

Too often standards are considered an impediment to technical advancement; not so. Standards are the handmaiden of innovation. They permit unbiased appraisal of what will be to what is. This author was one of a team of chemists that developed silicone elastomers. One of the key steps in the development of these artificial compounds was the comparison with natural rubber. This was done by the use of standard physical tests. From the results obtained, it was possible not only to match the properties of naturally occurring elastomers, but to improve the artificial elastomers such that they had properties never before available in a rubber-like compound.

Directly or indirectly, safety for human beings is the single most important driving force tor standards.

It is possible to solve engineering problems through the wider application of technological advancements. Professionals are crossing the artificial boundaries of disciplines: bio-engineering, mathematical synthesis, and astroengineering. The uniformity and discipline brought about by the intelligent use of standards has been a major tool in achieving these technological advancements. "Give me a place to stand and I can move the earth." That place, that function, is the platform of standards that make the application of technology possible now and in the future. This is one reason that standards are constantly updated.

U.S. MANUFACTURING UP TO STANDARD?

A comparatively low tech product such as an automobile has nearly 400 suppliers and 15,000 component parts. Without standards for even the simplest component, such as a screw, could such a vehicle be made to operate? Yet, there are engineers whose attitude towards standards, at best, can be described as scorn. The United States is the single largest producer of manufactured goods in the entire world. Yet, the U.S. reputation for quality is almost gone. It seems that as knowledge and understanding of standards has declined, so, too, has quality.

America cannot attain quality with slogans. "Quality circles, zero defects, and quality is job one!" are great, but worthless words, without an understanding of statistical quality control, without the development and application of quality standards, and without the understanding and drive of the engineer on the job. The understanding of standards and the standardization process as revealed in hundreds of locations is akin to osmosis. Some knowledge trickles through, but much hits a mental barrier. By comparison, Germany and Japan instill a belief in standards and quality in their engineers that is unmatched in the United States.

> Standards are the handmaiden of innovation because they permit unbiased appraisal of what will be to what is.

Some universities are teaching the principles of standards. Some corporate locations have training sessions. But, by and large, the attitude in foreign competitors verges on a cult. The god of that cult is standards. Scrap rates of two or three percent may be acceptable in a U.S. manufacturing plant. Visit a major manufacturing facility in almost any European country and talk about scrap rates. Mundane, low tech, basic standards; very unappealing, but cost-efficient. There is no glamour in standards and that makes it all the more important that the engineer recognize their importance, just as the doctor recognizes that a low tech stethoscope is every bit as important as a glamorous cat-scan.

Manufacturing and processing must be made a challenge once again. Cost reduction, quality control, scrap reduction, recall reduction, alternative materials, and processes are not catch words. They are worthy of the engineering profession's best efforts. Engineers do not realize industry standards are among the best tools to achieve efficiency goals.

DIALOG

One of the engineer's constant gripes is that others in the organization do not cooperate. Why is purchasing slow? Why do they get the wrong materials or wrong quality? Why can't they get the right part sizes? Well, sounds like the pot calling the kettle black. If engineers would specify standard parts, standard materials, and standard components, purchasing will be more efficient, more cost-effective and, in the long run, make the organization more competitive. Today, what kind of dialog takes place between the procurement and the engineering functions? Most importantly, what is the role of standards in such a dialog? Once there is an understanding and a commitment to standards on the part of engineers, there will be the right level of cooperation in procurement.

What is made must be sold, or rot. How many engineers even consider that the judicious use of standards is one of the best marketing tools ever devised? Standards are the language of trade. It is possible to define grades, sizes, performance. It is possible to describe just what can be produced. It is possible to define exactly what is needed. It is possible to ensure that what is sold is what is delivered. Does this facilitate trade? Absolutely! Standards are not only essential for domestic commerce, they are vital to world trade.

ACTIVE STANDARDS USERS

A major gap exists between the theory that engineers are taught standards and the reality that they appear to have little or no knowledge about standards. The purpose of these discussions is not to assess blame, but rather to delve for solutions. However, is it fair to ask engineers to use and understand standards when in the United States, at least, it would appear easier to count grains of sand on a beach?

Why does it take 423 organizations in the United States to write the standards this country needs, when it is done by five in Canada, three in Britain, two in Germany, and one in France? Why, in the United States, are there seven physical sizes for the paper on which standards are printed? There are almost 60 different designation systems containing duplicate acronyms and duplicate numbers. There are draft, proposed, tentative, recommended, and many other adjectives associated with standards. There are at least eight methods for revising standards. There exists mass confusion on what organizations promulgate standards. Some standards are without designations, even without attribution. Organizations stop issuing standards, yet their standards are mindlessly called out as requirements. Standards are transferred from one organization to another with change in designation, but without public notice.

Is it any wonder then that the engineering profession has difficulty learning what is going on? Is it any wonder that too often the untrained engineer faced with disarray, throws up his hands. The standards community has an obligation to generate standards, and to work with its constituent bodies and its users to rationalize the way it does business. In so doing, it must work also with academia to establish methods of indoctrination that will bring young engineers up to speed and involved in the standards process.

How many engineers understand that government standards are in many instances not standards, but are regulations or procurement documents? Do engineers realize that trade associations and professional societies generate standards in addition to strictly standards bodies? Does it make a difference? Do they understand the meaning of a national standard, an international standard? How involved are they in the generation, revision, review, or balloting of standards? For standards to succeed engineers must contribute and be part of the process; their professional input counts.

PLANTING SEEDS

The leadership of U.S. corporations is no longer in the hands of the technically qualified; attorneys and financial people lead the corporations and key government agencies. It cannot be said, however, that there are no engineers and scientists in position of influence. Thomas Edison, Herbert Hoover, Henry Ford, General Hoyt Vandenberg, Frank LaQue, F. Stanley Crane, and Miles Claire are not just names. These people were movers and shakers and, most importantly, they were active in standards development, use, and policy decisions. For similar support now and in the future, these seeds must be planted among the young.

Product liability suits are escalating at an ever increasing rate. The environment of the engineering marketplace has changed. But, badly designed products, unsafe products, dangerous processes, faulty construction, and poor procedures do exist. They exist in the workplace, on the road, in the home, in hotel atriums, and in the air. Whose fault is it?

As a nation, the United States seems willing to pay millions for cures and almost nothing for prevention. For every dollar spent for medical research, nearly 22 are spent for treatment. Half the amount spent on prevention makes the failure unnecessary. Standards are effective, cost efficient, and are based on the cumulative experiences of thousands of expert man hours. Unfortunately, in too many instances, this body of knowledge is either unknown or ignored.

The field of standards delivery is on the verge of tremendous progress. Even now, there are organizations that can deliver the vast majority of standards overnight. There are extraordinarily good indexing systems that make possible the location of some of the most obscure standards. On-line search capability is here; in the next five years, enormous strides will be made to increase the point of use capability. But for all the technological advances in the delivery mechanism, what good will it be to anybody, if the insight, training, and willingness to use these prosaic documents simply is not there? Standards do better, regardless of the barriers.

Curriculums are crowded. Students are bending under the load. Professors have little time. Text material is lacking. Many reasons can be given for the lack of proficiency in standards. But, alternatives must be explored and means set forth whereby standards will, once again, play a more meaningful role in the lives of engineers and scientists.

Manage the situation or lead it. Managers administrate. Leaders innovate. Managers maintain. Leaders develop. Managers believe in systems and forms. Leaders believe in people. Managers control. Leaders trust. Managers do things right. Leaders do the right thing. Let us do the right thing.

The United States is the single largest producer of manufactured goods in the entire world. Yet, U.S. reputation for quality is almost gone. It seems that as knowledge and understanding of standards has declined, so, too, has quality.

ENDNOTE

1. This article is taken from a paper presented at the symposium on Standards Education held on July 17, 1988.

13 The Case for Reforming Our Standards-Setting System

Ralph Nader and Peter Maier[1]

CONTENTS

The lack of a voice for consumers and small-business representatives at national standards-setting organizations has led to anticompetitive practices and a stifling of innovation. Engineers could help redress the balance.

Shivering consumers and kid-off factory workers who suffered through last winter's cold snap and natural gas shortage probably don't know that an abuse of standards-setting authority worsened their misery. Home gas heating bills and the number of industrial shutdowns could have been considerably lower had not the gas industry used its grip on gas appliance standards to stifle an important energy conservation innovation.

The suppressed device is the automatic vent damper. It prevents heat from escaping up the flue when a home gas furnace is off, reducing gas consumption by about 20 percent.

In 1969 a small firm called Save-Fuel asked the Z-21 gas appliance standards committee sponsored by the American National Standards Institute to write a damper standard for existing gas furnaces. Save-Fuel had no real choice but to seek a Z-21 standard because such a standard is a prerequisite for the testing-laboratory approval that's so vital to the salability of any gas appliance. ANSI, although it is a private organization, coordinates many of these standards in the United States.

Unfortunately for both Save-Fuel and consumers, its very structure makes the Z-21 committee inhospitable to a new firm manufacturing gas-saving devices. It is dominated by representatives of gas utilities and of gas equipment makers—who have a financial stake in excluding from the market newcomers whose gas-saving products would lower gas consumption and compete with equipment made by present manufacturers. Indeed, an industry trade group, the American Gas Association, serves as secretariat for the committee and pays its administrative and technical expenses.

Not surprisingly, the Z-21 committee showed little interest in developing a vent damper standard. It stalled for fully eight years before finally adopting one in April 1977. And that action came only after congressional hearings, an expose on the "60 Minutes" television program, and a bitterly cold winter had focused public attention on the committee's role in keeping vent dampers off the market. (The standard completed its journey through ANSI's approval process in November 1977.)

The vent damper case is but one example among many of trade product standards abuse. These abuses result in higher prices, fewer meaningful products for buyers to choose from, enormous waste, and avoidable hazards. They show the urgent need for fundamental reform of the standards-setting system.

Of course, trade product standards also have beneficial effects. They serve as the language of industrial commerce, allowing producers and buyers to define product specifications and thus to avoid costly misunderstandings. As nearly all industrial nations have discovered, product standards are vital to the efficient manufacture and distribution of complex modern products—everything from pipe, to plywood, to electrical components.

Another benefit, though largely unrealized in this country, is the stimulation of competition among producers by giving buyers product information to use for comparison shopping (for example, performance ratings for household appliances).

Despite this positive impact, widespread abuse of trade product standards harms the public and presents cause for serious concern. A July 1977 report by the U.S. Department of Commerce to the Office of Management and Budget reached a similar conclusion: "While our voluntary standards system has with very little public funding made a substantial contribution to our industrial growth, it would not be realistic to deny that standards have, in a significant number of cases, had an adverse impact upon competition and the consumer."

PATTERNS OF ABUSE

Aside from the Save-Fuel incident, another example of "adverse impact upon competition" involved low-water boiler cutoff devices. In that case a dominant manufacturer of boiler controls used its close ties to a standards committee of the American Society of Mechanical Engineers to spread the word—deliberately and incorrectly—that a small competitor's innovative product did not comply with the standard. According to testimony before the antitrust subcommittee of the Senate Judiciary Committee in 1975, the dominant firm, McDonnell & Miller, obtained a letter from ASME that was presented in McDonnell & Miller's promotional literature as an official society ruling against the boiler device (see *New Engineer,* July/August 1975). However, the device's design was such that the matter discussed in the letter was irrelevant. After prodding, ASME provided a corrective letter, but the damage had been done, since McDonnell & Miller was not bound to use it. By this maneuver, which sent the small manufacturer's sales into a tailspin, McDonnell & Miller was helped to retain its 85 percent share of the annual $10 million boiler control market. Other instances of anticompetitive uses of standards have been reported for products ranging from plywood to water backflow valves.

A second pattern of abuse is reflected in standards that boost dollar sales while benefiting only the producer. Excessive lighting level standards set by the Illuminating Engineering Society (IES), for example, cause the nation to light its buildings far more brightly than necessary for visual safety, efficiency, or comfort. Lighting experts estimate that U.S. lighting can be safely reduced by one-third; the savings in the nation's total electric lighting bill would be more than $3.5 billion annually.

The widely followed IES lighting level standards have increased dramatically in recent decades and now stand as the world's highest A good example of their wasteful excess is the IES school lighting standard, also adopted by ANSI. For certain classrooms, such as study halls and lecture rooms, IES recommends 70 footcandles—approximately the equivalent of ten ceiling globes, each with a 100-watt incandescent bulb, in a ten-by-ten-foot room. This school lighting levels more than twice as high as the 1957 standard's 30 footcandles. The graph shows the IES school lighting levels' startling increases over the past 60 years.

IES standards for offices are also excessive. Their illumination levels are roughly twice as high as those recommended by the Federal Energy Administration and by a General Services Administration study. Similarly, IES library lighting levels are double those that Keyes Metcalf, a top library lighting consultant, has recommended as adequate for the substantial majority of library users.

Proper vision does not require high general lighting levels. Ophthalmologists generally agree that lower lighting levels will not harm the eye; they consider the 1957 standard for school lighting adequate for most people in most reading situations. Nor are the high IES lighting levels necessary for most people to see well. Beyond 20 to 30 footcandles, increased illumination provides only minimal improvements in the eye's ability to perceive detail.

The reason for IES's wasteful standards is that the Society is essentially a trade association for the lighting industry. IES describes itself as a "scientific educational organization," but in fact it primarily serves the industry's candlepower pushers.

Fully 57 percent of IES members, and most of the IES "sustaining member" financial angels, sell electricity or lighting products and services and thus have a direct economic stake in keeping lighting levels high. IES obtains well over half its income from the lighting industry. The industry largely controls the IES technical committees where the critical work of writing the standards takes place. Sellers or installers of lighting equipment and electricity preside over half these committees. The roster of IES national officers also shows undue industry influence; almost half represent sellers of lighting equipment or electricity.

A third type of abuse is when lax standards allow the sale of dangerous products. Standards developed by Underwriters' Laboratories (UL), for example, facilitated the installation of hazardous aluminum branch-circuit wiring in more than two million homes from 1965 to 1972. This wiring soon earned a reputation as a fire hazard, due to two fundamental differences from traditional copper wiring: first, unlike copper oxide, the aluminum oxide that forms on wire surfaces at connectors is a poor conductor, and aluminum's expansion coefficient is one-third greater than copper's. These two weaknesses often lead to loose, overheated connections that can cause fires to begin within walls. Fire departments across the country have reported aluminum wiring as the cause of several hundred fires and hazardous electrical failures—and of at least 12 deaths.

It was on the basis of these reports that in September 1975 the U.S. Department of Housing and Urban Development banned the use of aluminum wiring in mobile homes. In November 1975 the federal Consumer Product Safety Commission followed suit and preliminarily determined certain sizes of aluminum wiring to present an unreasonable risk of death or injury. Unfortunately, the industry's legal challenges to the commission's jurisdiction over household wiring have stalled the 1975 action. But in October 1977 the commission renewed its attack by filing a lawsuit against 26 aluminum wiring producers. The lawsuit seeks to require the producers to inform consumers about the hazard and to repair some of the dangerous wiring systems.

The long string of weak UL standards began in 1966 when UL gave the go-ahead with only minimal restrictions for wiring certain household outlets and switches with aluminum. By 1969, however, enough fire departments and homeowners were complaining about fires and smoking wall sockets to wake even a sleeping giant like UL, which then cosponsored a survey with the National Electrical Manufacturers Association to study the problem.

Despite its knowledge of the problem in the late 1960s, UL did not significantly upgrade its standards until 1971. The most important of those changes amounted to a long-overdue prohibition of "old technology" connection devices. This prohibition supposedly took effect in September 1971, but in fact UL privately authorized the electrical industry to ignore it and to continue wiring homes with the same UL listed connection devices that the organization had just declared unsafe. Not until almost a year later, when new connection devices became available and the industry's profits were no longer endangered, did UL reconcile its private and public policies and require them.

UL's current aluminum wiring standards, though an improvement over earlier ones, remain inadequate to cope with aluminum wiring's dangerous properties. For example, the standards allow use of an indium coating to protect against poor connections, even though the coating literally disappears after only a few years of use. The Consumer Product Safety Commission has criticized the UL standards as deficient in 13 important respects.

Another instance of inadequate safety standards was publicized when, in 1973, the Federal Trade Commission charged the American Society for Testing and Materials and a group of plastics manufacturers with misrepresenting the fire hazard characteristics of polystyrene and polyurethane plastics. According to the FTC complaint, the plastics makers used ASTM standards to justify claims that their products did not burn or were self-extinguishing. Yet, the commission charged, the plastics frequently spread flames and generated smoke, heat, and flammable or toxic gases more quickly than conventional materials such as wood or cotton.

Lighting levels specified in standards set by the Illuminating Engineering Society have in some cases more than doubled in the past 20 years. This is hardly surprising, since IES is primarily an industry trade association.

Widely used for construction and home furnishings, these plastics have been involved in numerous fatal fires, including the 1970 dance hall fire in France in which 145 teen-agers died. ASTM and plastics makers had known of these serious fire hazards at least six years before the FTC complaint, but they had neither warned the public nor halted the misleading claims.

The FTC concluded that ASTM had been "dominated or unduly influenced" by the industry and had become "infected with commercial intent." In 1974, the commission obtained a consent order from the plastics manufacturers in which they agreed to stop misrepresenting their products, to notify the public of the fire hazard, and to undertake a $5 million research effort to find remedies to the problem. Unfortunately, ASTM itself was not included in the final consent order, partly because FTC's jurisdiction may not cover non-profit organizations.

ABUSES STEM FROM UNFAIR STANDARDS-SETTING PRACTICES

To prevent the kinds of abuses just cited, trade product standards-setters must adhere to basic principles of fairness and due process.

Unfortunately, today's standards-setting procedures fall short of the mark. They fail to afford all significantly affected interests an effective role in shaping standards, and they lack sufficient safeguards against the misuse of standards by parochial interests.

One of the most striking flaws is domination of the standards-setting process by industries directly affected by the standard. To supplement the examples already cited in standards-writing for gas furnaces and lighting levels, the American Nuclear Society provides a clear illustration of the extent of this domination. Take, for instance, the society's ANS-50 Nuclear Power Plant Systems Engineering committee, which sets nuclear reactor safety standards often subsequently adopted by ANSI. Fully 22 of the 26 ANS-50 committee members represent companies involved in producing or using nuclear reactors.

The ANS-50 committee's working groups are similarly industry dominated. The 11-member ANS-58.3 group, which wrote ANSI standard N182 regarding protection of reactor safety systems from physical hazards, has only one non-industry member; so does the ten-member ANS-55.2 group that wrote ANSI standard N199 regarding the processing of liquid radioactive wastes.

These companies participating in standards-writing usually are large corporations well aware of standards' marketplace clout. Economist David Hemenway concluded in a 1975 study of standards groups: "[I]n the United States, major firms tend to dominate the standards-writing process. This gives them great power in determining not only what standards are created, but exactly what those standards will say."

Large corporations' major role in standards-setting is reflected in ANSI's Board of Directors, which reads like a Fortune 500 honor roll. At least half the board's members represent giant corporations such as AT&T, Exxon, IBM, U.S. Steel, and Xerox, while few represent small businesses.

In contrast to the major role of large corporations, standards-setting committees generally have few representatives from among the consumers who buy the ultimate finished products in retail stores. Most have none. This holds true not only for committees setting standards for industrial equipment and materials, but also for those setting standards for final consumer products. For instance, ANSI's A197.3 standard for household garbage disposal plumbing requirements was written by a 20-member committee with *no* consumer representatives. (The home-appliance industry had 16 members, with three government officials and a private testing laboratory comprising the balance.) A similar dearth of consumers exists in the ANSI-ASME F2.1 Food, Drug, and Beverage Equipment standards committee. Food processors and equipment manufacturers comprise 100 percent of both its Executive Committee and its Beverage Subcommittee.

Nor do "consumer councils," relied upon by ANSI and Underwriters' Laboratories as a major source of consumer participation, improve the situation. Most of ANSI's consumer council members, and many of UL's, do not even represent consumers. Instead, they represent manufacturers and industrial or commercial intermediaries such as Whirlpool, DuPont, and the American Bankers Association. Moreover, the councils' reviews come after a proposed standard has been written, too late to significantly affect its content.

TOWARD A FAIR AND DEMOCRATIC STANDARDS SYSTEM

Engineers occupy an ideal strategic position to help reform the standards system. As professionals who routinely use standards in their jobs and who often help write them, engineers can detect emerging problems or abuses earlier than most citizens. They can be particularly attuned to the societal costs and to the personal frustration that results from the stifling of technological innovation and from other standards failings. And, several engineering societies themselves write standards. Engineers should therefore take the lead in obtaining standards reforms.

If they are to participate actively in standards-writing, consumers will need the help of technical experts. Perhaps engineers could set aside time to provide this, or give financial support for public-interest organizations.

A truly fair and democratic standards system would include the following procedural safeguards.

- **Full participation by consumers, small businesses, and other Important underrepresented Interests.**

Consumers, small businesses, and other important groups now severely underrepresented must be included as active participants in the standards-writing process. Their vocal presence can help prevent narrow special interests from abusing trade product standards. Had the Z-21 gas appliance standards committee included a substantial number of consumer members, for example, it probably would have been more sensitive to the pressing need to conserve natural gas and to lower home heating bills.

As a practical matter, financial assistance must be provided to representatives of consumers and small businesses. Unlike industry, these relatively impecunious interests often cannot afford to prepare for and attend committee meetings. Whether obtained from standards groups or the government, the financial aid should include both reimbursement for travel expenses and a per diem consulting fee.

To participate effectively in standards-writing, consumer representatives will also need access to advice from technical experts. By fulfilling this need, the engineering profession will gain an opportunity to make its long-overdue entrance into the public interest field. Engineering firms or individual engineers could devote a portion of their time to advising consumer representatives on technical issues arising from their participation in standards-writing (where the standards are unrelated to the engineer's sources of income).

An analogous arrangement is already found in the legal profession, where a number of law firms and individual attorneys donate a portion of their time to public interest cases. Alternatively, engineers could tithe a small percentage of their income to support public interest engineering groups who, as part of their consumer-related activity, could then advise the consumer standards representatives. Engineering societies could help establish and coordinate these public interest mechanisms, thereby helping to fulfill the professional engineer's expressed ethical obligation for civic service.

- **Procedures to prevent the affected industry from dominating standards-writing efforts.**
    Affected industries should be limited to no more than one-third of a committee's membership, with independent representatives such as government officials (especially where government standards apply) and consumers filling the other slots. All committee chairpersons should be independent of the affected industry. This will help assure neutrality in the chairperson's decisions, which greatly influence committee work.

Other important safeguards include:

- **Open standards-writing committee meetings, minutes, and records.** Secrecy fosters abuse. Public participation in and scrutiny of standards-writing will inhibit misuse of standards.
- **Preference for performance standards.** Standards that establish performance requirements rather than exact product specifications—for example, strength instead of thickness—are less likely to stifle innovation.
- **Written technical justifications for all standards, including minority technical opinions.** These justifications will give the user some idea of the premises and research that underlie the standard, and will assist government agencies in deciding whether to adopt it.
- **Effective mechanisms for appealing adverse standards decisions.** Appeal mechanisms provide a way to reverse or modify unwise decisions made by standards committees.

No trade standards group currently provides all these procedural safeguards. Indeed, most groups fail several parts of the fairness test. The 1977 Commerce Department report mentioned earlier concluded, "The procedures of the majority of the private standards-development organizations ... do not include minimum due process safeguards." Given the resulting abuses, there is ample reason for the federal government to require standards groups to follow specified fair standards-setting procedures.

Widespread governmental adoption of trade product standards further justifies regulation of standards-setting procedures. Federal, state, and local governments routinely incorporate privately developed standards in statutes and regulations, often at the urging of standards groups. This elevates standards to a mandatory status enforceable by law. Yet, too often the standard is adopted after only cursory review of its content because the governmental body lacks the time or expertise to study it closely or to develop a standard on its own.

It was this consideration that led the National Fire Protection Association (NFPA), a major standards group, to publicly endorse the concept of federal regulation of private standards-setting procedures. NFPA President Charles S. Morgan testified in April 1977 before the Senate Subcommittee on Antitrust and Monopoly: "[W]e support the concept of regulation and oversight by the federal government of the procedures by which standards are developed within the private sector to assure that such activities embrace procedural fairness and are free from illegal practices."

Morgan testified at hearings on Senate Bill S.825, the Voluntary Standards and Accreditation Act of 1977, introduced by Senators James Abourezk, Hubert H. Humphrey, and Birch Bayh. Title I of S.825 directs the FTC to write and enforce procedural rules for trade standards groups—for example, requirements for balanced membership on standards-writing committees and for effective appeal mechanisms. Last July, Title I received an important boost when the Department of Commerce endorsed it.

Title I is a modest enough reform that merely adjusts the standards system for smoother running. Nonetheless, it offers important improvements and would at long last introduce some fairness into the standards system.

Besides procedural reforms, we also need government institutions capable of setting standards independently when private standards prove insufficient. These government standards would provide sound alternatives from impartial sources to compete with private standards. For example, public school systems should have access to independent analyses to determine how much light their students need for reading, rather than to have to follow the self-serving recommendations of the lighting industry. Government standards and testing programs would also save taxpayers money in more efficient government purchases of supplies, equipment, and services.

Engineers should support these reforms and work to set them in place before still more engineering innovations are suppressed and more consumers harmed.

ENDNOTE

1. When this piece was written Peter Maier was a staff member of Ralph Nader's Center for Study of Responsive Law. Mauer and Nader have both written and testified about product standards issues. Copyright © 1978 by the Center for Study of Responsive Law.

14 Nuclear Power
More Than a Technological Issue

Ralph Nader

CONTENTS

> *Nuclear power is not merely a technical issue. Responsible engineers must recognize the human ramifications of their technologies. They must become sensitive to the myriad of human issues in the nuclear power question and to the overwhelming legal problems and burdens that are developing under the stresses this technology is placing on our society and world.*

Although the national debate on nuclear power is intensifying, some engineers and scientists view the nuclear issue as merely a technical one. Technologists such as Ralph Lapp and Norman Rasmussen condenscendingly dismiss all doubts. If ordinary people will just leave nuclear power to the "experts," they feel, all the problems will be solved or will disappear.

Of course, this technical elitism has been severely undercut by more than 2000 scientists and engineers who recently stated that the risks of nuclear power were "altogether too great" and urged a national energy policy based on energy conservation and nonnuclear power [1].[1]

But although this scientific controversy exists, the debate on nuclear power cannot be based on technical issues alone. Nuclear power is fraught with, if not dominated by, issues that are institutional and political rather than technical. There are technical issues, to be sure; but even these are ridden with controversy and unresolved problems.

ONE TECHNICAL QUESTION

Reactor safety is one example of an unsettled technical question. In August 1974, the Atomic Energy Commission (AEC) released a draft report which concluded that the chances of a major nuclear reactor accident were equivalent to the chances of a meteor striking a large city. From the fanfare with which the report was released, it is evident that the AEC and the nuclear industry hoped the report would lay to rest all concerns about reactor safety. But these hopes were not realized. The severe weaknesses of the report have contributed to greater skepticism over nuclear power.

At the outset, it should be emphasized that the report, the *Reactor Safety Study* (RSS) [2], only presumed to cover nuclear reactors themselves. The report did not cover the transportation of radio-active materials by truck, rail, and barge; the disposal of radioactive wastes; the risks of sabotage, theft, or terrorism; fuel reprocessing plants; or uranium mining processes and wastes.

The RSS attempted to predict probabilities and consequences of major reactor accidents. It assessed the chances of a meltdown accident as one in 17,000 per reactor per year. A meltdown accident would occur if the reactor generated heat faster than that heat could be removed. The most widely accepted scenario for this accident would begin with a piping rupture, allowing the water that cools the uranium fuel to escape.

The RSS concluded that most meltdown accidents would result in insignificant radiation exposure to the public. The chances that 70 deaths will be caused by a reactor accident are one in a million per reactor per year, according to the RSS. The worst accident that the RSS considered was predicted to occur once every billion years per reactor. This accident would cause 2300 immediate deaths, 5600 immediate injuries, and 6.2 billion dollars in property damage.

REACTORS ARE UNSAFE

The most extensive and severe criticism of the RSS comes from a joint review by the Sierra Club (SC) and the Union of Concerned Scientists (UCS), a group of professionals in the Boston area. The SCUCS critique [3] was organized by Henry Kendall, professor of nuclear physics at MIT.

The SC-UCS review found significant weaknesses in the conclusions of the RSS, with regard to both probabilities and consequences of reactor accidents. To predict reactor accident probabilities, the RSS used reliability estimating techniques developed by the aerospace industry. The SC-UCS pointed out that the aerospace industry later abandoned these techniques as a means of providing exact reliability estimates. As many as 35 percent of the failures in the Apollo program, for example, were deemed "incredible" before they happened.

The SC-UCS tested the validity of the RSS estimating techniques by applying them to a reactor accident that has already occurred. The result was an accident probability prediction of one in a billion-billion. The fact that this accident has already occurred casts a doubt on the RSS methods, to say the least. It seems that an accurate assessment of reactor accident probability might not be available until after an accident has occurred, which will be too late.

The SC-UCS review found that the RSS underestimates accident health consequences by at least a factor of 16. This finding is generally supported by the Environmental Protection Agency [4] and the AEC Regulatory Staff [5]. With the SC-UCS corrections, the most severe accident considered by the RSS could cause 36,800 *immediate* deaths and 90,000 *immediate* injuries. These totals are substantially equivalent to the results of accident consequence studies performed by the AEC in 1957 and 1965, which the RSS claimed to refute [6].

When the SC-UCS corrects the RSS for errors in probabilities and consequences, they find that the risks from nuclear power plants are similar to nonnuclear risks. Where the RSS found nuclear risks similar to the dangers of being struck by a meteor, SCUCS finds nuclear power as dangerous as other technological hazards. With the correction factors, the probability of a nuclear plant catastrophe becomes similar to the probability of catastrophes from air crashes, fires, explosions, dam failures, and toxic chemical accidents.

ZERO CREDIBILITY

A more recent review of the RSS by the American Physical Society (APS) [7] has destroyed whatever credibility the RSS might have retained after the critiques of EPA, SC-UCS, and the AEC Regulatory Staff. This latest review was conducted by a panel of 12 scientists for APS. Like the

SC-UCS review, the APS study also had no confidence in the RSS accident probabilities. The APS study assessed the calculated probabilities of accident "branches," or hypothetical accident sequences, in these words: "[B]ased upon our experience with problems of this nature, involving very low probabilities, we do not now have confidence in the presently calculated absolute values of the probabilities of the various branches." [8]

The APS panel chose not to review the consequences of the most catastrophic RSS accident, as this had already been covered by other critiques. Instead, the study looked at a "small" accident, which the RSS predicted would cause 62 immediate deaths, 310 genetic defects, and 310 lethal cancers over the long term. The APS study identified impressive faults in the RSS analysis of long-term health effects. The RSS assumed that persons downwind of a reactor accident would receive radiation only through the first day following the accident. In reality ground-deposited radioactivity would irradiate people over a large area for an extended period. Correction of this factor alone increased lethal cancers and genetic defects by a factor of 25.

The APS study also found that RSS had neglected the selective and much larger irradiation of special tissues, notably the lungs and thyroid. APS corrects for these errors and others and concludes that the "small" accident would cause 10,000–15,000 lethal cancers and 3000–20,000 genetic defects over the long term [9].

Until the RSS, reactor safety researchers had held that the consequences of a meltdown accident would be severe, but that the probability of such an accident was very low—one in a million or less. The RSS concluded that a meltdown was much more likely to occur, but the consequences of an accident might not be severe. Professor Frank von Hippel, a member of the APS panel, assessed the RSS in this manner: "…. two omissions in the calculations made there [the RSS] resulted in an underestimate of the average number of deaths from the reference accident by one or two orders of magnitude. Thus it appears that the APS group may have had the dubious honor of restoring the *status quo*: A reactor meltdown accident with containment failure would indeed be very serious." [10]

The critiques by the APS and other independent groups highlight the many uncertainties in nuclear reactor safety. These uncertainties cannot be swept away by the probability estimates of the thoroughly discredited *Reactor Safety Study*.

TOO LITTLE INSURANCE

The strongest business refutation of the RSS is the Price-Anderson Act, which provides for limited liability nuclear insurance. This Act requires each utility operating a nuclear plant to obtain the maximum available private insurance, which at present is $125 million. The federal government provides $435 million more in liability insurance. Were a nuclear accident to occur, liability of the reactor operator and the federal government would end at $560 million. This figure is less than one tenth the property damage alone of the worst RSS accident. More serious accidents are possible. A 1965 AEC study found that a reactor accident could cause 45,000 immediate deaths, 100,000 injuries, and $17 billion in property damage.

The Price-Anderson Act provides insufficient compensation for nuclear accidents, by any measure. But, if the chances of a reactor accident are really as remote as the RSS states, there will be no hardship for the nuclear industry to assume full liability for accidents it claims will never happen. If, on the other hand, reactor accidents are more likely than the RSS predicts, the public must be protected by unlimited liability. The question which must be put to the nuclear industry is: If nuclear reactors are so safe, why is limited liability necessary? To this, the industry has no satisfactory answer.

ASME, and ME in particular, has often been accused of catering to industry, especially to the nuclear industry. Nuclear power has been affirmed in recent articles by Seymour Baron ("The LMFBR: The Only Answer," Dec. 1974, p. 12); by W. E. Cooper and B. F. Langer ("Nuclear Vessels Are Safe," Apr. 1975, p. 18); and by Mike McCormack ("A National Energy Policy," May 1975, p. 18).

But what about the other side? This article is the result of our efforts to give both sides of the story, to present a balanced view of this controversial issue. It is a documented and detailed discussion of points made by Mr. Nader at the recent Joint Legislative Forum in Washington, D.C. (See "Nuclear Power: A Viable Alternative?" June 1975, p. 67.) ASME comments appear at the end of the article.

PLUTONIUM THEFT AND MISUSE

The Price-Anderson Act is by no means the only institutional issue facing nuclear power. If the nuclear establishment has its way, plutonium will soon be used as reactor fuel. This will subject an imperfect and unstable world to the dangers of a substance which is both extremely toxic and weapons grade material. Plutonium will subject workers and the environment to radioactive exposure from an industry that has not yet demonstrated it can handle the substance. But worse, plutonium will provide an attractive target for any group bent on sabotage or terrorism.

Experts within the nuclear establishment are concerned about plutonium theft and manufacture of an illicit nuclear weapon. An April 1974 AEC task force assessed measures to prevent plutonium theft and concluded that: "Even though safeguard regulations have just been reviewed and strengthened, we feel that the new regulations are inadequate and that immediate steps should be taken to greatly strengthen the protection of special nuclear materials." [11]

The AEC, in an environmental statement, admitted that present safeguards against theft would be inadequate if plutonium were introduced as reactor fuel [12]. The threat of an illicit nuclear weapon was reinforced by a public television program in which a 20-yr-old student designed, on paper, a workable nuclear weapon. A member of the Swedish Defense Ministry advanced the opinion that the weapon had a good chance of exploding. The only missing ingredient would have been plutonium. It should be noted that were a terrorist group unable to make a nuclear weapon, the group could still create havoc by releasing plutonium in a populated area.

DON'T EXPORT NUCLEAR POWER

Developing nations which may import reactor technology lack the technical infrastructure to adequately manage nuclear power, which is nowhere near safe in the U.S. itself. Moreover, there is probably no single item that developing nations need *less* than nuclear power. These nations cannot afford to tie large amounts of power production to a single central power station. When the plants shut down, a significant part of the country's power grid will be inoperative. In addition, many developing nations have substantial amounts of natural and renewable power that would be more appropriate. Clearly, a developing nation would want nuclear power largely for prestige and possible weapons manufacture.

The example of the nuclear weapon exploded by India indicates the ability of a country to divert plutonium from reactors to weapons production. The assurances of the State Dept. that U.S. reactors abroad will be subject to international safeguards are not at all convincing. International Atomic Energy Agency (IAEA) safeguards are little more than bookkeeping methods. The agency could detect a diversion of nuclear material after it occurred, but it requires minimal security measures. Even if diversion were detected, sanction would have to be approved by the United Nations Security Council. If the Security Council could agree to act on a diversion, it might be hard-pressed to devise adequate sanctions.

Exporting nuclear power also means exporting sabotage targets. A General Accounting Office investigation in October 1974 concluded that nuclear plants in this country were susceptible to sabotage [13]. The same would be at least as true for reactors abroad. Unfortunately, reactor exports to areas of strife such as the Middle East, Pakistan, Korea, and Argentina are either in progress or planned. The sabotage threat makes reactor export to these areas even more senseless.

A THREAT TO FREEDOM

Civil liberties will also be affected. The question here is: If adequate safeguards against plutonium theft can be devised, can they be implemented without unacceptable losses of liberties? The AEC had proposed, for example, to establish a federal plutonium police force and wide-scale background security clearances as part of a safeguard system [14]. One thoughtful observer has evaluated what such measures might mean to society:

> A basic objection to theft-preventive safeguards is that they would require individuals to distort their assessment of their own role in society. For example, civilian employees of the nuclear power industry who would have to comply with stringent new security regulations might come to believe that they were more like soldiers than civilians in light of the background checks that they would undergo to secure employment and in light of the limitations on their off-the-job activities that they would have to observe to retain employment.
>
> Post-theft recovery measures would create a situation approaching civil war, with the government arrayed against the perpetrators of the nuclear threat and with innocent citizens caught in the middle. [15]

Nuclear power threats to civil liberties have already surfaced, even without the use of plutonium fuel. The Virginia Electric Power Company (VEPCO), for example, has requested its state legislature to allow VEPCO to establish its own police force, with power to arrest persons anywhere in the state and to obtain access to confidential citizen records. VEPCO claimed such authority was required by AEC regulations, but the AEC denied that such sweeping private police authority was necessary.

There is a double danger for a society that chooses to use plutonium as an energy source. The first is that civil liberty infringements which undermine society will be necessary. The second is that a thief or saboteur will slip through any security system, no matter how carefully it is enforced, thus rendering the defense of the nation impossible.

Dr. Donald Geesaman, biophysicist and plutonium specialist at the University of Minnesota, would remind us that:

> Reality is more inclusive than the *remove numerata* of the systems analysts. There are Klaus Fuchs and Lee Harvey Oswalds. There are heroin thefts at the New York Police Department and gambling hanky-panky in the highest echelons of AEC security personnel ... There is terrorist violence at Khartoum and Munich and Tel Aviv; the attack on Princess Anne and the Hearst kidnapping. There are resurgent nationalism and alienated minorities, the hopeless, the poor, the vicious, the pathological, and most important of all, the brilliant, for raw human intelligence spreads across all human classifications; and "where there's a will there's a way." [16]

A LITANY OF PROBLEMS

There are other nuclear power institutional issues which because of limited space can only be mentioned:

- Nuclear power is an economic issue. Predictions of nuclear electricity so cheap it would not have to be metered have been shattered by skyrocketing nuclear plant costs and unreliable plant operation.
- Nuclear power is a subsidy issue. Even if utilities could produce electricity more cheaply for buyers they could only do so through subsidies and economic distortions throughout the nuclear fuel cycle. There is Price-Anderson, which reduces drastically the industry's insurance premiums. There is uranium enrichment, provided at reduced cost by government plants. There is fuel reprocessing, which will be subsidized by government floors for reused nuclear fuel. And, there are future costs which can only be estimated. The nuclear

industry has practically no idea of the costs to decommission large plants. Add subsidies for safeguards and waste disposal and research—to name a few—to the rising demands for direct financial aid and tax reductions to utilities and the picture of nuclear costs becomes clearer.

- Nuclear power is an environmental and health issue. The power plant is supported by the nuclear fuel cycle of mining, milling, enrichment, fabrication, reprocessing, waste storage, and transportation in between. These steps impose the dangers of occupational radiation hazard, environmental degradation through radiation releases, and catastrophic accidents. Some steps pose all three dangers.
- Nuclear power is a moral issue. If no solution can be found for the disposal of radioactive wastes, guardianship may be required for the quarter-million years for which the wastes can cause cancer, leukemia, or genetic damage. Thus the power we consume now may leave lethal garbage for thousands of generations to come.
- Nuclear power is an appropriations issue. For over 20 years, federal energy research has been almost totally devoted to nuclear power projects. If the breeder reactor and other nuclear projects continue to gobble up their present 60–70 percent of the energy research budget, development of new nonnuclear energy sources will be significantly forestalled.

THERE ARE ALTERNATIVES

The last point raises the alternatives to a nuclear future. Nuclear power proponents would have us believe that the future requires continuing exponential growth in energy demand, with the only energy sources being imported oil, dirty coal, and nuclear power. These underlying assumptions must be challenged.

In the first place, there is an enormous potential for reducing energy use by cutting out waste. The Federal Energy Administration (FEA) estimates that per capita energy consumption could be cut 30 percent with no significant effect on lifestyle or industrial output [17]. Other experts believe that at least 40 percent of U.S. energy use is waste.

On a per capita basis, the U.S. consumes two to three times the energy of the industrial Western European nations. Hysterical nuclear proponents who claim that energy conservation will lead to economic chaos are refuted by Sweden, Denmark, and Switzerland. These countries get along on 50 percent or less of the per capita energy consumption of the U.S., but they all have higher per capita gross national products [18].

The country should begin to think of conservation as an energy source. A recent American Institute of Architects study estimates that energy-efficient buildings alone could save more energy in 1990 than nuclear energy could supply; and the capital requirements to save that energy would be less than to provide it with nuclear power [19].

On the supply side, one should not ignore oil as an alternative. Much publicity has been given to recent pessimistic reports on oil reserves, but little attention has been focused on the qualifications that accompany the pessimism. The latest U.S. Geological Survey's (USGS) estimates are that only a 40- to 60-yr supply of domestic crude oil remains, at current production rates. But the USGS also admitted that its conclusions relied on data from the American Petroleum Institute (API) and the American Gas Association (AGA); that its analysis was done for pre-1974 oil prices; and that significant discovered and potential resources, including oil shale, had been excluded [20]. One should realize that the API and AGA have a vested interest, along with a historical record, of drastically underreporting reserves in order to justify higher prices. Also, as prices at 1974 levels and beyond are taken into account, much more oil will become economical to recover.

With all factors considered, a more realistic conclusion is that at least 100 years of oil remain at present consumption levels. This conclusion should certainly not be an excuse for wasteful energy consumption. But it can keep the country from being stampeded into nuclear power reliance.

Persons who still remain convinced that oil reserves are scanty should refer to professional geologist M. King Hubbert. Dr. Hubbert's reserve estimates are even more pessimistic than those of the USGS. Of nuclear power, he says, "I've gradually come around to look at the hazardous aspects and it scares the hell out of me." [21] Despite his low oil reserve estimates, Hubbert believes that dramatic conservation and development of solar energy are necessary to "phase out nuclear as fast as we can." [21]

Other sources include geothermal power and all the solar technologies—heating and cooling; conversion of urban, plant and animal waste to fuel [22]; wind power; and solar electricity. Even the FEA, which is becoming the Administration's nuclear power promoter, believes that solar power could replace the equivalent of 20 percent of our present energy needs by 1990 [23]. Some of these sources are in everyday use right now, in this country and abroad. If nuclear power can be prevented from usurping the federal energy research budget, these sources will provide more and more of our energy in the future.

WHAT IF ...

So the choices for even the immediate future clearly are not limited to nuclear vs. coal. But even if they were, nuclear would lose, for two major reasons. The first reason is that the problems of coal power are largely technical in nature—safer mines, reclamation of mined land, and removal of pollutants from the coal before or after burning. The problems of nuclear power, on the other hand, are institutionally serious as well as technically unresolved. Nuclear power tempts saboteurs, terrorists, and hostile nations; it will strain civil liberties; it will require massive measures of government-financed corporate socialism; it will burden future generations; and it will require incredible stability in human nature and human institutions if its present radioactive inventory and future garbage are to be controlled.

The second reason nuclear power must be rejected is its social ramifications as a technology. Even proponents of nuclear power acknowledge that to be a viable energy alternative, it must be free of catastrophic accident. Were a catastrophe to occur, from any plant or in any step in the nuclear fuel cycle, what would be the reaction of the citizenry? After all, they have been told for years "it will never happen" by government officials and an industry grown smug from its own propaganda. If a catastrophe occurred, there would be such an outcry that nuclear power plants all over the country might be shut down forever. Clearly this country cannot let such a frail technology become a major energy source, now or in the future.

It is clear, then, that nuclear power is not merely a technical issue. Responsible technologists recognize the human ramifications of their technologies. Engineers and scientists must become sensitive to the myriad human issues in the nuclear power question and to the overwhelming legal problems and burdens that are developing under the stresses this technology is placing on our society and the world. Only when all these issues are recognized can one engage in a rational and reasoned policy deliberation. When only some of these issues are conceded, one realizes that the decision on nuclear power must be made by an entire enlightened population, not by delegated authority. Just as war is too important to be left to the generals, so nuclear power is too important to be left to the so-called nuclear "experts."

ASME POSTSCRIPT

We feel it would be inappropriate for ME to publish Mr. Nader's article without comment because it might then be assumed by some that it has the endorsement of ASME. The reasons it does not have such endorsement are many. They have all been published in the open literature but have not received the widespread attention from the media that has been accorded the antinuclear factions. A detailed rebuttal of the points made by Mr. Nader would take too much time and space for inclusion here, but the following should be noted:

1. The article purports to prove that engineers do not recognize the human ramifications of their technology, but are interested only in the technology itself. The fact is that *all* safety studies are aimed at protecting the general public and when the technical problems have been solved the human issues will disappear.

2. The article refers to the August 1974 draft of WASH-1400 instead of the final version issued in October 1975 (WASH-1400, NUREG-75/014). It was prepared with consideration of the arguments in the Sierra Club, Union of Concerned Scientists, and American Physical Society publications and concluded that the probability of failure was even less than that given in the draft version.

3. The article refers to the UCS news release of August 6, 1975 where 2000 scientists and engineers signed an anti-nuclear statement, but ignores the pro-nuclear statement signed by about 30,000 technically knowledgeable persons in a declaration addressed to Mr. Zarb (news release dated Nov. 15, 1975).

4. The article accepts the fallacious and often rebutted argument that the Price-Anderson Act proves that government and industry are not sufficiently convinced of nuclear safety to back it with adequate insurance. Actually, insurance companies limit the amount of coverage they will provide in accordance with the established principles of insurance underwriting. Each company determines how much of its assets it can risk losing in one major event. The amount of coverage insurance companies provide the nuclear industry is the largest commitment of funds covering any single hazard and actually represents a vote of confidence in the industry. The Price-Anderson Act increases the dollars in this pool but does not constitute a subsidy to the nuclear industry.

5. Several statements in the article are given as fact but are really unsubstantiated conjectures. Examples are:

    a. "As many as 35 percent of the failures in the Apollo program, for example, were deemed "incredible" before they happened." What is the statistical definition of "incredible" as used in the Apollo program?

    b. "Plutonium will subject workers and the environment to radioactive exposure from an industry that has not yet demonstrated it can handle the substance." The nuclear industry has been handling plutonium quite safely for several years.

    c. "A 20-yr-old student designed, on paper, a workable nuclear weapon." The production of this nuclear weapon was deemed feasible on the false assumption that the plutonium could be handled by him and reduced to a form suitable for use in a bomb without danger to himself.

    d. "Moreover, there is probably no single item that developing nations need less than nuclear power." Actually nuclear power is the most attractive means for providing electric power in places where the shipment of fossil fuel is unfeasible or uneconomic.

    e. "Post-theft recovery measures would create a situation approaching civil war, with the government arrayed against the perpetrators of the nuclear threat and with innocent citizens caught in the middle." The security measures to prevent the misuse of stolen plutonium are no greater than those needed to prevent sabotage in any other dangerous activity.

    f. "Predictions of nuclear electricity so cheap it would not have to be metered have been shattered by skyrocketing nuclear plant costs and unreliable plant operation." Plant costs have increased significantly because of the obstructionist tactics of people such as Mr. Nader, and the reliability of nuclear plant operation is quite comparable and often better than that of fossil-fueled plant operation.

    g. "There is Price-Anderson, which reduces drastically the industry's insurance premiums. There is uranium enrichment, provided at reduced cost by government plants. There is fuel reprocessing, which will be subsidized by government floors for reused nuclear fuel." The industry actually pays its own insurance premiums and pays for its own fuel costs. We are already in the second decade of building water-cooled plants with no government subsidy at all.

h. "Nuclear power is an environmental and health issue." The lack of danger from low levels of radiation has been completely proven by the study, "The Effects on Population of Exposure to Low Levels of Ionizing Radiation," *Report of the Advisory Committee on the Biological Effects of Ionizing Radiations*, Division of Medical Science, National Academy of Sciences, National Research Council, Nov. 1972.

i. "If no solution can be found for the disposal of radioactive wastes, guardianship may be required for the quarter-million years for which the wastes can cause cancer, leukemia, or genetic damage." There are several possible solutions to the problem of waste disposal. The decision as to which one should be used can and should be delayed until the problem becomes more immediate and thereby use can be made of the best possible solution when it is needed.

j. "For over 20 years, federal energy research has been almost totally devoted to nuclear power projects." Studies of the latest budget requests show this to be completely untrue.

k. "There is an enormous potential for reducing energy use by cutting out waste." This is true but it has been shown that reduction of waste is only one of the many means necessary for solving the energy crisis.

l. "With all factors considered, a more realistic conclusion is that at least 100 years of oil remain at present consumption levels." This is an optimistic conclusion at variance with many other estimates that have been made.

m. "The problems of coal power are largely technical in nature." The same is true for nuclear power.

n. "Clearly this country cannot let such a frail technology become a major energy source, now or in the future." Nuclear power has been developed on a firmer and more solid foundation than any other industry.

In summary, Mr. Nader's article does not have the objectivity or technical accuracy usually required by this magazine.—*ASME Committee on Nuclear Power—Codes and Standards*.

ENDNOTE

1. Numbers in brackets designate References at end of article.

REFERENCES

1. Union of Concerned Scientists, News Release, Aug. 6, 1975.
2. WASH-1400, *Reactor Safety Study Draft*, U.S. Atomic Energy Commission, Washington, Aug. 1974.
3. *Preliminary Review of the AEC Reactor Safety Study*, Sierra Club-Union of Concerned Scientists, San Francisco-Cambridge, Nov. 1974.
4. *Comments by the Environmental Protection Agency on Reactor Safety Study*, EPA, Washington, Nov. 1974.
5. *Review of the Reactor Safety Study (WASH-1400) Draft of August 1974*, U.S. Atomic Energy Commission Regulatory Staff, Dec. 2, 1974.
6. The 1957 study was WASH-740, *Theoretical Possibilities and Consequences of Major Accidents in Large Nuclear Power Plants*, U.S. Atomic Energy Commission, Mar. 1957.
   The 1965 study updated the 1957 study by using for its model a more powerful reactor, which reflected the size of reactors being built in 1965.
7. *Report to the American Physical Society by the Study Group on Light-Water Reactor Safety*, Apr. 28, 1975. Referred to below as APS.
8. APS, pp. I-3 and I-4.
9. APS, p. V-68.
10. "Reactor Safety," invited Talk at the Spring Meeting of the American Physical Society, Apr. 30, 1975, Frank von Hippel, Center for Environmental Studies, Princeton Univ., p. 1.

11. "A Special Safeguards Study," AEC, reprinted in *Congressional Record*, S 6622, Apr. 30, 1974.

12. WASH-1327, *Generic Environmental Statement Mixed Oxide Fuel—Draft*, U.S. AEC, Aug. 1974, pp. V-6. Referred to below as GESMO.

13. Letter from U.S. General Accounting Office to Dixie Lee Ray, Chmn., U.S. AEC, Oct. 16, 1974.

14. GESMO, p. V-7 and elsewhere.

15. Ayres, R. W., "Policing Plutonium: The Civil Liberties Fallout," *Harvard Civil Rights—Civil Liberties Law Review*, Vol. 10, No. 2, Spring 1975, pp. 441–442.

16. Geesaman, D. "Statement in Support of a Nuclear Moratorium for the State of Minnesota," before the Minnesota Pollution Control Agency, Dec. 12, 1974.

17. U.S. Federal Energy Administration, brief filed with the State of California Public Utilities Commission, Apr. 2, 1975, p. 2.

18. Holdren, J. P., "Too Much Energy, Too Soon," University of California, Berkeley, *New York Times*, July 23, 1975, p. 31, and *Energy Reporter*, Federal Energy Administration, May 1975.

19. "Energy and the Built Environment: A Gap in Current Strategies" and "A Nation of Energy Efficient Buildings by 1990," American Institute of Architects, Washington, D.C., 1975.

20. "Geological Estimates of Undiscovered Recoverable Oil and Gas Resources in the United States," *Geological Survey Circular* 725, U.S. Dept. of the Interior, Washington, D.C. pp. 1, 2.

21. Prouty, D., "'Convert' Stresses N-Power Hazards," *Denver Post*, Feb. 4, 1975.

22. For an evaluation of the potential energy supply from plant waste, see "Energy Potential from Agriculture Field Residues," Famo L. Green, General Motors Corp., Detroit, June 1975.

23. *Project Independence Final Task Force Report: Solar Energy*, U.S. Federal Energy Administration, Washington, D.C., Nov. 1974, p. 1–9.

15 William Stieglitz

Ralph Nader, Peter Petkas, and Kate Blackwell

The standards were, in my opinion, a hoax on the American public, creating an illusion of improved safety which did not, in fact, exist. I could not be a party to this.

William Stieglitz

William Stieglitz received his copy of the first federal safety standards for automobiles along with the rest of the public on February 1, 1967. This was unusual because Stieglitz himself had been in charge of drawing up the standards as consultant to the Undersecretary of Commerce for Transportation. But ten days before the standards were issued he was abruptly assigned to other duties. When he saw them again in final form they were vastly different from those he had proposed.

Stieglitz took the regulations to his Washington hotel room and began to study them. It was approximately 5 P.M. At midnight he wrote out his resignation. That resignation, which Stieglitz personally handed to his superior the next morning, did more than any other single act to focus public attention on the deficiencies of the federal auto safety standards. As word of his departure carried to the press, newsmen rushed for an explanation. Stieglitz told them that he could not "in good conscience continue to serve."[1]

The next day press reports on the new standards featured Stieglitz's resignation prominently along with his comments on the failure of his former agency to follow sound safety policy. The *New York Times* commented in its typically restrained language that "Mr. Stieglitz's resignation adds another dimension to the furor that has grown since the twenty safety standards were issued. Heretofore, the criticism of the regulations came from outside the agency."[2] To the press and the public, the act of resigning was a powerful criticism in its own right. There were other critics, but Stieglitz's credentials as a long-time safety expert and above all as the man formerly in charge of the standards gave weight and substance to the criticism.

Stieglitz had not anticipated the publicity that attended his exit from the National Highway Safety Bureau. He had prepared no public statement and credits the press with being the real whistle blowers. "My only actions were to resign from a position that I found totally untenable on the basis of my professional convictions, and in reply to questions asked me by members of the press, to state the reasons for my resignation. The whistle blowing was done by the press in their coverage of my resignation and the reasons therefore, and by my professional peers who supported my actions."[3]

But if Stieglitz had not planned it, the effect was the same. Nor was his experience so different from others who have acted in accord with their standards of professional integrity and, in so doing, have exposed the failure of their organizations to achieve those standards. Stieglitz had no new information to divulge, but he exposed something equally serious: the refusal of the National Traffic Safety Agency to heed the advice and urging of one of its own chief experts. He also made public an example of professional responsibility that was to challenge and encourage other engineers. Finally, he became the focus, however inadvertently, of the criticism of the first auto safety standards.

If Stieglitz was surprised at the attention his resignation commanded, it may have been in part because it was a natural—he says "inevitable"—thing for him to do. For over thirty years he had worked in aviation safety and more recently had developed an expertise in automotive safety. Thirty years earlier he had promoted the idea that design errors lie behind many driver errors that lead to crashes. Throughout his career in private industry, where he has spent most of his life, and in the private consulting business he built later, he was concerned with designing vehicles that were "crashworthy." that protected, as far as possible, the occupants in crashes.

In February, 1971, Charles O. Miller, director of the Bureau of Aviation Safety of the National Transportation Safety Board, writing in the MIT *Technology Review*, pointed to Stieglitz as a pioneer safety advocate, one of the earliest to point out the design challenge to professional engineers. "In 1948," Miller wrote, "William Stieglitz, who is still quite active in automotive as well as aviation safety, gave the landmark paper in what today has become known as system-safety engineering. 'Safety,' he said 'must be designed and built into airplanes just as arc performance, stability, and structural integrity... A safety group must be just as important a part of a manufacturer's organization as stress, aerodynamics, weights, and so forth.'"[4]

Born in 1911 in Chicago, Illinois, Stieglitz studied electrical engineering for two years at Swarthmore College and graduated from MIT in 1932 with a bachelor of science degree in aeronautical engineering. He went to work in the aircraft industry as a specialist in aerodynamics and flight testing and was soon deeply involved in safety problems. While chief of the Fleet Wings division of Kaiser Aircraft, he lectured on aviation safety at Princeton in 1941–42. In 1944 he went with Republic Aviation, where he spent the next twenty years as a design engineer.

In 1946 Stieglitz organized a design safety group at Republic, the first such group to be formed as an integral part of the industry. In 1950, on loan from Republic, he worked with the Air Force on accident prevention and flight safety research. From 1950 to 1955 he was a member of the subcommittee on aircraft fire prevention for the National Advisory Commission for Aeronautics, the predecessor of NASA. During 1952–55 he was also a member of the NACA committee on aircraft operation problems.

Meanwhile, he was broadening his field to include automotive safety engineering, applying to automobiles the same ideas on "crashworthiness" he applied to aircraft design. In 1959 he testified before the House Subcommittee on Interstate and Foreign Commerce, chaired by Representative Kenneth Roberts, which was then holding the first congressional hearings to consider automobile safety. He submitted a paper on crash-worthiness written in 1950. The paper was addressed to aircraft but, as he was to show, applied equally to automobiles. For example, he cited aircraft crashes in which "many of these fatalities could have been prevented by stronger safety belts and seats."[5] He said the same design factors pertained to the automobile—to dashboard design, access of controls to the driver, and shiny chrome and high gloss paint that can blind the driver.

"Quite frankly, from personal observation of automobiles," he told the subcommittee, "it is my conviction that the interiors of these cars are designed, not from the standpoint of effective human utilization, but ... from the standpoint of appearance and style. I think unquestionably there are a large number of needless accidents resulting from improper consideration of the human being in design."[6]

By the time Dr. William Haddon, head of the new federal traffic safety agency, persuaded him to become his chief engineering consultant, Stieglitz had built a reputation as one of the foremost safety experts in the country.

In the fall of 1964, Stieglitz resigned as manager of Republic's design safety and reliability division to set up a private consulting practice on aviation and automotive safety in Huntington, New York. Two years later, he was contacted by Dr. William Haddon, who had been appointed to head the new National Traffic Safety Agency. Dr. Haddon came to New York and asked Stieglitz to accept a consulting position and take charge of drafting the standards. Stieglitz was not enthusiastic. In a subsequent meeting with Haddon and Alan Boyd, then Undersecretary of Commerce for Transportation, he suggested several people he considered better qualified for the post. Haddon and Boyd objected to all of them. They wanted Stieglitz.

Stieglitz's reluctance was partly based on the fact that the job involved certain personal sacrifices. "It meant abandoning the consulting practice which I had spent two years building up, and a way of life that I enjoy."[7] But there were more serious misgivings.

"In the discussion prior to my accepting the position, I called attention to my earlier testimony before committees of both the Senate and the House of Representatives on the then pending traffic safety bills. One of the measures being considered involved merely extending to all vehicles the standards of the General Services Administration for automobiles purchased by the federal government. I had opposed this vigorously in my testimony, stating that in my judgment these standards were inadequate. I therefore told Dr. Haddon and Undersecretary Boyd that I could not and would not be a party to issuing these same standards as the initial federal motor vehicle safety standards. I was assured that whether or not this were done would be my responsibility."[8]

Stieglitz had indeed made his position absolutely clear during hearings before the Senate Commerce Committee and the House Committee on Interstate and Foreign Commerce in March and April of 1966. He criticized the General Services Administration standards as "woefully and totally inadequate to provide the level of safety that they are intended to provide." He then attacked the bill that would apply these standards to 1968 cars driven by the public, standards "which, on the industry's recommendation, are written around what they had in the 1965 cars." If that were done, he said, "we have improved nothing. We have applied a label, we have spread some whitewash, we have said, now it is safe. But we have not changed a thing. We cannot get safety this way, gentlemen. We cannot protect people by just giving a blessing to what already exists, and which we know hurts people or kills people. There is no reason we cannot write these standards."[9]

Haddon and Boyd still wanted Stieglitz, wanted him to write the standards. Finally, having been assured, to some extent at least, that the standards would be his responsibility, Stieglitz agreed. "I felt obliged to accept. Had I refused the position and then found the standards to be inadequate in my engineering judgment, I could only blame myself for not having tried."[10]

Stieglitz directed the first two phases of the initial safety standard–setting process, the preparation of the advance notice requesting suggestions for standards, and of the proposed standards published in the Notice of Proposed Rule Making. But already a fundamental difference was developing between Stieglitz and his superiors at the agency. The industry's strong argument was that the proposed safety features, as Stieglitz and his staff drafted them, could not be developed for the 1968 models. They wanted standards to reflect what they had already planned. Henry Ford threatened that the proposed standards would close down the industry, and House minority leader Gerald Ford demanded an explanation. Stieglitz was willing to acknowledge that the industry needed time to institute some of the new features. But he argued that the standards themselves should require building at least some crashworthiness performance into vehicles in which forty-eight thousand people were dying yearly and millions were being injured.

During the last phase of preparing the final standards, Stieglitz found the ground slipping from under him. Others in the agency were leaning toward industry's argument of what was "practical." They were demanding what he considered serious weakening of the standards. He objected strenuously. Finally, ten days before the final standards were issued, he was assigned to other duties and not consulted again.

When he read the standards in his hotel room on the night of February 1, he found that of the initial twenty-three standards, three—relating to tires, wheel rims, and headrests—were dropped on the grounds that more technical data were needed. Of the twenty standards issued, all were reworded, most were softened, and six were substantially compromised and scheduled for future rulemaking. The date for most to go into effect was moved forward four months.[11]

"They offered no advance in safety but were basically a rubber stamp of what would have been done in most 1968 automobiles without standards," he says. "From this point of view, they were, in my opinion, a hoax on the American public, creating an illusion of improved safety which did not, in fact, exist. I could not be a party to this. Under the circumstances I had no alternative to resigning. It was not an agonizing decision."[12]

When their differences became public, Haddon explained that Stieglitz's recommendations were "preposterously impossible."[13] In hearings before the Senate Commerce Committee in March, 1967, Stieglitz explained his position and stressed the problem of lead-time. "In my opinion, what should have been done on many of these standards was not to weaken the standards but to change the effective date to provide a legitimate lead-time requirement of industry.... I am perfectly willing to say that some of the proposed standards could not possibly have been met for 1968 automobiles. I didn't think they could be when I proposed them. But I felt that things had to start moving forward, and this lead-time block had to be broken."[14]

Rather than abandon the provisions which had value but truly could not be incorporated in the short time available between issuance of the standards and the new car production beginning in July, 1967, Stieglitz argued that two-part standards should be issued, with the first step applicable to 1968 vehicles and the second step applicable to 1969 vehicles. In this manner, the agency would not be merely endorsing existing industry practice from year to year but rather would be assuring orderly priority for safety advances. As had already been documented in the 1966 hearings leading up to the passage of the safety legislation, without mandatory requirements there is little priority or urgency in the auto companies to build safety features into motor vehicles.

Stieglitz documented his skepticism about industry arguments for lead-time. He cited as an example the Chrysler request for twenty-five to seventy-nine weeks to change their brake hoses to a type using a different reinforcing braid in order to comply with the proposed standard. Chrysler said it would require one engineer and one mechanic two weeks to work out the front wheel brake hose installation (even this Stieglitz stated was excessive). Further, to accommodate the various Chrysler models there would have to be eleven different installations, or a total of twenty-two weeks (two weeks for each of the eleven installations). Stieglitz chided Chrysler for implying that in all its corporate divisions there was only one engineer and one mechanic capable of routing a brake hose.

Stieglitz also criticized the agency for unnecessarily bowing to industry pressure. He cited as an example the proposed motor vehicle safety standard 101, "Control Location and Identification." As proposed, the controls listed had to be within the operational reach of a fifth-percentile adult female driver restrained by an upper torso restraint with enough slack to permit "a five-inch movement of her chest." The standard as finally issued required controls to be within the reach of "a person," with a "reasonable degree of slack" in the upper torso restraint.

As Stieglitz pointed out, the final standard contained numerous loopholes. The seat position was not defined, thus allowing compliance with the seat in the most forward position (even if the subject could not drive the car in such a position). The manufacturer could interpret "reasonable degree of slack" very broadly, even if the amount of slack made the restraint virtually ineffective. But most important, the "person" in the standard could be a six-foot, five-inch male with extra long arms. The standard was clearly meaningless.

The agency, according to Stieglitz, deleted the reference to fifth-percentile female because it claimed it did not have sufficient data on arm reach. Stieglitz cited a study providing data on the measurements of Air Force female pilots and flight nurses obtained in 1946. The agency rejected this information as non-representative of the public. Stieglitz also cited a 1961 Society of Automotive Engineers publication, "Human Body Size and Passenger Vehicle Design," by Ross McFarland and Howard Stoudt of the Harvard School of Public Health. The authors said their data were approximations of the general driving population interpolated from several studies (including the 1946 Air Force information) and that: "Although the adjusted and weighted values in Table 1 [of the aforementioned publication by McFarland and Stoudt] are tentative, we believe they do, for most practical human engineering purposes, approximate sufficiently closely the true body measurements of this group to enable them to be used with confidence."

The agency refused to base its standard on these careful approximations. This difference in the arm reach of fifth-percentile Air Force flight nurses and the estimate by McFarland and Stoudt is two-tenths of an inch. The arm reach of a tall man—who would qualify as a "person" in the agency's standard—is many times greater than these values. Stieglitz concluded: "Thus in order to avoid a possible discrepancy of a fraction of an inch, the standard permits controls to be nine or ten inches beyond that of even a tall woman. I do not believe that the standard as written can be claimed to contribute appreciably to safety."

Stieglitz assesses the influence of his action on the agency as "negligible." "Many standards that I have objected to in February of 1967 are still basically in effect and have yet to be revised to a point where they are meaningful." But he does see one major change. "In one respect I do find the bureau acting in consonance with what I then argued for; that is, establishing standards on a basis of safety requirements and setting an effective date that allows adequate lead-time rather than weakening the standards to endorse what would be immediately achievable. I cannot say that this was the result of my influence, although it does represent a position I had taken before my resignation."[15]

The response from many fellow engineers has given Stieglitz tremendous personal satisfaction, however, and he credits his peers with helping him blow the whistle. He cites in particular the support of Frederic Salinger, an engineer in Seattle, Washington, and the Washington Society of Professional Engineers which, in response to a motion by Mr. Salinger, passed and forwarded to Senator Magnuson a resolution supporting Stieglitz's position and urging Senate committee, hearings into the circumstances. "Judging from letters and comments I have received," he says, "it may be that my actions gave encouragement to other engineers who also feel strongly about their professional responsibility."[16]

Stieglitz has worked actively to encourage safety committees in industrial plants, like the one he started at Republic Aviation in 1946. He has also encouraged physicians with special knowledge in treating crash victims to support safety measures. In an editorial written for the *Medical Tribune* on June 23, 1965, Stieglitz urged physicians to support the then pending bills to create a federal auto safety agency and set federal standards. "It is essential that the legislation be supported by all those interested in automobile safety," he wrote, "but especially by those with professional knowledge in crash injury and accident prevention."[17]

His experience has not been without some hardship. "From a personal standpoint it unquestionably was expensive economically," he says. "I had not only to start over, but was faced with the attitude of some clients expressed as 'You walked out on us once. Can we be sure you won't do it again?' Against this, of course, were new clients who had heard of me through the publicity attending my resignation, and who had approved of my stand. Nevertheless, it took considerable time to rebuild the practice to the level it had reached prior to my going to Washington."[18]

When asked whether he would take the same action again, Stieglitz responded that "there is no question in my mind that under similar circumstances I would do the same thing, and I sincerely believe that in the same situation I would do it in the same way. I wish to stress the words 'the same situation.' No two cases are ever the same, and I think that the manner in which one performs an action of this type must depend on the particular circumstances."[19]

To others who find themselves in similar situations, Stieglitz says, "There is no question that there are probably economic penalties which in some situations may be prohibitive. One's responsibility for supporting a family may be so great that blowing the whistle publicly is impossible. Further, there is the pressure exerted by retirement plans, for one's investment in such a plan could make resignation almost impossible. This would be particularly true for older men.

"None of these factors, however, should inhibit a man from privately calling the situation about which he is concerned to the attention of someone who can act, and will protect his source of information. The cost of violating one's sense of professional integrity must be weighted against possible economic loss, in determining one's course of action. Peace of mind is intangible, but very important."[20]

Reprinted with permission from Grossman Publishers, from *Whistle Blowing*, edited by Ralph Nader, Peter Petkas, and Kate Blackwell. Copyright © 1972 by Ralph Nader.

ENDNOTES

1. *Washington Post*, February 3, 1967, p. 11.
2. *New York Times*, February 3, 1967, pp. 1, 15.
3. Whistle Blowers Conference, January 30, 1971.

4. *Technology Review* (MIT alumni journal), February 1971, pp. 30–31.
5. "Motor Vehicle Safety," Hearing of the House Subcommittee on Health and Safety, Committee on Interstate and Foreign Commerce, July, 1959, p. 234.
6. Ibid., p. 227.
7. Conference.
8. Conference.
9. "HR 13229 and Other Bills Relating to Safety," Hearing of the House Committee on Interstate and Foreign Commerce, April 1996, p. 898.
10. Conference.
11. *New York Times*, February 3, 1967, pp. 1, 15.
12. Conference.
13. *New York Times*, February 3, 1967, pp. 1, 15.
14. "The Implementation of the National Traffic and Motor Vehicle Safety Act of 1966," Hearings of the Senate Committee on Commerce, March 1967, p. 158.
15. Conference.
16. Conference.
17. Conference.
18. Conference.
19. Conference.
20. Conference.

16 Whistle-Blowing
Not Always a Losing Game

Karen Fitzgerald

CONTENTS

Five engineers in nuclear power, aerospace, and air-traffic control recount their experiences in following their consciences.

Exposing errors or unethical conduct in any occupation is risky, but when engineering judgment is involved, the risks of "blowing the whistle" acquire an added dimension. A technical decision cannot always be categorized as strictly right or wrong—unlike situations in which an organization is falsifying documents or overcharging for a product. Consequently, the engineer must be convinced of being right and then wait, sometimes years and even decades after lives are lost or millions of dollars are spent, to be proved right or wrong. Frequently, the whistle-blower's career is destroyed in the meantime.

In the following cases, which date from the 1970s and 1980s, the whistle-blowers have by now been vindicated to a degree for their actions, though the verdict may not be unanimous. And the careers of the first group may have even benefited by blowing the whistle.

NUCLEAR POWER, FEBRUARY 1976

Three engineers quit General Electric Co.'s nuclear division to protest alleged inadequate testing and unsafe designs, not only at GE but throughout the nuclear industry.

Though engineers Greg Minor, Richard Hubbard, and Dale Bridenbaugh expected that their decision to resign their jobs might get some attention in the newspapers, they thought it would play itself out in a few days. The day after they quit, however, they realized their lives had irrevocably changed when they were thrust in front of hot lights, cameras, and reporters at a press conference in Los Angeles.

"We're technical people," said Minor. "Our lives changed because we had to become political people—not because we wanted to, but to keep the debate focused on the technical issues."

The three engineers were able to continue practicing engineering and have fared much better than most whistle-blowers. Nine months after resigning from the San Jose, Calif., GE division, they started a successful consulting firm on nuclear power, MHB Technical Associates, also in San Jose.

Each of the GE engineers came independently to his decision to resign, but agreed to do it together publicly in order to have more of an impact on the debate over nuclear power, which at that time was raging in California because of Proposition 15, an initiative requiring that certain safety problems be resolved before more nuclear plants were licensed.

RUDE AWAKENINGS

With a bachelor's degree in mechanical engineering, Bridenbaugh went to work for GE in 1953 and was a start-up supervisor in 1960 at the world's first commercial boiling-water reactor at Dresden Nuclear Power Station, in Morris, Ill., near Chicago. After the plant had been in operation a short while, Bridenbaugh said failures began occurring that the designers had not foreseen; pumps stopped working, heat exchangers leaked, and control rods started warping, bending, and cracking. "The reactor began falling apart," he said. "The designers really didn't understand what they were dealing with."

The final straw came when Bridenbaugh, as manager of nuclear reactor performance improvement, reviewed reactor containment response to possible accidents. A computer simulation showed that 19 GE plants in operation in the United States might not survive a serious accident: the release of pressure during an accident would throw up a violent swell of water that the containment could not withstand, possibly resulting in the release of radioactivity into the atmosphere.

Refusing to believe the simulation results, Bridenbaugh's manager wanted to keep the plants operating; he told Bridenbaugh that if they had to shut down, it would be the end of GE's nuclear business. In his last major task at GE, Bridenbaugh and other GE engineers met with the utilities to convince them they had to spend millions of dollars on plant modifications, but the utilities representatives refused, saying that they did not have the authority to make the decision.

Minor began working in GE's San Jose division in 1960. He received a master's in electrical engineering from Stanford University in 1966, and took part in designing the routing of the electrical cabling at the Brown's Ferry plant in Decatur, Ala. It was the first one to incorporate safety requirements increasing the separation of redundant cables. "I thought this was the best plant we had designed," he said.

But the cables were sealed with flammable polyurethane foam, and in 1975 when technicians performed a routine check for air leaks with a candle, the sealant caught fire. Because of about half a dozen mostly human errors, the fire raged for seven hours, damaged 1600 cables, and disabled most of the emergency core-cooling system. The core came close to being uncovered, but the crisis ended when an auxiliary pump was used to dump water into the reactor. "To me it was a disaster," said Minor. "I felt we were very, very lucky we hadn't had a major catastrophe." Richard Hubbard, who began working for GE in 1960 with a degree in electrical engineering from the University of Arizona, was also disturbed by the Brown's Ferry accident. But as manager of manufacturing quality assurance for the nuclear division, he had already seen what he regarded to be fundamental problems with the company's attitude toward safety. While trying to eliminate a serious vibration in the reactor core, GE engineers stumbled upon a miscalculation in the core-cooling system flow rates and realized that the water would not flood the core as quickly as expected during an accident. When told of the problem, "a GE vice president said, 'It's important when you look under a rock, the angle that you look,'" Hubbard recalled. "He meant to look straight under the rock—don't look around and find other problems."

When the engineers quit, they made a point of not criticizing any person or the company specifically, but focused on the issue, which for them was that there were fundamental safety problems that the industry and the Nuclear Regulatory Commission (NRC) had not addressed adequately. They felt that there was too much reliance on theoretical models and not enough prototype or field testing. "The rate at which nuclear power was developed outstripped our knowledge and our understanding of the consequences and the side effects," said Minor. "All we were saying was 'Let's slow down and look at the side effects.'"

TESTIFYING IN WASHINGTON, D.C.

After their resignations, phone calls began flooding in from around the world for interviews and information about nuclear power. The engineers were asked to meet with commissioners of the

NRC, who, the engineers believed, seemed more interested in undercutting their position than in understanding the technical reasons for it. Two weeks after resigning, they testified before the Joint Committee on Atomic Energy in Washington, D.C., where they again encountered skepticism and harsh questioning. Physicist Edward Teller and others accused them of being paid by the Soviets to speak against nuclear power.

GE disputed the engineers' views on nuclear power, and today still believes that their resignations were part of a preplanned publicity campaign to influence the Proposition 15 vote. "The resignation letters presented no fresh views or arguments but repeated the emotional claims of an antinuclear group of which they and their families were apparently members," GE spokesman Hugh Hexamer told *Spectrum* in a written statement.

Among their consulting firm's first projects was campaigning for initiatives similar to California's Proposition 15 that were on the ballot in six other states that fall. They also began a safety assessment of the Diablo Canyon Nuclear Plant in Avila Beach, Calif., for the Center for Law in the Public Interest, which was supporting intervenors in the NRC licensing process.

That year, the Union of Concerned Scientists hired them to write a formal critique of the WASH-1400 risk assessment study, which estimated the probability of a core meltdown. The Swedish Government in 1977 asked them to perform a similar risk assessment for Sweden's nuclear plant at Barsebäck.

In 1978, the firm was asked by movie actor and producer Michael Douglas to devise a technically accurate scenario for a feasible nuclear reactor accident for *The China Syndrome*. For two weeks after it opened, MHB spent most of its time defending the sequence of events portrayed in the movie. After two weeks, the Three Mile Island (TMI) accident occurred in Pennsylvania, with many of the same precursors as the movie's scenario, but it went a step further to a partial coremelt. The press now began asking them to explain what happened at TMI.

The TMI accident gained credibility for MHB's views, and in addition to doing safety studies for intervenors in the licensing for the Shoreham nuclear plant, Brookhaven, N.Y., and others, the firm was hired by state attorneys general and utility regulatory agencies concerned about the economics of nuclear power and the prudence of rate increases.

The firm recently conducted a study of advanced reactors for the Union of Concerned Scientists, putting the engineers in the news again—and back in the position of being pitted against at least some in the nuclear industry. While MHB found that many of the features of the new reactors improved safety considerably, the engineers are concerned that the designs introduced new vulnerabilities not yet adequately explored, and that, despite these weaknesses, the nuclear industry and the Department of Energy are promoting them as inherently safe.

It seems that the three engineers are still blowing the whistle. Undeterred by the difficulties they encountered, they all say they would do it again. "The only regret I have," said Bridenbaugh, "is the hard feelings that developed between the people I worked with for many years who apparently don't understand why I had to do this."

AEROSPACE, JANUARY 1986

Engineers at Morton Thiokol Inc. warn against the launch of the space shuttle Challenger because low temperatures predicted for the next morning might stiffen O-rings. The launch proceeds as scheduled, and seven astronauts die in an explosion caused by the O-rings' failure to seal rocket booster joints.

Like the GE engineers, mechanical engineer Roger Boisjoly was one of three who spoke out against a management decision at his company, Morton Thiokol Inc. (the aerospace division is now Thiokol Corp.), Brigham City, Utah. However, in his case, Boisjoly found little advantage in numbers. After six years with the company, he lost his job, and despite a 27-year career in the industry, could not find another. His two former colleagues have fared somewhat better—both have retained their jobs at Thiokol, even though apparently derailed from the fast track.

Boisjoly explains the uneven outcome by what transpired in the hearings of the President's commission investigating the shuttle disaster. The company instructed the engineers to give only "yes" and "no" answers and to volunteer nothing. Boisjoly spoke first and "bared all," revealing that the engineers warned that the cold temperatures—a low of 18°F (−8°C)—predicted overnight before the morning launch might render the booster O-rings so stiff that they would be unable to seal the gases properly. He and Thompson presented evidence of a past launch at 53°F in which one of two redundant joints had not sealed.

Thiokol's upper management initially would not approve a launch below 53°F, Boisjoly told the commission, but they later buckled under to pressure from the National Aeronautics and Space Administration (NASA), which had already postponed the launch four times. NASA argued that no launch criterion had ever been set for the booster joint temperature. Allan McDonald, the only Thiokol manager at the Kennedy Space Flight Center where the launch took place, fought against the launch to the end.

Although McDonald and engineer Arnold Thompson volunteered information during the commission hearings, Boisjoly believes that their 25 years with the company, compared with his six years, gave them more job security. Furthermore, Boisjoly was singled out for harsh punishment because of his outspokenness, he said. He saw his ultimate transgression as challenging testimony at the commission's public hearing by the company's general manager, who said that no unanimous engineering position against the launch had existed the night before. "That simply was not true," Boisjoly told *Spectrum*. "By telling the commission that, I put myself further and further in the quicksand. When I left that meeting, I knew I was at the top of the hit list."

GROWING ISOLATION

Shortly afterward, the company took Boisjoly off the failure investigation team and sent him back to Utah, while Thompson remained on the team in Washington. Boisjoly was assigned to work on the redesign of the booster seal, and although told he was very important to that redesign effort, began to realize he was being left out of meetings. When questioned a couple of months later by the President's Commission looking into whether the Thiokol engineers' jobs had been affected, U. Edwin Garrison, president of Morton Thiokol's aerospace operations, said he had given an official order to isolate Boisjoly in order to minimize friction with NASA.

But Boisjoly believes the real reason was that management did not want a thorough redesign. "I truly believed them when they told us as engineers that we were going to have a clean sheet of paper to redesign these joints, to do the job right." He said he was "devastated" when he realized that management itself had devised a redesign only marginally different from the original in a strategy to fend off outside criticism of the failed design. All management wanted the engineers to do was make the design work, he added. "I really cared about the program," he said. "I had devoted my whole being to doing the best possible job I could do, and when I found this out, it just destroyed me."

Furthermore, he and four other engineers who testified before the commission were feeling heat from co-workers who viewed the group as troublemakers out to hurt the company. Formerly close associates turned away when meeting them in the corridors and would not speak to them.

Boisjoly soon began experiencing stress-induced symptoms, including double vision and pains in his chest and shoulder. After going on sick leave for six months, he was diagnosed in the fall of 1986 as having post-traumatic stress disorder, qualifying for long-term disability benefits. But when the benefits took effect in January 1987, the company terminated his job.

Since that time, Boisjoly has given more than 100 talks on his experience to students, professional societies, and businesses. Anticipating that he might have trouble finding another job, he studied for a professional engineering license so that he could become a consultant. In prior job searches, he had to decide among two or three offers, having had experience at companies including Hamilton Standard Electronics Systems Inc., Atlantic Research Corp., and Rockwell International Corp., but this time he received only one job interview after sending out 150 resumes. In March of last year, he started a consulting business in Mesa, Ariz., to provide expert technical testimony in legal cases.

Although this past year his fees match the salary he received in industry, Boisjoly is bothered by his income's instability. He filed a lawsuit against Morton Thiokol for compensation for "ruining his career," but the case was dismissed by a Federal judge in Utah. "People who are branded whistle-blowers have no rights," Boisjoly said. "The Whistle-blower Protection Act was passed in 1989, but it only deals with Federal employees. I was trying to make the law with my lawsuit."

Arnold Thompson and Allan McDonald are still at Thiokol. Thompson is now working on the manufacturing side of the shuttle program, instead of engineering analysis, and McDonald is off the shuttle program altogether, with the title of vice president of special projects. Thiokol refuses all requests to talk about the Challenger incident. Asked to comment on Boisjoly's case, the company told *Spectrum* in a written statement, "It is our sincere belief that these articles serve no legitimate purpose, and continue to cause unwarranted suffering among our employees, the astronaut families, NASA, and the nation."

AVIATION, MARCH 1981

A Federal Aviation Administration engineer appears on the television show "60 Minutes," charging that lives have been lost because of the agency's mishandling of collision avoidance system development.

Jim Pope had a clear vision of what he wanted to accomplish when he went to work as a mechanical engineer for the Federal Aviation Administration (FAA) in McLean, Va., in 1966. He had dealt with the agency through its district and regional offices in a five-year stint as chief of aviation safety in Nebraska's Department of Aeronautics in Lincoln in the early 1960s. Afterward, he experienced firsthand what he felt were inefficiencies in the agency's certification process when he started a company to sell a device he invented for preventing wheels-up landings.

"When I tried to get certification for my landing gear control, I found that the FAA officials didn't seem to care about the cashflow problems that people on the outside were experiencing while waiting for certification," he told *Spectrum*. "I joined the FAA to get that agency moving in a positive direction of service to the aviation community."

But eventually, Pope's mission mushroomed when he concluded that the agency had lost any sense of its original purpose—making flying as safe as possible. The FAA's technical decisions, Pope believes, have needlessly cost the United States 1000 lives in airplane accidents since 1975.

In 1971, after heading R&D efforts at the FAA, Pope became chief of the industry and government liaison division in the office of general aviation (nonairliner craft). "People in the small business aviation community across the country learned in a hurry that our office—and I in particular—was acting as a catalyst for getting them a timely response," he said. "We got a lot of attention, and we got a lot of things done."

Over the next two years, Pope circulated in meetings among the top leaders of the agency, and began questioning the agency's decisions to spend millions of dollars for R&D on two systems that he felt were unnecessary.

One was a collision avoidance system that would prevent mid-air crashes through ground radar. In 1972, after two years of work on a system that the agency believed would be on-line in 20 years, the FAA was approached by Honeywell Inc., Minneapolis, Minn., with an airborne collision avoidance system (ACAS), which allowed planes to communicate directly without the ground as go-between. An onboard box would create an egg-shaped envelope of RF energy around each craft, and when two envelopes overlapped, the system would warn the pilots of a potential collision 45 seconds beforehand. If no action was taken 25 seconds from a crash, the system would give complementary commands for evasive action to both pilots.

Congress directed the FAA to evaluate the Honeywell ACAS, as well as two others by RCA Corp. and McDonnell-Douglas Corp. in a four-year, US $12 million program. The conclusion of the study, according to Pope, was that the Honeywell system was the best and the least expensive at US $1000 for small aircraft and US $7000 for airliners, and as stated by an FAA executive committee report of Dec. 16, 1975, that it "meets all the objectives of the agency."

But the FAA told Congress that ACAS caused too many false alarms and that it was concerned about the system's compatibility with the current air-traffic control system. Pope charges that the FAA lied to keep its own development program alive.

"ACAS conflicted with their NIH [not-invented-here] mindset," said Pope. "They wanted control of every airplane from the ground, so they weren't going to certify it."

With the backing of his boss, Pope wrote what he called "hardhitting" letters to FAA administrator John L. McLucas, trying to convince him of the merits of selecting Honeywell's ACAS. In response to these letters, Pope said, and a run-in with the agency over its microwave landing system (MLS), another technology he said was already available from industry, the next FAA administrator, Langhorne Bond, eliminated the office of general aviation in 1978, fired Pope's boss, who supported Pope's positions, and transferred Pope to the FAA's Seattle office.

The FAA pursued the development of a ground-based system called the discrete address beacon system (DABS), in which air-traffic control would send warnings and evasive action when necessary. The agency also worked on an interim technology it told Congress would be ready in 1978, called the beacon collision avoidance system (BCAS). The active version of BCAS, in which aircraft would have their own receivers so they could interrogate transponders on other aircraft directly, is the basis for the agency's current TCAS (traffic alert and collision avoidance system). Unlike ACAS, BCAS would not allow the two closing craft to communicate and so each would have to take independent evasive action.

Pope claims that a 1975 study on midair collisions conducted by Mitre Corp. found BCAS to be dangerous, creating interference problems in high-density areas. In a study the next year, Mitre examined the 494 midair collisions that occurred in a prior nine-year period and concluded that improvements in the air-traffic control system could have prevented 118 of them; BCAS would have prevented 120; DABS would have avoided 190; and ACAS came out on top, preventing 228. Nonetheless, problems with DABS moved FAA administrator J. Lynn Helms in 1981 to make the BCAS technology the focus of collision avoidance efforts, though Helms renamed it TCAS, Pope charges, to make it appear to be a new technology. TCAS, which airlines began installing this year, costs about US $150 000 per aircraft.

When Pope was sent to Seattle in 1979, his family remained in McLean, Va., where he had just built a house. After arriving, he found he had little to do, and his boss eventually told him that the job had been fabricated to get him out of FAA headquarters. After a long and fruitless struggle to get reassigned to FAA headquarters, in 1981, Pope went public with his story, appearing on "60 Minutes" and testifying before Congress' Subcommittee on Transportation, Aviation, and Materials.

Pope said the FAA then began a campaign of harassment to build a case for terminating his job. His management reprimanded him for insubordination, failure to carry out orders, and improper use of duty time. When Pope began experiencing stress-related symptoms, including three kidney stone attacks, his doctors recommended that he take sick leave and return to his family in Virginia. A few months later, the agency fired him.

Pope contested his firing as a violation of laws prohibiting retaliation against Federal employees who testify before Congress. In order to prevent a public hearing, Pope believes, the agency converted his termination to retirement due to disability.

FAA associate administrator for regulation and certification Anthony J. Broderick commented on Pope's charges in a written statement to *Spectrum*: "In a perfect world, the [TCAS and MLS] projects may have moved faster and in a direct line from conception to production. But aviation research and development is not filled with black and white choices… Mr. Pope would have your reader believe that the development of the collision avoidance system was victimized by small-minded bureaucrats, and as a result a needed safety innovation was unconsciously (sic) delayed. To say this, is to be less than accurate.…"

In 1985, Pope found employment at NASA's Goddard Space Flight Center in Green Belt, Md., working as an engineer on the Cosmic Background Explorer (COBE) and later the space station. But he left NASA in 1988 after a heart attack. He is now writing a book documenting his experiences at the FAA.

To Probe Further

A good overview of whistle-blowers in a variety of professions (including discussion of the Boisjoly and Pope cases) is *The Whistleblowers: Exposing Corruption in Government and Industry* by Myron Peretz Glazer and Penina Migdal Glazer, Basic Books Inc., 1989.

For information on the remedies available to government and corporate whistle-blowers, contact Sarah Levitt, in take coordinator at the Government Accountability Project, 25 E Street, N.W., Suite 700, Washington, D.C. 20001; 202-347-0460.

17 Knowing How to Blow the Whistle

Tekla S. Perry

CONTENTS

Speaking out about an employer's unethical practices may bring public esteem to the whistle blower, but the many possible pitfalls must be considered.

An engineer working on a project for an employer knows of a technical condition in the design that appears to be dangerous or morally reprehensible. Should the engineer speak out? When? To whom?

Asked these questions most engineers would say, "Yes, right away, to management." But what if management appears unwilling to heed the warning? What if, in addition, it provides no satisfactory arguments that could assuage the engineer's concern?

Such cases are rare in the engineering environment—and rarer, perhaps, in electrical engineering than in mechanical or civil engineering. But they can occur.

In fact, engineers who have taken critical stands publicly—those who told of shortcomings in San Francisco's Bay Area Rapid Transit system, for example—sometimes find themselves honored as heroes of the profession. In cases where no one informs the public and flawed products lead to injuries, the public is quick to ask why someone didn't blow the whistle. That reaction followed disclosures of design troubles with the fuel tank of the Pinto car and the rear cargo door of the DC-10.

In today's questioning society, the implication for working engineers who see themselves as highly ethical seems to be that they should no longer always accept management's decisions passively; rather they are expected to keep a whistle in their pockets. Though most engineers will never find themselves in a situation that suggests whistle blowing, the pressure to be prepared bears on the entire profession, and engineers increasingly crave information on how to act if they ever have to.

But such advice is hard to give. There are dilemmas of such complexity that even experts will disagree on the best course of action. *Spectrum* found this to be true when, several months ago, it polled both experts and members on how to handle five complex cases. The results, published in the June issue [p. 53], showed most engineers will concede that certain situations require that as ethical engineer act, but they disagreed widely on how far such as engineer should go.

This month, we propose to attack these dilemmas differently. Putting aside the question of whether a case warrants action, we will assume that an engineer has suddenly become one of those rare individuals who must decide whether to blow the whistle or not. He wonders how he might best do it and what the consequences will be. Addressing such concerns, *Spectrum* has combed the literature, asked experts, and asked some whistle blowers themselves what they learned from their experiences.

BE AWARE OF THE RISKS

In spite of wide disagreement on many points, there is at least one area of consensus: the whistle blower takes serious risks. Some are:

- He or she may have been premature in concluding that management is the "bad guy." Management's motives may have been misjudged, when it, in fact, is seeking to work things out in its own way.
- Potential human damage may occur from the whistle blower's action—to the company, and its shareholders, its clients and to his colleagues. Should he go through with his act?
- Whistle blowing engenders hostility, right or wrong. It puts people on the defensive—not only company leaders, but also fellow employees who had access to the same information; there is strong motivation within a company to prove that whistle blowers are either wrong, vindictive, or publicity seekers. A company functions as a team, one manager pointed out, based on mutual trust and interdependence. A whistle blower, even if he is vindicated legally, loses this trust though he may keep his job.
- Whistle blowing is expensive—financially, psychologically, and professionally. Clifford Richter, who reported a violation of a Nuclear Regulatory Commission rule [see "Whistle blowers speak—Was it worth it?," p. 58], says, "Even if you are proven correct, you feel like a professional leper."

On the other hand, Federal and state governments are increasingly recognizing the need for knowledgeable employees to speak out in certain instances and are offering various legal guarantees to obviate managerial reprisals. So are taxpayers who see millions bilked because no one blows the whistle. And so is enlightened management in private industry. Some companies are encouraging outspoken employee criticism by setting up formal procedures to air complaints.

So when should an engineer blow the whistle? Only when all other remedies have been exhausted, most social and organizational scientists and consumer specialists say. Only when all the risks have been weighed. Only when the whistle blower is ready to follow an exacting sequence of steps. Many experts have agreed on these steps but they are best summarized by a perhaps surprisingly cautious Ralph Nader, who listed seven considerations a whistle blower ought to keep in mind:

1. He must have complete and accurate knowledge of the matter.
2. He must know exactly what public interests the questionable practices threaten.
3. He must estimate the probability of the organization responding positively to his concern.
4. He must be aware of possible violations of rules or laws when contacting outside parties.
5. He must know the best way to blow the whistle.
6. He must be prepared for the likely response to whistle blowing.
7. He should understand what whistle blowing should achieve in the situation.

KNOW THE FACTS

In the first area, gaining complete and accurate knowledge of the matter, experts point out that information can be fact, rumor, or something in between. Documented evidence is needed to convince an impartial observer. In many situations, some managers warn, the issue is one of policy rather than one of right and wrong. If the evidence is not so clear, writes Peter Raven-Hansen, a lawyer who has represented several whistle blowers, a court test might be applied: would reasonable investors take the information into account in making investment decisions?

In the second area, determining exactly what the objectionable practices are and what the public interests harmed are, experts say the engineer should consider the seriousness of the threat to the public or determine whether responsibility to the public interest overrules responsibility to his or her own career, to colleagues and family.

TRY AN INTERNAL ROUTE

The third concern is how far the objection can be taken inside the organization. Before considering going public, the experts agree, an engineer should go first to an immediate supervisor, then up through all available channels in the organization. This should be done for several reasons, says Mr. Raven-Hansen. First, it might work. Second, attempting to use internal channels will minimize the breach of company loyalty and perhaps prevent the question of motivation from being raised. Third, from a practical standpoint, many laws under which a whistle blower may appeal require that internal channels be used before public disclosure.

There are exceptions to trying internal remedies first, notes Sissela Bok, a lecturer on medical ethics at the Harvard-MIT Division of Health and Sciences. There may be no time to go through routine channels, or the institution may be so corrupt that steps would be taken to silence the whistle blower in internal channels.

MAKE A DECISION

The fourth consideration, weighing the probability of violating any rules and whether the action is nevertheless justified, is easier settled if speaking out is mandated by law, say the advisors, as it is when the issue involves a Nuclear Regulatory Commission license application or Environmental Protection Authority regulation, for example.

The question of exactly how to blow the whistle once the decision to act is made—anonymously, overtly, by resignation prior to speaking out, or in some other way—brings into consideration not only the potential whistle blower's actual motives, experts warn, but also the different possible motives of others—such as revenge or personal gain.

In addressing the sixth area, that of the likely responses from sources inside and outside the organization to the whistle blowing, the engineer should consider whether he or she is articulate enough to arouse a public response and also considered knowledgeable enough about the topic of concern to be believed, say Dr. Bok and others. Perhaps the engineer needs someone to corroborate the information at issue or to speak on his or her behalf.

Experts say the final question, what should be achieved by whistle blowing in any particular situation, requires that the potential whistle blower have an end in mind—correcting a defect or stopping the sale of a product, for instance. Speaking out can degrade the company's public image and perhaps the reputation of the engineering profession. Richard DeGeorge, professor of philosophy at the University of Kansas, in a keynote address last year to the National Conference on Engineering Ethics, said that the engineer is morally obligated to blow the whistle only if there is very strong evidence that doing so will have a positive result.

STUDY THE ALTERNATIVES

But suppose internal procedures have been exhausted and the engineer has decided it is time to speak out. Those who have done so indicate that the action is not simple and clear-cut.

"It happens to people slowly," says Morris Baslow, who pressured his employer to report data to the Environmental Protection Authority that appeared unfavorable to a client company [see "Whistle blowers speak," p. 58]. "The pressures get stronger and stronger—they do one little thing, then another."

See a lawyer before taking any action, the experts agree, or as soon as possible. If the dispute goes to litigation, it will be decided on its legal, not necessarily its technical, merits. A legal review will show the engineer whether his or her evidence is sound and not merely hearsay. It can also help block countercharges of slander. In this connection it is wise to keep personal accusations against supervisors and others—even if justified—out of the complaint. Focus only on the technical facts.

As to how the critical information should be released, doing it anonymously may appear to be the safest method, but many experts advise that information given anonymously loses credibility, and the anonymity probably will not last long in any case.

Attorney Raven-Hansen suggests a format. The whistle blower should summarize the information, its significance, and what should be done. An engineer cannot assume that the numbers will speak for themselves. They may be sufficient for internal disclosure, but the public needs words. The source of the information should be given and, if possible, the technical facts verified by outside sources, such as a professional society.

PROTECT YOURSELF

After making the information public, the whistle blower should anticipate attempts by the company to focus on him as a troublemaker instead of on the facts. Retaliation can take many forms—being passed over for choice assignments, demotion, or even dismissal.

If retaliation begins, advisors say, the employee must be prepared to document the chain of events, because although up until now engineers could be fired at will, the laws are changing and the whistle blower may be able to sue for reparations.

WHISTLE BLOWERS SPEAK—WAS IT WORTH IT?

Spectrum asked eight of the protagonists in well-publicized whistle-blowing cases if they would take the same path again and, if so, what they might do differently. These were:

- Frank Camps, who was senior principal design engineer at the Ford Motor Co. during the testing of the Pinto in the early 1970s. In 1973, Mr. Camps began a series of letters to Ford's management expressing concern about the procedures used in safety testing and charging violations of Federal law in this testing.
- A. Grace Pierce, who in 1975 was a director of medical research for Ortho Pharmeceutical Corp., a division of Johnson & Johnson. She would not test a new drug containing high levels of saccharin, a suspected carcinogen, on human beings while a form containing much less of the substance was available.
- Morris Baslow, a marine biologist who was a senior scientist at Lawler, Matusky, and Skelley Engineers. He pressured the company to include in its testimony before the Environmental Protection Agency data potentially unfavorable to the company's client, Consolidated Edison, on the effect of power plant thermal effluents on fish in the Hudson River.
- Clifford Richter, who, as Director of the Department of Medical Physics at the Ellis Fischel State Cancer Hospital, reported in 1978 a violation of a Nuclear Regulatory Commission rule to that agency, as required. (Seeds of iridium-192 were accidentally left in a cancer patient's body when the patient was discharged.)
- Gregory C. Minor, who with Dale G. Bridenbaugh and Richard B. Hubbard, resigned as an engineering manager with General Electric Co.'s Nuclear Energy Systems Division, citing deep concern about the safety of U.S. nuclear power plants, and in 1976 testified on those concerns before the U.S. Congress' Joint Committee on Atomic Energy.

- Peter Faulkner, who was a systems application engineer for the Nuclear Services Corp. when he became concerned with engineering deficiencies throughout the industry and presented a paper critical of the industry to a Senate subcommittee in March of 1974.
- Virginia Edgerton, an IEEE member who, as senior information scientist working for the Criminal Justice Coordinating Council of the City of New York, warned her immediate superior, then New York City's Criminal Justice Steering Committee, that the configuration of two computer facilities might be faulty and lead to degradation of the on-line police car dispatching program. The IEEE backed Ms. Edgerton in this case. She declined to comment for this article.
- Holger Hjortsvang, an IEEE member who, along with Max Blankenzee and Robert Bruder, was involved in the building of the Bay Area Rapid Transit (BART) system in San Francisco and went to BART's board of directors with the conclusion that the automated train control, developed under a contract with Westinghouse Corp., had grave defects (later proven by dangerous failures after the system was in operation). Mr. Hjortsvang also declined to comment for this article.

WOULD THEY DO IT AGAIN?

There was no other way I could have gone. I did not want to compromise my standards as certain persons in the Ford Motor Co. compromised theirs.

Frank Camps

I don't think I had any choice. When you feel strongly, you're committed.

A. Grace Pierce

There is no question that I would do it again, because I couldn't just sit back and let it [harm to the environment] happen.

Morris Baslow

It led to a lot of harassment, but I was required by the hospital's license, as a radiation safety officer, to cooperate with regulatory agencies. And my report did not identify the patient or any doctors, so in no way can it be said I did it with malice. If I hadn't reported the mistake and it was later discovered that I knew about it, the NRC would never trust me to be a radiation safety officer anywhere.

Clifford Richter

Considering the circumstances in effect at that time, we did it the way it had to be done. Internal mechanisms wouldn't have worked to satisfaction.

Greg C. Minor

I'd do it again even if I'd suffered more than I had because the technology we're dealing with [in nuclear power] isn't like that in the Ford auto or the Lockheed 1011, having limited effects, but a new technology being rushed into with insufficient testing, with widespread biological risks.

Peter Faulkner

WORDS OF WISDOM FOR OTHERS

Document everything you do, thoroughly and completely, and keep a log from the day you begin, listing the moves you make and all the people you contact. I did, and that is why I survived.

Frank Camps

I think speaking out is one hope for the future. If people care about their jobs and their dignity and their country, they may have to speak out. If every person did every small thing possible, perhaps we wouldn't have so many ills in the future.

A. Grace Pierce

Do everything you can to avoid it, but I hope that if you had to, you would do the same thing I did. Perhaps I should have done it earlier—it took me 21 months from asking the company to report the situation to blowing the whistle. And I should have gone out and gotten a lawyer immediately.

Morris Baslow

We have to do this [speak out] to be professional, but we also need the support of our professional societies. As it stands now, even when you are vindicated in the courts, proven correct, you are treated like a leper by some members of your profession. People think that if you hadn't done something wrong you wouldn't have ended up in such a position and you wouldn't have been fired.

Clifford Richter

What is needed is an improvement in internal procedures so someone in such a situation can use them—but such procedures aren't in place yet.

Greg C. Minor

Don't expect any breaks—a whistle blower has to make up his mind that he's going to get his throat cut. Cover your ass, because they'll try to get you any way they can. Have the facts straight, have documentation. Get an ombudsman to play devil's advocate, and don't stop until you've achieved your objective. If I did it again, I would go for the jugular—I would have moved faster and been more aggressive. I was right, but I didn't know just how right I was.

Peter Faulkner

WHERE ARE THEY NOW?

Frank Camps was demoted to a position that did not involve vehicle testing and eventually resigned. He is currently working as a consultant.

A. Grace Pierce was demoted and stripped of her research projects. She is currently in private practice.

Morris Baslow was fired in 1979 when after two years of trying to persuade his employer he threatened to go to the EPA directly. He then did supply documents to various hearing bodies. He is currently an independent consultant, involved in work for Cornell University.

Clifford Richter was dismissed in 1978. After a lengthy legal battle, he was reinstated at his 1977 salary in April of this year. But, he says, although he is physically back on the job, his former responsibilities have not been returned to him. He is completing a book about his experiences entitled "Why Cancer Wins."

Greg C. Minor, Richard C. Hubbard, and Dale G. Bridenbaugh are partners in their own nuclear energy consulting firm, MHB Technical Associates.

Peter Faulkner is running a children's summer sports camp and winter ski program.

Virginia Edgerton is a private consultant in New York.

Holger Hjortsvang is retired.

T.S.P.

According to traditional interpretations of common law, explains Alfred G. Feliu, a lawyer engaged in research for the educational Fund for Individual Rights, unless explicit provisions in a contract specify otherwise, every employment is an employment "at will" and can be terminated by the employer or employee at any time for any reason, even if the reason is unjust. Employees covered by collective bargaining agreements are protected in most cases, for such agreements commonly specify that the worker cannot be fired unless there is just cause. These employees have access to formal grievance procedures. But few engineers fall into this category.

CHECK FEDERAL LAWS

Government employees who report wrongdoing are protected by the Merit System Protection Board, created by the Civil Service Reform Act of 1978. Federal employees are also protected by the First Amendment to the Constitution—the right of free speech—and the right to petition Congress.

The Nuclear Regulatory Commission has gone further than other Government agencies in extending job protection. In an official policy statement last October, it guaranteed "that all employees have the opportunity to express differing professional opinions, that these opinions will be considered by the NRC in good faith, and that employees will be protected against retaliation in any form."

Employees do not see the Government's system for protection as perfect, however. A survey by the Merit System Protection Board indicated that 45 percent of Federal workers had observed waste or illegal activities in 1979–80 but that 70 percent of those also had did not report them because they feared reprisals or believed no action would be taken. Still, Government employees are better off than most engineers in industry.

Two parallel trends are spreading the umbrella of legal protection to cover previously ignored employees. First, Government regulatory agencies, aware that socially conscious employees in industry are helpful in enforcing agency regulations, have fostered legislation to protect employees who speak out when company action goes against regulations. Second, court rulings in some states have indicated that an employee might sue an employer if the employee was discharged for refusal to violate public policy.

Though the scope of whistle-blower protection and procedures for enforcement vary, a number of laws state that employers may not discharge or discriminate against employees with respect to pay, working conditions, or job privileges because the employee reports company noncompliance with laws. Several of these problems are enforced through the Department of Labor.

According to Seth Zinman in the Department of Labor's Office of the Solicitor, these statutes include the following acts: Occasional Safety and Health, Safe Drinking Water, Water Pollution Control, Toxic Substances Control, Solid-Waste Disposal, Clean Air, 1974 Energy Reorganization, Fair Labor Standards, and Federal Mine Safety and Health Amendments.

The picture may not be as rosy as it looks, however. The department of Labor currently has "an enormous number" of complaints pending, says Mr. Zinman, and whistle-blowing cases get no special priority. The backlog, which has been estimated at 3000, can only get worse, says Alan Westin, director of the Educational Fund for Individual Rights, as the Reagan Administration is cutting back on staff in Government agencies, and antireprisal action usually has low priority. He suggests that the Reagan Administration take a second look at this area, particularly in light of the general deregulation philosophy.

"One way to carry out a policy of deregulation in a country that wants clean air and water and a safe work place," Mr. Westin says, is to put money into whistle-blower support. "A dollar spent this way is worth a hundred put into outside regulatory enforcement, because the insider is always the first to know" about violations.

WHISTLE BLOWING AROUND THE WORLD

Engineers who dissent from an employer's position on ethical grounds find themselves in widely differing circumstances depending upon the country they are in. In some countries whistle blowing may lead to extreme retaliation or terror, whereas in others engineers can use legal or institutional mechanisms designed to hear, resolve, or prevent ethical confrontations.

Repressive regimes, with tight control of newspapers and the courts, can prevent the whistle from being blown. And formal or familial ties between business and the government may make blowing the whistle a very risky business. Moreover, public apathy can blunt the efforts of the whistle blower, for to be effective that action depends upon an aroused public bringing pressure for reform. Whistle blowing can be rendered ineffective where corruption is acknowledged openly—where a politician can say publicly, "I steal but I achieve."

In developing nations, one who speaks out faces serious obstacles if the public perceives industrialization as a good to be realized at virtually any cost and is prepared to tolerate high risks to health, safety, and even life.

There are mechanisms in other countries, however, that can aid or ameliorate the expression of ethical concerns. In Germany and the Scandanavian countries such mechanisms as "outsider boards" and boards for workers or the public have been instituted to promote more socially conscious corporate decision making.

Some political regimes have encouraged employee involvement in decision making with such formal mechanisms as the elected-workers' councils in England; worker participation in Sweden; codetermination, in which employee representatives are involved in management, in Germany; and self-management in Yugoslavia. These are forums for examining decisions and raising and addressing questions of unethical behavior, constituting an alternative to hierarchical management structures, which can make engineers powerless by overriding their ethical concerns.

The presence of engineering and professional unions in Germany and Australia also changes the context for ethical dissent. Unions can protect the engineer from retaliation and can challenge managerial decisions more effectively than an individual engineer with limited resources. Modifications of the employment-at-will doctrine can also protect an engineer from retaliation. These include the legal recognition of quasiproperty rights invested in employment by requiring a valid, documented case against an employee prior to dismissal or by requiring substantial compensation for termination.

Evidence of the wide divergence of social mechanisms and climates that can help or hinder an engineer's expression is being collected as part of a project designed to study the ethical dilemmas of engineers practicing in countries other than their own. The study was begun in 1978 by E.C. Jones, professor of electrical engineering and chairman of the technology and social change program at Iowa State University, and C.A. Smith, associate professor of philosophy at the University of Missouri at Rolla. *Spectrum*'s readers, particularly international readers, are invited to participate by forwarding information and ideas about actual whistle blowing, along with descriptions of whistle-blowing cases, to C.A. Smith, 225 Humanities and Social Sciences Building, University of Missouri-Rolla, Rolla, Mo. 65401.

C.A. Smith, associate professor of philosophy
University of Missouri at Rolla

THE IEEE MEMBER CONDUCT COMMITTEE: HOW IT WORKS

Suppose you, as a member of the Institute of Electrical and Electronics Engineers, believe that you are involved in an ethics clash at work. What support can you find in the IEEE?

The Member Conduct Committee, established by the IEEE Board of Directors in 1978, is composed of five members appointed by the Board for five-year terms. Its function is to accept and investigate submissions by IEEE members, which may be either complaints against other IEEE members or requests for support by the Institute. Complainants must allege specific violations of one or more provisions of the IEEE Code of Ethics. Requests for support must demonstrate adherence by the member to the Code of Ethics, which adherence involves a matter of ethical principle and has jeopardized or can jeopardize the member's livelihood, compromise the discharge of the member's professional responsibilities, or be detrimental to the interests of the IEEE or the engineering profession.

It is the responsibility of the committee to investigate all matters brought to it and to prepare in writing a "Report on Preliminary Investigation." In all cases this report is furnished to the member submitting a complaint or request for support. As the title suggests, each report is the initial step preceding any disposition by the Institute of a request or complaint. If, in the case of a complaint, the committee is unable to determine that a reasonable basis exists (1) for believing that the facts alleged, if proved, constitute cause for censure, suspension, or expulsion or (2) for believing that the allegations can be proved, the investigation is terminated. Should the committee make both determinations in the affirmative, the matter is presented to and decided by a separate hearing board and is thereafter reviewed and acted upon by the IEEE Board of Directors.

If, in the case of a request for support, the committee is able to conclude that member support is warranted, its "Report on Preliminary Investigation" is submitted to the Executive Committee with recommendations on the appropriate support. The Executive Committee may accept or reject, in whole or part, the determinations and recommendations of the Member Conduct Committee.

The 24 cases submitted to the committee to date have been nearly equally divided between complaints and requests for support. A few cases have been withdrawn by the initiating member. Five are under review. None of the complaints submitted has met the twofold test required by the procedures to warrant a hearing on the merits of censure, suspension, or expulsion of a member of the IEEE. Only two requests for support were submitted to the Executive Committee after investigation. One was endorsed; the other was not.

The Member Conduct Committee is unanimous in its view that it does not have the responsibility to "make a case" for the member submitting a complaint or request for support, for diverse opinions may flow from the same set of circumstances.

None of the cases the Member Conduct Committee reviews are as simple as they seem to be. There is nearly always at least one other side to the story, and there are considerations that the first presentation of a case ignores. Self-righteousness and misplaced indignation also tend to muddy the ethical waters.

This presents an extremely difficult problem for the Member Conduct Committee, since it is given, at least at the outset, only one view of the facts. Until it has obtained other views or information in some way and has arrived at what it believes to be a full version of the events, any judgment it makes may well be unjustified.

IEEE Member Conduct Committee

James F. Fairman Jr., chairman
2025 I St., N.W., Suite 1022,
Washington, D.C. 20006; 202-296-3795

Howard B. Hamilton

William W. Middleton

Raymond W. Sears

Robert H. Tanner

KNOW STATE POLICIES

If a whistle blower's situation is not covered by any Federal statutes, his job may still be protected. Since 1974, the interpretation of common law concerning "employment at will" has begun to be questioned. Most states, particularly in the South, says Mr. Feliu, still adhere to the common-law doctrine even if, as a recent Vermont ruling stated, "discharge was in bad faith, malice, and retaliation." Alabama, Mississippi, Kentucky, Arizona, North Carolina, Colorado, and Texas have recently reaffirmed the "employment at will" doctrine. But in California, New Hampshire, West Virginia, Oregon, Washington, Iowa, Nebraska, New Jersey, Connecticut, Michigan, and Illinois the State Supreme Courts have indicated that an employee may sue for wrongful discharge if an employer fires an employee for refusing to violate public policy. In New Jersey, the Supreme Court went on to say that "in certain instances a professional code of ethics may contain an expression of public policy." Not all of these courts have ruled in favor of an employee on these grounds, but they have indicated that such a situation could arise.

Whistle blowers in Michigan now have statutory as well as common-law protection—the Whistleblower's Protection Act of 1980, which protects "employees who report a violation of state, local, or Federal laws, or employees who prove by clear and convincing evidence that they were about to report a suspected violation, from dicrimination or discharge. Such legal changes allow a whistle blower to protect his or her career, but only upon proof that dismissal was a direct result of the disclosure.

Peter Faulkner, who was fired from his position as systems application engineer at Nuclear Services Corp. when he criticized the nuclear industry before a Senate subcommittee ["Whistleblowers speak," p. 58], is against all "statutory cushions." "Knowing that his action will lead him to get his throat cut ensures a high-quality whistle blower," he says. "If the law assures a safe landing and a warm towel, we will have all sorts of strange people coming out of the woodwork."

MANAGEMENT EFFORTS MAY HELP

To enable engineers and others to give their best technical opinions even when the opinions are controversial, a small number of corporations have an internal mechanism that gives employees a fair hearing. Though the majority of cases handled under these systems are employee complaints and grievances, they can be and are used for ethical questions. The IBM Corp. is considered a pioneer in institutionalizing such procedures; its programs have been used as a model by other corporations. According to an IBM spokesman, under the company's Open Door policy, an employee with a complaint is urged to discuss it with the immediate manager, the manager's manager, or the site manager. Then, if the employee is not satisfied or chooses to skip the first step, he or she can get in touch with a general or regional manager and can even go further by writing to the chief executive officer and describing the situation or requesting a meeting. IBM also has what it calls a Speak Up program to deal confidentially with written complaints. It has handled 200 000 letters since its inception, and the company says it receives approximately 18 000 such contacts annually from IBM's 341 000 employees worldwide.

SUPPORT FOR WHISTLE BLOWERS: WHAT DO OTHER ENGINEERING SOCIETIES DO?

The major engineering societies contacted by *Spectrum* have ethics codes but differ greatly in how far they are prepared to go to ensure adherance. The National Society of Professional Engineers (NSPE), like the IEEE, can handle complaints against members and support some members against an outside organization. The American Society of Civil Engineers (ASCE) and the American Society of Mechanical Engineers (ASME) have procedures for enforcing members' adherance to their codes but no formal mechanism of support for those involved in an ethical conflict. The American Institute of Chemical Engineers (AlChE) and the American Institute of Mining, Metallurgical, and Petroleum Engineers (AIME) make practically no attempt to enforce their codes.

The NSPE's Board of Ethical Review interprets its code of ethics and handles complaints against members. Its opinions on cases—about 12 annually—are published in the society's quarterly journal *Professional Engineer*. The NSPE maintains a legal defense fund to assist engineers in cases that broadly affect the engineering profession. It has not yet been used in a clear-cut whistle-blowing case, but, says Milton F. Lunch, NSPE general counsel, it could give such assistance.

The ASCE has a hotline service for questions on interpreting its code. The society's Professional Conduct Committee investigates complaints against members for code violations and initiates actions against members found guilty of crimes in a court of law. The Professional Conduct Committee acts as a grand jury and can issue a letter of reprimand, or recommend suspension or expulsion, in which case the charge is brought before the full Board of Directors. On the average, of the 30 to 40 cases considered each year, about five are heard by the Board and three members are suspended and one is expelled.

The ASCE has no mechanism to support a member charging a nonmember or an organization with ethical wrongdoings. "Our attorney has told us that when a member voluntarily joins he has said that he will follow a code of ethics, but we can't hold the entire profession to our standards," ASCE Managing Director for Educational Affairs Don A. Buzzell says.

The ASME's procedures for investigating and enforcing its ethics code seem relatively simple compared to the IEEE's [see "The IEEE Member Conduct Committee: how it works," p. 60]. If a member is accused of unethical conduct, the Society's Executive Director reviews the complaint. If the Executive Director finds it substantive, it is passed to the Board for

Professional Practice and Ethics, which can decide to investigate. If that board finds that the complaint is justified—which according to Donald R. Haworth, managing director for education, happens once every two years—a hearing panel made up of past ASME presidents is set up by the Board of Governors. That panel can censure, suspend, or expel a member judged guilty.

Members who seek support in a complaint against a non-member or an organization cannot be helped through any formal ASME procedures, but they are not turned away, says Mr. Haworth. Rather, they are referred to the vice president of the Board for Professional Practice and Ethics or a member of that board living in their area for informal advice.

The AIChE can, according to its Constitution, suspend or expel a member for violation of its code of ethics. It has a Committee on Ethics charged with developing and maintaining a code and considering all matters of professional ethics and reporting its recommendations to the AIChE Council. That committee is made up of past presidents, and, says AIChE Executive Director J. Charles Forman, "It's the most inactive committee in the institute and we want to keep it that way so as not to threaten our C-3 tax status."

The AIME has no mechanism to enforce its code of ethics. Adherance is left to the individual members.

T.S.P.

The Bank of America Corp. is also viewed as a model company when it comes to handling employee complaints. Its Open Line program, begun in 1972, is a close copy of IBM's Speak Up. Open Line receives about 100 letters and 10 phone calls each month, says the program's coordinator, Constance Johns, and it has handled 13 000 questions or complaints since its inception.

The Bank of America uses a modified ombudsman approach to employee complaints: the Employee Assistance Department, formed in 1978, which can try to resolve the problem on an employee's behalf. The Assistance Department is used by approximately 25 000 employees annually, the bank says, and all contacts are kept confidential. In addition, there is a system for impartial review of complaints if the employee and an assistance officer cannot settle the problem with discussion up the corporate ladder.

The Polaroid Corp. has staff people who will help other employees process a grievance through company channels if the employee has first discussed the problem with an immediate supervisor without receiving satisfaction. The Polaroid Employees Committee, formed in 1941, is made up of 38 elected representatives, a chairman, and a vice chairman. Though employees have no collective bargaining agreement or striking power, the committee can seek impartial arbitration of their grievances. According to the committee's chairman, Paul Kelley, the panel takes approximately 25 complaints annually to the president of the company but has only gone to arbitration five times, if that, in its 40 years of existence.

These corporations instituted their programs voluntarily because they saw listening to employee grievances as being in their best interests. But, says Mr. Westin, such internal procedures are not common, and they can work badly.

A bill introduced in the House of Representatives' Judiciary Committee's subcommittee on crime last year (though not yet reintroduced) could make such programs more prevalent. It would put the responsibility for product safety in the hands of corporate leaders and subject them to fines or imprisonment for permitting the production or sale of dangerous products or services (H.R. 4973; rewritten, H.R. 7040).

Such a law would force the restructuring of corporations into clearer lines of responsibility, says Dr. De George. And perhaps employees who pointed out errors to their supervisors could expect rewards instead of reprisals.

FOR FURTHER READING

Alan F. Westin discusses whistle blowing's past and future, focusing on 10 first-person accounts in *Whistle-Blowing! Loyalty and Dissent in the Corporation*, McGraw-Hill Book Co., New York, N.Y., 1981.

When whistle blowing is morally appropriate and when it is not is addressed by Sissela Bok in "Whistleblowing and professional responsibility," *New York University Education Quarterly*, Summer 1980, pp. 2–10. Peter Raven-Hansen also discusses how to make this distinction and how to handle the disclosure in "Dos and don'ts for whistleblowers—planning for trouble," *Technology Review*, May 1980, pp. 34–44.

David Ewing of the Harvard Business School discusses the legal aspects of employees rights in the book *Freedom inside the Organization*, E.P. Dutton, New York, N.Y., 1977.

18 The Whistle-Blower's Dilemma

Jean Kumagai

Speaking out may be the ethical thing to do, but too often it comes at a steep price

When Salvador Castro, a medical electronics engineer working at Air-Shields Inc. in Hatboro, Pa., spotted a serious design flaw in one of the company's infant incubators, he didn't hesitate to tell his supervisor. The problem was easy and inexpensive to fix, whereas the possible consequences of not fixing it could kill. Much to his surprise, though, nobody acted on his observation, and when Castro threatened to notify the U.S. Food and Drug Administration (FDA), he was fired. "I was shocked," Castro says.

Castro's case is far from unique. Indeed, it's the rare whistle-blower who manages to expose wrongdoing and remain on the job. The vast majority suffer a fate similar to Castro's—they end up being harassed, fired (often on trumped-up charges), and blackballed from their professions. The financial and emotional strain can snowball further, breaking up marriages, draining bank accounts, and taking a toll on physical and mental health.

"I've interviewed hundreds of whistle-blowers over the years, and hardly any have been successful in both not suffering reprisals and leading to a change in the situation," says Brian Martin, an associate professor in science, technology, and society at the University of Wollongong, in Australia, who has written a how-to for whistle-blowers [see "To Probe Further"]. "Even if you've got everything going your way, it's still hard to be successful."

And yet, an open society relies on those who are willing to come forward and reveal wrongdoing. Think of Roger Boisjoly, the Morton Thiokol engineer who tried to avert the *Challenger* disaster and later testified about how his company ignored problems with the shuttle's booster rockets. Or perhaps the most famous whistle-blower of all, Deep Throat, who exposed criminal activity within the Nixon administration. The act of speaking out is even built into certain codes of professional ethics. The IEEE code, for example, states that engineers shall "protect the safety, health, and welfare of the public and speak out against abuses in those areas affecting the public interest."

How then can the ethical engineer do the right thing and not sacrifice his or her career?

Everyone who works with whistle-blowers agrees that there are certain basic steps that potential whistle-blowers can and should take to protect themselves—and that very few actually take such steps, much to their detriment. When Martin found that the people he interviewed were making the same mistakes over and over again, he decided to lead off his book with a chapter on "seven common mistakes" whistle-blowers tend to make.

Mistake number 1: trusting too much. "Most whistle-blowers believe the system works," Martin says. "So when they find a lapse in their organization, their instinct is to go to their boss or through the regular grievance process. And then they're shocked when bad things start to happen." Dina Rasor, principal investigator for the Military Money Project, observes, "Whistle-blowers tend to have a real strong sense of right and wrong." Her organization, which looks into fraud and waste at the Pentagon, is run under the auspices of the National Whistleblower Center, a nonprofit advocacy group in Washington, D.C. "They're the ones who believed as kids that if you throw a ball through a window and you just tell the truth, you won't get spanked. Most of us learn to ignore that message. Whistle-blowers don't."

Among the other mistakes Martin cites are that people don't collect enough evidence of the problem they're trying to expose, don't build support among colleagues and others, and don't wait for the right opportunity to come forward. "My advice to most people is, 'Don't do it—until you're done investigating, preparing an escape route, and weighing your options,'" he says.

That last piece of advice is especially important. "People think the right thing to do is just speaking out. But there are many different ways to do the right thing. It may be best to wait and collect more information. You also have to look at the consequences, for yourself, your family, your colleagues."

"I hate the term whistle-blower," says IEEE Fellow Stephen H. Unger, a computer science professor at Columbia University in New York City. He has a long-standing interest in engineering ethics, and as chairman of the IEEE Ethics Committee in the 1990s he helped develop a set of guidelines for engineers faced with ethical dilemmas. "It conveys the wrong impression, of someone running around, being noisy and disruptive, behaving in an erratic way. Which is the very opposite of all the engineer whistle-blowers I'm aware of. They did everything they could to avoid publicity, to avoid making waves. Engineers are very quiet people."

His basic message for any engineer who's contemplating speaking out is to "make sure you're right. Check and recheck whatever calculations you've made, talk to people on the other side so that you understand their case, and be able to back off at any time if you see your case is weak."

"Don't exaggerate at all," Unger adds. "You could be 99 percent right, but if you make one little mistake, they'll focus on that to discredit you."

Because of the many bad things that happen to whistle-blowers, Dina Rasor likens the act to "setting your hair on fire for one glorious minute." She has two words of advice for would-be whistle-blowers: remain anonymous. "If there's any way to get the information out—through a nonprofit, or a trusted reporter, or a friend—without identifying yourself and having your fingerprints all over it, that's preferable to going public. Then the fraud becomes the issue, and not you."

A common tactic used against whistle-blowers is to dig up—or manufacture—personal or professional problems. When Rasor first began investigating Pentagon fraud back in the 1970s, "people who didn't like what I was doing spread rumors that I was a lesbian, or that I was 'living in sin' with a man. At the time, that was scandalous stuff," Rasor recalls. "I was in fact living with a man—my husband."

Some people find the idea of leaking information sneaky or cowardly, she adds. "But if you're doing it because there's some horrible fraud going on, it's the smart thing to do. If whistle-blowers could get up and be protected, I'd say come forward. But the reality is they can't." An insider is also in a much better position to keep the investigation going, she points out. Once the person's identity is known, any further access to critical evidence usually evaporates.

A number of organizations now exist to help whistle-blowers publicize their messages without having to put their careers on the line. In the United States, the National Whistle-blower Center, the Government Accountability Program, and the Project on Government Oversight are three such groups. Rasor runs a Web site, http://www.quitam.com, which educates whistle-blowers on filing suits against contractors or others who have defrauded the U.S. government.

If used carefully, the Internet can also be a boon to whistle-blowers; anonymous remailers let people send e-mail that can't be traced to its source, and Web sites make it easier both to publicize wrongdoing and to offer advice to whistle-blowers. [For more about staying anonymous online, see "The Illusion of Web Privacy" in this issue.]

Martin, for one, believes the climate for whistle-blowers is gradually improving. Over the last few decades, he notes, media coverage and public attitudes toward whistle-blowers have improved. He adds, though, "Problems tend to arise in an organization when people are too afraid or too powerless or too cynical to speak out. Whereas if more people are willing to speak out, then it's less likely a problem will occur in the first place."

As for Salvador Castro, he sued Air-Shields for wrongful termination, and his case has been tied up in the Pennsylvania courts for nearly eight years; the company has tried three times to have the case dismissed but hasn't succeeded yet. The IEEE, of which Castro is a Life Member, has promised to file an amicus curiae brief on his behalf should his case go to trial.

In the process, Castro has had a crash course in labor law and whistleblower protections. Before his dismissal, for instance, he'd never considered Pennsylvania's "at-will" employment laws, which allow companies to fire workers "for a good reason, or a bad reason, or no reason at all," he says. Had his employer been polluting a stream rather than designing defective medical devices, he might still be on the job; the Federal Water Pollution Control Act and other environmental legislation make it illegal to fire someone for blowing the whistle, but the FDA has no such protection.

Meanwhile, his old employer has changed hands twice since firing him; most recently it was acquired by Germany's Draeger Medical. Air-Shields, which didn't respond to IEEE Spectrum's interview requests, recently offered to settle out of court; Castro declined. "This will set a precedent for all engineers in Pennsylvania," he says. "The next guy who figures he can fire an engineer for doing the right thing will think twice."

Although he has worked only sporadically since his firing, Castro has no regrets about his actions. "I'd do it again in a heartbeat," he says. Nor has his long fight gone unrecognized. In 2001, the IEEE Society on the Social Implications of Technology presented him its Carl Barus Award, given for outstanding service in the public interest. And in December 1999, the FDA finally forced his former employer to recall the incubator and correct the defect Castro had brought to light four years before.

TO PROBE FURTHER

Two guides to blowing the whistle are Brian Martin's *The Whistleblower's Handbook: How to Be an Effective Resister* (Jon Carpenter Publishing, 1999) and *Courage without Martyrdom: A Survival Guide for Whistleblowers* (Fund for Constitutional Government, 1997), by Tom Devine of the Government Accountability Project (http://www.whistleblower.org). Martin's Web site, http://www.uow.edu.au/arts/sts/bmartin/dissent/, also has useful links about whistle-blowing.

Stephen Unger's *Controlling Technology: Ethics and the Responsible Engineer* (Wiley, 1994) examines ethical and moral conflicts that engineers face on the job and offers practical advice on how to deal with them.

The IEEE Ethics Committee's "Guidelines for Engineers Dissenting on Ethical Grounds" is available at http://onlineethics.org/codes/guidelines.html.

Part III

Raising the Bar

It is all too easy to show the world is filled with efficient and economical structures which are ugly and oppressive. The engineer can no more easily create a beautiful structure by minimizing only materials and costs than can a poet create a beautiful poem by using a few words as possible to express an idea. Something else is needed and that additional component is imagination. Here I mean this overworked word to imply a non quantifiable talent for putting things together in unique ways that work, i.e., the creation of something that is beautiful and permanent.

David Billington (1985)

… it is crucially important to allow our students to become competent not only as scientific and technical problem-solvers … but as dreamers and visionaries as well.

Langdon Winner (1990)

Part III offers creative, concrete, and sustainable engineering solutions. In an age of designs generated by committees or computers, or by what Rosalind Williams (2002) called the "dematerialization of engineering," some think that technologists are losing their creativity and imagination. In 1966, for example, Donald Frey, then vice president of Ford Motor Company, described that "some of the most reactionary people in industry are engineers …," a fact which results in "new departures that require creative thinking and innovation … wind[ing] up in the file marked N.I.H.—Not Invented Here." Many decades later, astute critics within the engineering profession were remarking that the "… industry and in particular the engineering industry, has begun to focus on the lack of 'creative thinking' and innovation in graduates," thus calling for "unblockers in engineering education" in order to explore and further enhance students' creative potential (Baillie, 2002). Both creativity and imagination are optimistic concepts connoting resolution, originality, and a future orientation. With that being said, Part III of *Ethics, Politics, and Whistleblowing in Engineering* comes full circle to cover novel solutions.

More often than not, a layperson's contact with engineering is founded less upon understanding per se and more on the belief that the future will consist of "new ways of being in the world" conceived in terms of technological innovation. Engineers, on the other hand, are famous for considering facts hard and values soft. At the same time, though, they also believe in specific professional mind-sets without question.

Even so, as shown in Part I of *Ethics, Politics, and Whistleblowing in Engineering*, since the Second World War, engineers have become increasingly more reflective about the impacts of their practice in society and the environment at large. The engineer's perception that his or her work lags behind some sort of social concern must be a part of the "engineering culture" and must become the intellectual soil in which the engineering imagination grows. Moreover, albeit still prevalent among a large segment of the population, the notion that affected communities cannot, or should not, have an active say in the direction that technological affairs are taken is being debunked by both engineers and nonengineers (Part II). As nonengineers contribute their imagination, engineering projects are influenced by the intersection of social values and the complexity of the natural world.

David Noble shows in enlightening detail in his popular *America by Design* that the engineers' undergraduate curriculum was intended to fit the agenda of a developing corporate colossus (Noble, 1979 [1977]).[1] In addition, industries diminished the capacity to explore because they often fund university projects that blur the distinction between academia and the corporate world. The Energy Biosciences Institute and the Dow Chair in Sustainable Chemistry are striking examples of the hybridization of academia, the state, and the world of business at the University of California (UC) at Berkeley. The former committed $500 million to UC Berkeley, the Lawrence Berkeley National Laboratory, and the University of Illinois at Urbana-Champaign for research on biofuels; the latter brought together Dow Chemical's "commitment to sustainability" with an operational expense of $10 million to be used by Berkeley's College of Chemistry (Scheiber, 2009).

Being taught by the industry and being involved in practical, hands-on projects as a student are different in an important way. The politics and infrastructure of technological innovation have been part of an ongoing debate, with some arguing that a "triple helix of university–industry–government relations" be the appropriate institutional matrix of research in science and engineering (Etzkowitz and Leydesdorff, 2000). Nevertheless, those models fail to account for how the needs and the knowledge of community geography, culture, and social expectations must become an integral part of engineering solutions and product design.

Engineers themselves are influenced by an alleged lack of "engineering leadership" as they generate technological innovations. Poor public perception of engineers, as shown in Part II, has sustained the view that engineers are not good enough to lead technological affairs. However, the core of the problem lies more with the absence of a clear link between the people who design and the people who use engineering projects. Although more corporate involvement does not create engineers who are more socially aware and/or environmentally responsible, this must not be taken to mean that university research shall remain unresponsive to real-world problems.[2] On the contrary, engineering colleges and departments of engineering *must* intervene in the process of delivering life-sustaining technologies in the most vulnerable communities in America as well as outside the United States.[3]

In his "Engineering Ethics and Political Imagination," quoted in the epigraph of this introduction, the political scientist Langdon Winner (1990) recounts the following anecdote from his experience in teaching ethics to engineering students:

> After the class had gone over the facts in the readings, my colleague [an aeronautical engineer] turned to the group of about thirty science and engineering undergraduates and asked a question that I would probably never have had the nerve to raise. "Well," he said, "how does it feel to be on a conveyor belt being turned into slots?" There was a long embarrassing pause. Finally, one of the students, a fellow who had been very quiet most of the term, raised his hand and answered. "I don't want to be put into any slot," he said confidently. Then he thought for a moment and added, "I want to find a good slot!"

The student at that point understood the link between his education and the conveyor belt job. Like many undergraduates training in engineering, for example, he benefited from practical experience in making the (difficult) transition from academia to the workplace.[4] Practical experience and social skills, such as the ones obtained by students at Howard University or in the Franklin W. Olin College of Engineering, prompt thinking about "engineering as a people field, rather than just sitting at a computer."[5] Likewise, the responsibility gained through involvement in engineering projects which are socially attuned to design is both inspiring and empowering for budding technical experts.[6]

"Raising the Bar" gives examples of how some engineers imagine the future by incorporating people who possess different engineering perspectives and by making sustainability meaningful from an engineering perspective.[7] These engineers give special attention to developing technical trends, environmental sustainability, and renewable energy.[8] More specifically, these technical experts have recently begun to make sustainability meaningful from an engineering perspective. Finally, we are reminded how the process of reimagining the place of technology in society inherently involves engineers reinventing themselves.[9] Engineers may reassess their epistemic beliefs about technology and experiment with their identity when faced with social or political challenges. *New, reflective, holistic, peace, adaptive, green, humanitarian, local/global, focal, women, minorities,* and *citizen* are some of the descriptors that have been put next to the word *engineer,* and their existence, as well as their plurality, is in fact indicative of the various dimensions of the changing engineering identity.

Part of that changing identity is the inclusion of women and other engineering minorities. Three main points of view dominate discussion and material written on women and other minority engineers. Increasingly, academics and others are developing a feminist perspective on women in engineering. Much of the feminist perspective literature parallels the general difficulties women face in our society at large with nothing particular to engineering. Second, a significant part of the literature quantifies engineering minorities and their professional involvement and reveals a lack of female perspective. Engineers turn to those in the minority in their field when faced with low productivity or decreased enrollment. In addition, women, in particular, are thought of as resources because their social skills are supposedly better than those of their male counterparts. In engineering today, communication skills are given greater importance, but with the increased attention given to women, it does not seem that women in the engineering field will ever be simply engineers. Women continue to be referred to as "women engineers" and are suddenly valued more when their gender proves advantageous. The challenge is to include minorities in engineering who, by virtue of different experiences in life, contribute to an engineering profession that interacts with its publics.

Since the 1970s and before,[10] engineers have not shied away from dialogue regarding the meaning of sustainability and systems thinking—upon which a lot of the sustainable development literature is based. The transition from problem-solving to active problem definition is very relevant to the broader and more systems-oriented training in technology currently referred to as "sustainability science and engineering." Sustainability is also about redefining what makes engineering an intellectually rigorous enterprise; sustainability, one may argue, is about adding quality to the "existential pleasures of engineering."

Studying what engineers say about their future, as well as reviewing the many solutions and successes in the field, generates optimism. Part III of *Ethics, Politics, and Whistleblowing in Engineering* will inspire a "new engineer" who will apply an intensity to engineering design that is driven by inside knowledge and outside detachment. The hope is that young, technically minded people, women and men alike, will place more importance on protecting their fellow humans and the environment. This book provides a starting point for engineers to create their own engineering image unbounded by present stereotypes. The future holds a "new" breed of practitioners: inspired by heroic engineers of the past, in touch with their various publics, and motivated by the challenges of an increasingly engineered world.

ENDNOTES

1. See, specifically, the chapter entitled "Technology as People" (David, 1979 [1977], pp. 224–256).
2. According to a report written by Chris Langley and Stuart Parkinson of Scientists for Global Responsibility, "[t]he power and influence of some corporations, and the increased pressure on researchers to bring in funding from business, means that academic departments are increasingly orientating

themselves to commercial needs rather than *to broader public interest* or curiosity-driven goals. This is a trend especially evident in biotechnology, pharmaceutical, oil and gas, and military partnerships" (Langley and Parkinson, 2009; emphasis ours).

3. A notable example of such intervention is the inauguration (2006) of the *International Journal for Service Learning in Engineering.*

4. Ironically, as Winner recounts, the student did find his "good slot"; he became a programmer for inter-continental ballistic missiles.

5. See "Redesigning First-Year Engineering: Experiment Lets Students Do More," by Brooke A. Masters, and "Re-engineering Engineering," by John Schwartz.

6. See "Students Discover the Soul of Engineering by Serving the Disabled," by William Celis 3rd and "Ground Swell on Wheels: Appropriate Technology Could Bring Cheap, Sturdy Wheelchairs to 20 Million Disabled People," by Ralf D. Hotchkiss.

7. See "Engineering Professionalism and the Imperative of Sustainable Development," by Mark Manion (Chapter 21).

8. See "Engineering Students Help San Diego Region Secure $154 Million in Solar Bonds," by Andrea Siedsma.

9. See, for example, Downey et al.'s (2007) article. See also Wisnioski's (2012) book.

10. See *Principles of Systems* (Forrester, 1968) and, particularly, *World Dynamics* (Forrester, 1970).

REFERENCES

Baillie, Caroline. Enhancing creativity in engineering students. Engineering Science and Education Journal 11, no. 5 (2002): 185–192.

Billington, David P. Engineering in the liberal arts: The new art of structural engineering. Paper presented at Southern University Conference March 9, 1985.

Downey, Gary Lee, Lucena, Juan C., and Mitcham, Carl. Engineering ethics and identity: Emerging initiatives in comparative perspective. *Science and Engineering Ethics* 13, no. 4 (2007): 463–487.

Etzkowitz, Henry and Leydesdorff, Loet. The dynamics of innovation: From national systems and "mode 2" to a triple helix of university–industry–government relations. *Research Policy* 29, (2000): 109–123.

Forrester, Jay W. 1968. *Principles of Systems.* Portland, OR: Productivity Press.

Forrester, Jay W. 1970. *World Dynamics.* Portland, OR Productivity Press.

Langley, Chris and Parkinson, Stuart. Science and the corporate agenda: The detrimental effects of commercial influence in science and technology. Scientists for Global Responsibility, September 2009. http://www.sgr.org.uk/publications/science-and-corporate-agenda. Accessed February 3, 2011.

Noble, David. *America by Design: Science, Technology and the Rise of Corporate Capitalism.* Oxford: Oxford University Press, 1979 [1977].

Scheiber, Jane. College of Chemistry. mygreeneducation.com, October 8, 2009. http://www.mygreeneducation.com/sustainable-green-chemistry-at-uc-berkeley. Accessed February 21, 2011.

Williams, Rosalind. *Retooling: A Historian Confronts Technological Change.* Cambridge, MA: The MIT Press, 2002.

Winner, Langdon. Engineering ethics and political imagination. In *Broad and Narrow Interpretations of Philosophy of Technology,* edited by Paul T. Durbin, 53–64. Dordrecht: Kluwer Academic Publishers, 1990.

Wisnioski, Matthew. *Engineers for Change: Competing Visions of Technology in 1960s America.* Cambridge, MA: The MIT Press, 2012.

19 Dead Poets and Engineers

Domenico Grasso

The year 1989 saw the debut of the feature film *The Dead Poet's Society*. This critically acclaimed movie is a story of inspiration: English teacher John Keating (played by Robin Williams) motivates and inspires his students to a love of poetry and to seize the day.

That same year, I began as a young assistant professor of engineering at the University of Connecticut. As luck would have it, Professor Samuel Pickering, the true-life inspiration for the Keating character and founder of the Dead Poet's Society, was on the English faculty at Connecticut. Because I was so inspired by the movie, I contacted Professor Pickering and asked him to deliver a seminar to the School of Engineering, hoping that his vision and excitement for the importance of poetry in enhancing the human condition would have relevance to engineering—a profession also ostensibly committed to the enhancement of human society.

His reaction to my request was astonishing. He felt that he had nothing to say that would interest engineers and respectfully declined my invitation. I was not only unprepared for this response, but incredibly disappointed at this latter-day manifestation of C.P. Snow's "Two Cultures."

The discontinuity in communication between sciences and the humanities (the "two cultures") and how this chasm can be an obstacle to solving the world's complex and integrated problems was the subject of Snow's Rede Lecture delivered at Cambridge University in 1959. Indeed, this rift in our sensibilities led Robert Pirsig, the author of *Zen and the Art of Motorcycle Maintenance*, to observe that what is "wrong with technology is that it's not connected in any real way with matters of the spirit and of the heart. And so it does blind, ugly things quite by accident and gets hated for that."

My encounter with Professor Pickering left me dismayed but undeterred in my search for a bridge connecting the two cultures. I eventually left Connecticut and moved on to Smith College where I became the founding director of the first engineering program at a women's college in the United States and, just as significantly, one of the few engineering programs at a liberal arts college in the United States.

What I have discovered since my interactions with Professor Pickering is that many engineers, in fact, **do** want to hear what poets have to say, dead or otherwise. And indeed there is a bridge being built to connect the two cultures, the humanities/arts and sciences/engineering. But, in my experience, the construction appears to be proceeding at a much faster pace from the science/engineering shoreline. In fact, the overarching philosophy of both the Smith College and the University of Vermont (my current affiliation) engineering programs is to achieve a true unity of knowledge rather than the factious paradigm characterized by Snow's "Two Cultures."

From an academic or intellectual perspective, the "Two Cultures" is an interesting concept to contemplate and debate. However, the implications of this concept can be pernicious when played out in a profession such as engineering, a profession that is dedicated to harnessing the powers of nature to benefit humanity. In preparation for this profession with such noble goals, students have unfortunately devoted the vast majority of their studies to understanding and manipulating the natural world and have often paid scant attention to understanding the motivations and intentions of the human spirit or of our connection to the natural world. This has been the state of engineering through most of the 20th century, a century that has seen meteoric growth in the role that technology plays in our everyday lives. A testament to the power of engineering thought and action.

My former colleague from Smith College, writer Kurt Vonnegut (who studied engineering), thought that this bias in many an engineer's intellect was worthy of comment in his first novel,

Player Piano. In that novel, one of his characters laments, "If it weren't for the people, the god-damn people always getting tangled up in the machinery ... the world would be an engineer's paradise."

To what has this cultural schism led? A "developed-world," where between one-half and three-quarters of our industrial inputs or feed stocks are returned to the environment as waste within one year. While engineers are instrumental in designing life-saving technologies and have taken men to the moon, we have also constructed enough dams in the northern hemisphere to—in one estimate—actually change the earth's center of gravity and hence the length of a day. This estimate was made before the largest dam in the history of humankind, the Three Gorges Dam in China, even came on line. We have also been able to develop energy and transportation technologies that have warmed our planet by increasing the CO_2 in our atmosphere with an annual influx of 3.2 billion metric tons of carbon. To be sure, the power of engineers is truly remarkable.

All too often decisions makers, with a superficial knowledge of science and technology, assign to engineers, who have only a passing exposure to the humanities and the arts, the tasks of providing the detailed designs for the life support systems that are the scaffolding for our civilization.

To better prepare engineers for a century that Peter Senge describes as having problems that come from yesterday's solutions, a recent National Academy of Engineering (United States) report entitled *Educating the Engineer of 2020* called for universities to revise their engineering curricula. The report is aimed at creating a more holistic engineering profession to better prepare engineers for the complex challenges of the 21st century, from climate change to cyber security. Higher education's goal, they argued, should be to create "technically proficient engineers who are broadly educated, see themselves as global citizens, can be leaders in business and public service, and who are ethically grounded" where "learning disciplinary technical subjects to the exclusion of a selection of humanities, economics, political science, language, and/or interdisciplinary technical subjects is not in the best interest of producing engineers able to communicate with the public, able to engage in a global engineering marketplace, or trained to be lifelong learners."

While there are few universities that have embraced fully this thinking, the rhetoric makes us hopeful that, at least in the United States, engineers are moving toward a more holistic perspective regarding their design products.

If we are to transcend the engineering thought of the twentieth century, the successful technological designs and solutions of the future must be ever more dependent on creative, imaginative, and intuitive skills. Daniel Pink, in his provocative book, *A Whole New Mind*, argues that "the era of left brain dominance is giving way to a new world in which artistic and holistic right-brain abilities mark the fault-line between who gets ahead and who falls behind." Not surprisingly, ambidextrous thinkers will be the best poised for success. Two centuries after the industrial revolution, we are starting to bridge the two cultures and develop both sides of the American engineer's brain.

Regrettably, the progress in integrated thinking that is unfolding in the United States is not necessarily being shared by the majority of engineers around the globe. Much is said of the earth being flat and the global economy that has been created by the new Information Age. This economy has seen much of the United States' manufacturing and a significant portion of our engineering services outsourced. As we recognize the need for more holistically trained engineers and move forward to re-form our engineering education, we continue to allow major portions of our products to be designed and manufactured by overseas engineers that have been trained (albeit technically rigorously) in a model that has been shown to have significant shortcomings and result in less than holistically sustainable solutions.

If leaders in the United States and overseas continue to proffer education strategies for aspiring engineers, the designers of our collective future, that are solely focused on science, technology, engineering, and mathematics disciplines to the exclusion of the humanities, social sciences, and the arts, we will not realize the full potential of 21st century innovation and global sustainability.

We have a responsibility to ensure that holistic approaches are the norm of engineering's global future. U.S. leadership concepts of "sustainability," "environmental justice," "holistic design," or "service systems engineering," cannot be economically supported or justified if our overseas colleagues are not our collaborators. A world in which global engineers designing exciting technologies, but disconnected from the richness of the human condition, society, and the natural environment, is a world in which we are all the poorer.

Perhaps Tsinghua University in Beijing should invite a "John Keating" to inspire its engineering students with poetry. This time, I hope he accepts.

Domenico Grasso is the Dean of the College and Professor of Engineering, at the College of Engineering and Mathematical Sciences, The University of Vermont, Burlington, VT 05405-0156; email: Domenico.Grasso@uvm.edu. He is Editor-in-Chief of the journal *Environmental Engineering Science* http://www.liebertpub.com/ees.

20 Engineering Professionalism and the Imperative of Sustainable Development

Mark Manion

CONTENTS

INTRODUCTION: THE ABET 2000 CRITERIA AND ENGINEERING ETHICS EDUCATION

Criterion 3 of the ABET 2000 criteria dictates that engineering students demonstrate: (a) an understanding of professional and ethical responsibility; and (b) an understanding of the impact of engineering in a global and societal context [1]. Hence, criterion 3 mandates the inclusion of professional ethics in accredited engineering programs. No less important, however, is the focus on the understanding of the social responsibilities of engineers, and the environmental and global impact of engineering. In fact, Criterion 4 places an emphasis on the larger, social impacts of engineering by requiring that "a major design experience ... must include ... the following considerations: economic; environmental; sustainability; manufacturability; ethical; health and safety; social; and political" [1]. From these criteria one gets a sense that the concepts of professionalism and sustainability seem to be central to carrying out the ABET requirements that engineering students demonstrate a heightened sense of engineering professional responsibility and a sharpened awareness of the impacts of engineering and technology on society and the environment.

Sustainable development was first defined by the World Commission on Environment and Development (WCED) (The Brundtland Commission) as development that "meets the needs of the present without compromising the ability of future generations to meet their own needs" [2, p. 8]. In effect, sustainable development requires, in the least, that industries minimize wastage, increase material and energy efficiencies and clean up the pollution they create. Moreover, it will be engineers who will be responsible for developing the new technologies that will accomplish such mandates.

The importance ABET places on educating engineers on the concept of sustainable development is reflected in the "ASEE Statement on Sustainable Development Education." The statement reads:

> Engineering students should learn about sustainable Development and sustainability in the general education component of the curriculum as they are preparing for the major design experience. For example, studies of economics and ethics are necessary to understand the need to use sustainable engineering techniques, including improved clean technologies. In teaching sustainable design, faculty should ask their students to consider the impacts of design upon U.S. society, and upon other nations and cultures. Engineering faculty should use systems approaches, including interdisciplinary teams, to teach pollution prevention techniques, life cycle analysis, industrial ecology, and other sustainable engineering concepts...

ASEE believes that engineering graduates must be prepared by their education to use sustainable engineering techniques in the practice of their profession and to take leadership roles in facilitating sustainable development in their communities [3].

Engineering schools have responded to the new demands placed on them by ABET and ASEE for more social impacts and engineering ethics education in engineering curricula in many ways—some successful, some not so successful. For example, theorists such as Davis [4], Rabins [5], Herkert [6], and Lynch [7] have identified numerous methods for integrating engineering ethics and science-technology-society studies into the standard engineering curriculum.

Engineering educators have also debated the best method of instruction to use in the engineering ethics classroom. Whitbeck, for example, argues that, generally speaking, engineering ethicists have become dissatisfied with the traditional "applied ethics" approach that starts with abstract moral principles and then reasons deductively from these first principles to determine how they would apply to particular instances [8]. Instead, ethics educators have shifted focus toward case-based reasoning that starts from particulars, and then argues inductively to first principles. For example, according to Harris, Pritchard, and Rabins, extensive discussion and analysis of case studies involving ethical issues in the technological field is the key to the learning process [9].

In order to advance the needed discussion and dialogue on the goals and purposes of engineering ethics education, This paper articulates some preliminary foundations for a philosophy of engineering ethics grounded in the two principles of professionalism and sustainable development. Whatever else is included in a course in engineering ethics, the themes of professionalism and sustainable development should play a crucial organizing role. One important reason why this is so is because these two grand themes answer, in the most complete and thorough way, to the demands of the ABET 2000 criteria and the fulfillment of such proclamations as the ASEE Statement on Sustainable Development Education.

ENGINEERING PROFESSIONALISM

The literature on the sociology and history of professionalism is vast, but one common theme running through all the theories about the nature and function of professions is that there exist crucial characteristics, essential attributes that any occupation must exhibit before it qualifies as a profession [10]. Different theorists have produced different sets of attributes; nevertheless, there seem to be a few attributes that almost all agree are the distinguishing characteristics of a profession [11]. They are: (1) mastery of a theoretical body of knowledge; (2) full decision-making autonomy in the workplace; (3) provides a crucial social need and is sanctioned by the public to provide that need; and (4) professional organization centered around a code of ethics.

Moreover, these attributes are interrelated and mutually effect and support each other. For example, since the professional has mastered a complex body of knowledge—medical science, engineering, chemistry, etc—that statistically few of us master due to its esoteric and complex nature, the professional is the only one in a position to use that knowledge effectively and with a sense of ethical responsibility. Therefore, professionals alone, it is argued, can make appropriate decisions regarding the application of such sophisticated and powerful knowledge. Hence, the need and importance of decision-making autonomy concerning such knowledge, residing exclusively in the hands of the professional, follow as a natural corollary. Moreover, such knowledge usually serves a crucial social function such as health and well-being, law and order, social justice, and technological development, etc. Needless to say, with such esoteric knowledge comes immense power. Given that society needs and depends on such professional services as health-care, justice, or technological development, and given that individual citizens become very vulnerable in their absence, there is always a potential for misuse and abuse of that power.

Therefore, the public must be assured that the professional will always act with a strong sense of responsibility toward the public good and a strong commitment to advance the public interest. This role is, in turn, played by the professional organization's ethic of public service, which is grounded, moreover, in an implicit social contract that exists between society and the professions.

CODES OF ETHICS AND THE SOCIAL CONTRACT

Well-established professions communicate their ethic of service through a code of ethics and a procedure for monitoring their members' compliance with the ethical principles enunciated in the code. The establishment and enforcement of a code of ethics, reaffirms, moreover, the social contract established between the profession and society at large.

It has been argued that the roots of the engineering profession's duties and responsibilities to society can be said to be derived from a "social contract" that holds, at least implicitly, between the engineering profession and society. For their part, the public provides engineers, through the tax base, with educational opportunities, and through legislation, with the means for licensing and regulating themselves. In addition, professionals are given a place of honor and are usually compensated with higher than average salaries. In return, engineers have a responsibility for protecting the safety and well-being of the public in their professional practice.

It is important to point out, moreover that professional codes have an important role to play in establishing the social contract between the engineering profession and society at large. In other words, the injunctions of the codes to protect the safety, health and welfare could be interpreted as standing as a "promissory note" to society, made by the engineering profession, to uphold their part of the contract.

In fact, there exist strong commitments to an implicit social contract between the engineering profession and the public at large, which can be found in almost all contemporary engineering codes of ethics. The emphasis on obligations to the public—the hallmark of the social contract model—seems to be accorded an increasingly prominent role in such codes. The Code of Ethics for the Engineering Council of Professional Development was revised in 1974 to include the so-called "paramountcy clause" which states; "engineers shall hold paramount the safety, health and welfare of the public in the performance of their professional duties." During the intervening decades, this provision has been adopted by almost all of the major engineering societies. This clause requires members to include ethical and social concerns not as something over and above the professional's duties to self and to his or her employer, and not as remedial, but as integral to the central requirements of the profession *per se*—integral to what the profession is all about.

Embedded in such prescriptions is the assumption that the goal of engineering is to uphold the values of public health, safety and welfare above *all* other values. On this view, values such as safety, health and welfare are overriding values that always "trump" other values such as market competitiveness, profit, efficiency, cost, productivity, consumption, benefit-burden trade-offs, etc.

In addition, professional engineering societies have made important, albeit preliminary commitments, to the environment in their codes of ethics. The IEEE Code of Ethics was one of the first to make such a commitment. Its first injunction to its members is to:

> Accept responsibility in making engineering decisions consistent with the safety, health and welfare of the public, and to disclose promptly factors that might endanger the public or the environment [12].

An even stronger commitment to environmental issues is found in code of ethics of the American Society of Civil Engineers, which entreats their members to maintain responsibility towards sustainable development. Canon 1 of the ASCE code states that:

> Engineers shall hold paramount the safety, health and welfare of the public and shall strive to comply with the principles of sustainable development in the performance of their professional duties [13].

In addition to this bold move by the ASCE, the American Association of Engineering Societies (AAES) has taken an active role in setting the pace for sustainable engineering. The AAES is presently involved in an active joint venture with the World Engineering Partnership for Sustainable Development (WEPSD), a global coalition of engineering organizations committed to long-term actions in support of sustainable development. The WEPSD vision statement is as follows:

> Engineers will translate the dreams of humanity, traditional knowledge, and the concepts of science into action through the creative application of technology to achieve sustainable development. The ethics, education, and practices of the engineering profession will shape a sustainable future for all generations. To achieve this vision, the leadership of the world engineering community will join together in an integrated partnership to actively engage with all disciplines and decision makers to provide advice, leadership, and facilitation for our shared and sustainable world [14, p. 7].

If one looks at the practice of engineering over the last 150 years, engineers have indeed embraced the goal of public safety as an automatic objective of most engineering projects. The hope is that the principle of sustainability will be integrated into the nature of engineering so that an engineer would no more design a project that is unsustainable than one that is unsafe.

The ASCE, AAES, and WEPSD philosophies not only stress the importance of sustainable development for sound engineering practice, but they make it unethical for engineers not to strive for such goals.

ENGINEERING AND THE AMERICAN BUSINESS ESTABLISHMENT

Engineers and ethicists have, however, been sharply divided over whether this interpretation of the so-called "paramountcy clause" is correct. Many claim the paramountcy clause asks that engineers promise something they cannot fulfill. Ought implies can, so no one can be held responsible for something that is not under their control. There is one influential set of arguments which try to demonstrate that one cannot derive the social and environmental responsibilities of engineers from the paramountcy clause alone, because engineers alone cannot be held responsible for the risks, harms and non-sustainability of technological processes, artifacts and systems.

Such arguments claim that there exist various organizational and legal constraints engineers working in large corporations are subjected to, and which do not permit the necessary decision-making autonomy required in order to hold engineers responsible for the societal and environmental risks of technology. Engineers are not the only ones responsible for decisions concerning technological development or technology policy—managers, corporate executives, government regulators, legislators, consumers, etc. all play their part. Hence, even if such ultimate values as expressed in the paramountcy clause were possible, engineers would be severely restricted in being able to honor them even if they wanted to. This point is no more evident than in the variety of harassments, firings, black-listings, etc. that constantly confront whistleblowers.

The "received view," as Michael Davis [4] calls it, sees the engineer—who lacks the full autonomy necessary to a be responsible and ethical engineer—in constant conflict with corporate management—who often ends up overriding engineering judgment concerning technological design in their incessant pursuit of the bottom line. Moreover, engineers are often chastised for failing to act on their moral responsibilities when confronted with demands made by corporate executives that are inconsistent with the moral ideals of the engineering profession.

If they mean to be professionals, engineers themselves will have to take moral responsibility for their work rather than unquestioningly accepting whatever orders come down to them from Government or corporate employers [15, p. 42].

The "received view" goes far back to the origins of the American engineering profession (Layton, 1986) and it is also reflected in the history of the Canadian Engineering profession as well. · As Millard puts it:

> Engineers were not independent professionals. Most of them were employees of large corporations and governments. This was the most important factor affecting their professional life. Emerging from the canal- and railway-building enterprises of the nineteenth century, the engineer was a creature of large bureaucratic organizations—the original "organization man" [16, p. 41].

On a more contemporary note, Deborah Johnson and other professional ethicists are quick to point out that engineers working in large organizations are unduly restricted in terms of autonomous decision-making. According to such researchers, even though engineers alone have the expertise to estimate the risks of the technologies they design, and hence are in a strategic position to best judge whether a certain technology is unsafe or counter to the welfare of society, they have little or no decision-making power over that technology. "Engineers do not have the final say concerning what is built, how it is built, or whether it is ultimately manufactured and marketed to be sold and used" [17, p. 23]. In fact, managerial decisions, not engineering decisions, often define the course and shape of the design, development and implementation of technological artifacts. As the "received view" goes, engineers simply lack the professional decision-making autonomy essential for them to act on such noble ideals that fulfilling one's "social responsibility" entails. As Johnson points out:

> Engineers need autonomy because they have special knowledge. If they are to use that knowledge in a responsible manner and for the good of society, they must have the power to do so. However, as they work for corporations with complex organizational structures and decision-making procedures, and as these large organizations need co-ordination of their various parts, it would seem that engineers must often do what they are told [17, p. 24].

Lack of professional autonomy leaves scant room for moral choice. As a result, it is often argued that:

> Engineers are a "captive profession" in a highly compartmentalized environment. Managers choose what to do, divide work into small jobs, and assign each job to one engineer or small group. Communication between engineers is kept to a minimum to assure management control. An engineer may need permission from his boss even to discuss a project with an engineer in another department or working group. Engineers identify options, test them, and report the results to managers. Managers combine these reports with business information they alone have. Managers decide. Engineers merely advise [4, p. 42].

According to such accounts, engineers working in large corporations are all too often used as mere "hired-hands" who design and develop technology with the sole purpose of advancing the market-competitiveness or the cost-effectiveness demands of the corporate client.

This view has not gone uncontested, however. Researchers such as Davis himself have, in fact, recently challenged the claim that engineers lack any decision-making authority in the corporate workplace. After an empirical investigation involving interviews and surveys of 60 engineering and management employees in 10 companies, Davis concludes that:

> Instead of the rigid hierarchical, and compartmentalized decision making process of the received view, we found a highly fluid process depending heavily on meetings and less formal exchange of information across even departmental boundaries. Managers seemed to have little control over what information would reach their engineers. Indeed, they seemed anxious to get their engineers to hook up with others on their own [4, p. 51].

There are numerous examples, however, throughout the history of engineering, some which are chronicled in Layton's book *The Revolt of the Engineers* [18], which indicates that there is some truth to the "received view;" namely that engineers merely "serve" the corporate client and have little or no role in decisions regarding questions about the ultimate ends for which the technology that they design and develop are used. In fact, despite its inherent flaws, an influential model of engineering ethics proposed by Samuel Florman is entirely consistent with the received view! [19,20]. We must develop beliefs and attitudes that can overcome such hindrances to sustainable engineering.

The greening of American Business and technology policy may very well be the solution to the seemingly intractable dilemmas the ethical engineer confronts when trying to reconcile business interests with the public interest. Such changing conditions in corporate American culture provide, perhaps for the first time, the opportunity for engineers to exercise what Stephen Unger calls one's "right to be an ethical engineer" [21]. This is another reason why engineers must be made aware of the role of sustainable development and its potential impacts on engineering theory and practice.

THE GREENING OF AMERICAN BUSINESS AND AMERICAN TECHNOLOGY POLICY

In 1992 American corporations and Governmental legislators and regulators received a wake-up call. The Rio Earth Summit highlighted the potential risks to ecology and long-term economic and social development created by current patters of industrialization, population growth, and social inequality. The issues were set out in various documents; one of documents produced was Agenda 21. Over 175 heads of state and government signed this document. Agenda 21 stresses the need for fundamental political, social, economic, and industrial change in order to conserve natural and biological resources, limit pollution, build strong and prosperous communities in all parts of the world, and addresses issues such as social justice and human rights violations [22].

The American government responded to Agenda 21 by creating the President's Council on Sustainable Development (PCSD), which was established by President Clinton through Executive Order No. 12852 on June 29, 1993. So far, the PCSD has created eight different task forces, which have produced recommendations thorough workshops, demonstration projects, case studies, regional round tables, public comment, published reports, as well as other courses of action.

Such activity on the global, as well as national, scale is beginning to have substantial effects on American business. The message for companies doing nothing is that they must respond with critical action. The message to companies already taking the environment seriously is to do more—and to pay greater attention to the issues of sustainable development and sustainable engineering. In 1990 DuPont's then CEO, Edgar Woolrad, identified the developing trend toward incorporating ecological sustainability into an already growing "Total Quality Management" (TQM) movement when he stated:

> The green economies and lifestyles of the twenty-first century may be conceptualized by environmental thinkers, but they can only be actualized by industrial corporations. Industry has a next-century vision of integrated environmental performance. Not every company is there yet, but most are trying. Those that aren't trying won't be a problem long-term, simply because they won't be around long-term. This is the new competitive reality [23, p. 1].

The words of Woolrad are reflected in the words of Maurice Strong, Secretary General of the UNSED Earth Summit, who remarked:

> Efficient enterprises are at the head of the movement to sustainable development. Corporations that are on the leading edge of a new generation of opportunities created by the transition to sustainable development will be the most successful in terms of profits and the interests of their shareholders. Businesses that are defensive, fighting yesterday's battles, will fall by the wayside and will be caught in the backwash of the wave of the future [24, p. 2].

In the pursuit of total quality, businesses and industry in the United States and around the world are noticeably becoming more responsible in terms of their environmental effects. They are, in effect, "going green." Indeed, it is becoming more and more evident to corporate leaders that becoming environmentally responsible is the next step in total quality management—some call it total quality environment movement (TQEM)—a step that is becoming more and more necessary in order to stay competitive and profitable [25]. In a survey conducted in 1991 by a major polling firm, 75 percent of the Fortune 200 firms surveyed claimed that the environment was a central strategic issue [24, p. 2].

Pollution prevention, sustainable, full life-cycle packaging and product development, green marketing, and measurement of environmental performance are presently all sectors of intense focus and investment in numerous large corporations [26]. Environmental performance is increasingly a determinant of a company's reputation, among employees, customers, and stakeholders alike [27]. A poor environmental reputation can harm recruitment, retention, and morale, damage sales, even threaten a company's "license to operate" [28]. In sum, businesses are not only going green, but, moreover, they are actually maintaining and improving their competitive positions by going green.

The greening of business and industry is happening at a fast rate and it is happening globally. The global characteristic of the movement towards more responsible corporate environmental performance is apparent not only in what individual organizations are doing, but also in the growth of international, environmental business networks. Presently, it is not longer a question of *whether* companies will focus on the call to sustainable development, or when, but of *how*, and how fast!

For the first time, the goals of the corporate world are beginning to be identified with the goals of environmentalists. In fact, if there is presently a conflict at all between the concerns of environmentalists to protect the environment and renew crucial natural resources, and the concerns of corporations for the bottom line, it is that the changes taking place appear to be occurring too slowly for the environmentalists, and too fast for corporate leaders. Both agree, however, that the changes are inevitable and good for all.

In sum, the various legal and social pressures placed on corporations from environmentalism and the philosophy of sustainable development have placed serious constraints on corporations to develop and design more environmentally sound and human friendly technological systems and artifacts.

In order to accomplish such a tall order, corporate executives will have to turn to engineers in particular, and to the engineering profession, in general. Such radical changes in management theory and business practices are requiring a new type of engineer with new skills, attitudes and personal qualities. As one researcher put it: "It is in the area of attitudes and values where corporations are making a dramatic shift in their expectations" [29, p. 395]. In other words, the requirements of sustainable development dictate that engineers and businesspeople develop a new set of values and attitudes. Technical knowledge is not enough. In addition, the new engineer must be educated about the social and ethical impacts of engineering and technology. This is because the philosophy of sustainable development requires that engineers and businesspeople value the environment alongside economic development; value future as well as current generations; value social equity as well as material growth; value the poor along with the rich. Thus, sustainable development relies on a change in corporate culture, supported by an adapting economic system with newly defined market values, and sustained by appropriately developed technology. It is the last requirement for successful sustainable engineering, the development of appropriate or "clean" technology, which should concern the engineering profession.

Designing technological processes, artifacts and systems that respect the call to "reduce, reuse, and recycle" is a perfect example of how engineers can take the lead in developing sustainable engineering. The inclusion of recycled wood products, metals, glass, plastics, concrete, etc. can benefit society and the environment in three ways. First of all, recyclables reduce waste and the growing problem of its storage and disposal. Second of all, recyclables contribute to the conservation of raw materials and resources that no longer have to be taken from the environment. Thirdly, recycling pays!

For example, the recycling of already contaminated metals for reuse in waste containment is projected to save the DOE and the nuclear industry the cost of two million tons of new metal resources. This also greatly reduces the overall poisoning of the environment because two million tons less toxic waste will enter the environment. Or, take the example of the Texas Department of Transportation (DOT) which has begun to use styrene and carbon black from used printer and copier cartridges as an asphalt additive. This additive actually improves pavement high temperature performance, and keeps lots of waste out of already overcrowded landfills. This is part of the Texas DOT's overall to use recyclables such as glass, shingles, and shredded brush in their roads and parking lots, instead of further clogging up the state's dumps

Other examples include new construction methods using fiber-reinforced wood, wood composite products, and polystyrene, in various types of structures, from homes to office buildings, to skyscrapers. This eco-efficient approach to engineering construction has reportedly helped Dow-Europe cut overall fuel costs by 30 percent, and the cost of raw materials by 40 percent in all building projects.

Sustainable manufacturing also pays, as examples from the chemical industry have shown. Two well-cited examples are Dow Chemical's "Waste Reduction Always Pays," and the 3M Corporation's "3P" program ("Pollution Prevention Pays"), which is reported to have saved the company hundreds of millions of dollars [23, p. 33–38].

Engineers of all disciplines are in a unique position to lead by example, whether by designing alternative fabrication processes that do not produce large amounts of waste, or by designing cost-effective and environmentally sound energy sources such as solar power, or designing environmentally sound materials for construction, etc. Driven by a philosophy of clean engineering design and a promise to uphold the principles of sustainable development through their profession and their codes, engineering will rapidly become one of most the sought-after professions in the corporate world.

In effect, the shift toward sustainable economics and business has created the perfect opportunity for the engineering profession to finally end its quarrels and conflicts with business and industry over the question of autonomy and control over decisions about technological design and development. Under such new conditions, management are now "all ears;" willing and eager to learn everything they can from engineers about how to design, develop, and deploy "clean" technologies, and praising and rewarding them if and when they design such technologies. Whether it be how to design more efficient energy sources through conservation measures and switching to renewable sources, or waste minimization through greater recycling and reuse of materials, or more "green" production, including full product life-cycle analyses, or more comprehensive technology assessment and better management of risk and resources, engineers are beginning to play leadership roles, whether inadvertently or not. All of these changes have the potential to catapult engineers to the forefront of the sustainable development revolution.

As one expert put it: "The environmental challenge will be one of the central issues of the 21st century" [30]. Moreover, as Mitcham and Duval argue, "Engineering perhaps more than any other profession has an opportunity to contribute to the nuts and bolts of just such [sustainable] development" [30, p. 35]. To meet these challenges, whole new dimensions of engineering theory and practice are emerging. Terms that point toward these new forms of engineering are: alternative technology, preventive engineering, sustainable engineering, design for environment, and green design. My point is that the engineering profession surely does not want to miss this opportunity to exert its influence in matters of business and economics.

Sustainable development and sustainable engineering are, at least as described above, beneficial and achievable visions. However, as Herkert correctly points out, some theorists are critical of such concepts to the extent that they become "co-opted and corrupted by those who would ignore [their] ethical implications and focus on the sustainability of markets to the neglect of environmental and human considerations [31, p. 48]. This is because, as Herkert puts it: "despite proclamations that engineers have an ethical responsibility to endorse the principles of sustainable development, questions of just distribution and other questions of equity (such as risk distribution) are often ignored when engineers consider sustainable development policies" [31, p. 49]. This is why it is crucial, as this paper argues, to never separate an engineer's commitment to sustainable engineering practice from his or her commitment to an ethic of sustainable development.

THE ENGINEERING PROFESSION AND THE IMPERATIVE OF SUSTAINABLE DEVELOPMENT

The unprecedented level of negative impacts due to turbo-industrialization has forced us to confront the necessities of sustainable development. This confrontation has, in turn, forced us to recognize the limitations of traditional ethical systems (utilitarianism, social welfarism, traditional deontology, etc.) and has stimulated the development of an alternative ethics that can aid us in understanding our duties, obligations, and responsibilities under conditions of sustainable living. The limitations of risk-cost-benefit-analysis (RCBA) is a case in point. The moral soundness of RCBA is no better than the utilitarian ethic that provides its moral foundations. However, utilitarianism's insensitivity and ignorance to questions of distributive, intergenerational, and ecological justice proves that we must formulate a new engineering ethic to meet the needs of sustainable engineering. This means that engineers need to reconsider the validity of their typically utilitarian way of responding to questions about the social impacts of technological innovations.

Engineers must also respect an ethic grounded in the philosophy of sustainable development. Moreover, this new ethic should not be called engineering ethics. The new ethic would be more far-reaching and would merge all of the so called "applied-ethics" disciplines such as engineering ethics, business ethics, environmental ethics, computer ethics, biomedical ethics, etc. with the philosophy of technology and philosophy of science. Call it an "ethics of technology" or "technology ethics."

One form the new technology ethics could take would be following what I call the imperative of sustainable development. Here are a few formulations of the imperative of sustainable development:

- "Act so that your effects of your action are compatible with the permanence of genuine life";
- "Act so that the effects of your action are not destructive of the future possibility of such life";
- "Do not compromise the conditions for an indefinite continuation of humanity on earth";
- "In your present choices, include the future wholeness of Humanity among the objects of your will."[1]

These imperatives as stated are indeed very general in nature and need explicit formulation, which, unfortunately, cannot be developed here. However, they at least provide a preliminary foundation for a philosophy of engineering ethics that goes beyond the traditional utilitarian ethics of many practitioners of engineering.

CONCLUSION

The cultural attitude toward the environment has been changing, and the engineer can play an active role in this transformation. In the changing climate of the greening of American business and Government regulation, corporate management and political leaders will have to rely on the engineer more than ever to invent the visionary technologies that will be required to fulfill the promise of sustainable development. Therefore, the move toward sustainable development is a prime opportunity for engineers to gain full autonomy over decision-making concerning technological development and technology policy. In order to help sustain development, the engineer must be able to reconcile the goals of engineering as a business as well as a profession. As one researcher put it:

> If we [engineers] do not meet this challenge [of sustainable development], then we will be left behind in the decision making process that will influence shape of this world. Engineering for a sustainable future will require engineers to engage more actively in political, economic, technical, and social discussion and processes to help set a new direction for the world and its development [33, p. 3].

In their new role, engineers could become true counselors to both corporate executives as well as to the citizenry in general. As counselors to corporate executives, they could play leadership roles in the cultural paradigm shift toward sustainable development. As counselors to the public, they could play leadership roles in enhancing public perception and understanding of risk, as well as helping to improve the quality of technology-intensive choices in public policy, particularly when it comes to developing sustainable technology. Engineers have traditionally seen themselves serving three "clients": their clients or employers; society at large; and their profession. If engineers can really integrate principles of sustainable development into their designs as well as their attitudes, they can actually fulfill their duties to all parties concerned.

In integrating such sustainable engineering techniques as full life cycle production, reduced pollution, energy efficient manufacturing, etc. they can help their corporate clients prosper and stay competitive. In addition, practicing such environmentally sound and humane engineering, then can regain the coveted trust and respect of the public at large which is so important to professional success, but which has been on the wane in recent decades. Finally, in following the principles of sustainable development, the will be serving and upholding the ideals of their profession. Moreover, following the principles of green engineering, they will be fulfilling their obligations to "self" in that individual engineers can be proud they are "doing good" by society and the environment. These are some of the reasons why the concepts of professionalism and sustainable development must play a role in any course in engineering ethics, and also why such a philosophy of engineering ethics centered around the concept of and sustainable development answers best to the demands of the ABET 2000 criteria.

This essay has attempted to provide the rationale for such a philosophy of engineering ethics. This essay has not, however, attempted to demonstrate what the content of a program in engineering ethics grounded in such ideals would look like. It is central to my present thesis, however, that this new philosophy of engineering ethics can be best inculcated into the culture of engineering through engineering education—experience and intuition are not enough. Engineering ethicists must work more closely with engineering scientists to ensure all facets of sustainable technology become a practical reality. While professors of engineering science can increase awareness by stimulating engineering students to build sustainable ideas into their designs, professors of engineering ethics would work to complement this by helping to transform the attitudes, values and philosophies of the "new engineer."

If the engineering profession can accomplish this grand challenge through engineering ethics education, and train future engineers to become leaders in business and social policy, as well as counselors to corporate executives and citizens alike, they can finally fulfill their professional ideal as benefactor of humankind, and no longer be cast as obedient servants to corporatism.

BIOGRAPHY

Mark Manion received his PhD in philosophy from Temple University. His dissertation focused on developing a theory of moral epistemology grounded in the writings of Ludwig Wittgenstein.

ENDNOTE

1. These formulations are from the philosopher of technology Hans Jonas [32, p. 44] who develops them in a different context.

REFERENCES

1. ABET, 1998. *Engineering Criteria 2000.* Available on the World Wide Web at http://www.abet.org/eac /EAC_99-00_Critcria.htm#EC2000.
2. World Commission on Environment and Development (1984). *Our Common Future.* New York: Oxford University Press.
3. ASEE Statement on Sustainable Development Education (1999). Approved by the ASEE Board of Directors June 20, 1999. Available on the World Wide Web at: http://www.asee.org/welcome/stalements /sustain.cfm.
4. Davis, Michael (1992). "Technical designs: time to rethink the engineers responsibilities?," *Business and Professional Ethics Journal* 11 (3): 41–55.
5. Rabins, Michael. (1998). "Teaching Engineering Ethics to Undergraduates: Why? What? How?" *Science and Engineering Ethics* 4: 291–302.
6. Herkert, Joseph. (1990). "Science, Technology and Society Education for Engineers" *IEEE Technology and Society Magazine,* September: pp. 22–26.
7. Lynch, William. (1998). "Teaching Engineering Ethics in the United States." *IEEE Technology and Society Magazine* 16 (Winter): 27–36.
8. Whitbeck, C. (1995). "Teaching Ethics to Scientists and Engineers: Moral agents and moral problems." *Science and Engineering Ethics* 1: 397–308
9. Harris, C. Pritchard, M. Rabins, M. (2000). Engineering Ethics: Concepts and Cases. Stamford, CT: Wadsworth Publishers.
10. Davis, Michael (1997). "Is there a profession of engineering?" *Science and engineering ethics* 3 (4): 407–428.
11. Greenwood, E. (1991). "Attributes of a profession," in *Ethical issues in engineering,* Johnson, Deborah, Ed. Englewood Cliffs, NJ: Prentice Hall.
12. Institute of Electrical and Electronic Engineers (IEEE), Code of Ethics, Approved by the IEEE Board of Directors, 1990.

13. American Society of Civil Engineers (ASCE), Code of Ethics, adopted 1996.
14. Ellis, Monica, Ed. (1994). *The Role of Engineering in Sustainable Development: Selected readings and references for the profession.* Washington, DC: The American Association of Engineering Societies.
15. Walter, Kenneth (1973). "Professionalism and engineer/management relations," *Professional Engineer* 43 January: 42–48.
16. Millard, Rodney (1988). *The master spirit of the age: Canadian engineers and the politics of professionalism 1887–1922.* Toronto: University of Toronto Press.
17. Johnson, Deborah (1992). "Do Engineers Have Social Responsibilities?" Journal of Applied Philosophy 9(1): 21–34.
18. Layton, Edwin (1986). The revolt of the engineers: Social responsibility and the American engineering profession. Baltimore, MD: Johns Hopkins University Press, 1986.
19. Florman, Samuel (1978). "Moral blueprints," *Harpers* 257, October: 32–38.
20. Florman, Samuel (1976). *The existential pleasures of engineering.* New York, NY: St. Martin's Press.
21. Unger, Stephen. (1994). *Controlling Technology: Ethics and the Responsible Engineer,* 2nd edition. New York: John Wiley and Sons, Inc.
22. Brown, Noel and Quiblier, Pierre, Eds. (1994). *Ethics and Agenda 21: Moral implications of a global consensus.* New York: United Nations Environment Programme.
23. DeSimone, L. and Popoff, F. (1998). *Eco-efficiency: The business link to sustainable development.* Cambridge, MA: The MIT Press.
24. Kinlaw, Dennis (1993). *Competitive and Green: Sustainable performance in the environmental age.* Amsterdam: Pfeiffer.
25. Bhat, V. N. (1998). *Total quality environmental management: An ISO 2000 approach.* Westport, CT: Quorum Books.
26. Post, James, Collins, Dennis and Starik, Mark, Eds. (1995). *Sustaining the natural environment: Empirical studies on the interface between nature and organizations.* Greenwich, CT: JAI Press.
27. Stead, W. and Stead, J. (1995). "An empirical investigation of sustainability strategy implementation in industrial organizations," In Post, James, Collins, Dennis and Starik, Mark, Eds. (1995). *Sustaining the natural environment: Empirical studies on the interface between nature and organizations.* Greenwich, CT: JAI Press: 44–67.
28. Greening, Daniel (1995). "Conservation strategies, firm performance, and corporate reputation in the U.S. electric utility agency. In Post, James, Collins, Dennis and Starik, Mark, Eds. (1995). *Sustaining the natural environment: Empirical studies on the interface between nature and organizations.* Greenwich, CT: JAI Press: 345–369.
29. Vasilica, G. (1994). "Personal view: Engineers for a new age: How should we train them?" *The International Journal of Engineering Education* 10: 394–400.
30. Mitcham, Carl and Duvall Shannon (2000). *Engineering Ethics.* Upper Saddle River, NJ: Prentice Hall.
31. Herkert, Joseph (1998). "Sustainable development: ethical and policy implications," CHEMTECH (November): 47–53.
32. Jonas, Hans (1973). "Technology and responsibility: Reflections of the new task of ethics," *Social Research* (40): 31–54.
33. Statement of the American Association of Engineering Societies on the Role of the Engineer in Sustainable Development (1994). In Ellis, M. Ed. (1994). *The Role of Engineering in Sustainable Development: Selected readings and references for the profession.* Washington, DC: The American Association of Engineering Societies.

21 The Problem of Knowledge in Incorporating Humanitarian Ethics in Engineering Education
Barriers and Opportunities

Jon A. Leydens and Juan C. Lucena

CONTENTS

INTRODUCTION

From its 19th century origins as an organized profession, US engineering has had a marginal connection to nonmilitary humanitarian efforts. Most humanitarian work took place among medical professionals in the battle fields; the International Red Cross/Red Crescent, founded in 1864, originated as a humanitarian response to battlefield conditions. During World War I, Herbert Hoover became one of the first engineers to deal with large-scale humanitarian problems as he headed the Commission for the Relief of Belgium in 1914 and directed the relief efforts of the Great Mississippi Flood of 1927 [1]. So significant was his leadership in these humanitarian efforts that the American Society of Mechanical Engineers (ASME) established the Hoover Medal in 1929 "to recognize great, unselfish, non-technical services by engineers to humanity" [2]. However, the engineering profession had yet to engage in humanitarian work in its own terms. It was not until the mid-20th century that engineering became more instrumental to international development policies and projects, many of which were humanitarian in character and driven mainly by US foreign policy.

Most recently, following the example of other professions, engineers have created national and international networks that bring humanitarian relief through the application of engineering knowledge and skills to communities in need [3]. National networks include student organizations such as Engineers for a Sustainable World (ESW) and Engineers without Borders (EWB). These networks are complemented by local organizations such as Engineers in Technical and Humanitarian Opportunities of Service (Iowa State University), Technology Assist by Students (Stanford University), Engineering World Health (Duke University), and Engineers for a Better World (Colorado School of Mines).

Related efforts have resulted in recent curricular changes at many institutions with strong engineering programs, such as Purdue University, University of Colorado, Michigan Technological University, MIT, Caltech, Cornell, and Colorado School of Mines [4–6]. Purdue's Engineering Projects in Community Service (EPICS) aims to "create partnerships between teams of undergraduate students and local community not-for-profit organizations to solve engineering-based problems in the community" [7]. The University of Colorado-Boulder's Engineering for Developing Communities™ (EDC) program educates "globally responsible engineering students and professionals who can offer sustainable and appropriate solutions to the endemic problems faced by developing communities worldwide" [8]. At Michigan Technological University (MTU), engineering students have four potential opportunities for humanitarian engineering: A minor in International Development, a student chapter of Engineers without Borders as of 2005, International Senior Design Projects, and a Master's Degree in International Peace Corps. The latter program allows students to obtain a civil or environmental engineering master's degree while working as an engineer with the Peace Corps [9]. Yet in neither EPICS, EDC, nor the MTU graduate program is there an explicit graduate-level engineering ethics component drawing on humanitarian traditions of practice and knowledge.

At Colorado School of Mines (CSM), the Humanitarian Engineering Ethics (HEE) initiative is predicated on the idea that many unexplored questions, particularly ethical ones, arise from the intersection between engineering and humanitarianism. However, engineering ethics education has to date mainly focused on individual and social responsibilities. In his comprehensive survey of US engineering ethics education, Herkert shows that the content of engineering ethics instruction has mainly focused on professional responsibility toward society—public safety and welfare, risk and the principle of informed consent, conflict of interest, and whistle blowing, among others [10,11]. In focusing on individual and social responsibilities, engineering ethics education has overlooked an important dimension of engineering practice: the role of engineers in domestic and international humanitarian activities. To address these issues and to complement our institution's existing undergraduate Humanitarian Engineering (HE) Program, which has a strong service-learning component, we have spearheaded an NSF-funded project to implement Humanitarian Engineering Ethics (HEE) education at the graduate level. The National Service-Learning Clearinghouse defines service learning as "a teaching and learning strategy that integrates meaningful community service with instruction and reflection to enrich the learning experience, teach civic responsibility, and strengthen communities" [12]. The significant capstone experience in the undergraduate HE minor is a senior design project focused on humanitarian engineering, defined as the "application of science and mathematics to improve the welfare of the less advantaged;" to date, over 100 students have been involved in 21 humanitarian engineering projects on four continents in nine countries [13]. We are currently exploring intersections between the HE program and the HEE initiative, including the idea of a HEE design experience.

This paper focuses on the methods and findings related to the barriers and opportunities involved in launching the HEE graduate minor. Although still under construction, the working idea for the HEE minor involves an introductory seminar team taught by faculty in engineering and the liberal arts, modules developed within new and existing engineering courses, and humanitarian ethics courses in the liberal arts.

Before implementing HEE, we need to better understand the opportunities and barriers inherent in general reforms of graduate engineering education and the specific reform involving HEE.

Our literature review has accentuated the importance of understanding both institutional culture and systemic barriers to ethics education and curriculum development [14,15]. HEE will also need to heed the organizational and institutional constraints that arose in other engineering education reform initiatives such as the Engineering Education Coalitions [16].

METHOD

To expand our knowledge base, we conducted semi-structured interviews with colleagues at our own and other institutions. Interviewees were chosen using purposeful sampling methods, which involve selecting participants most likely to provide rich, detailed descriptions of the issues under investigation [17,18]. Thus, study participants (interviewees) came from institutions with engineering programs and most are developing or have developed community-service or service-learning programs. Of the 15 study participants, 10 were engineers, and most of these were engineering faculty. The remaining interviewees included non-engineering faculty, an academic administrator, and an engineering and non-engineering student. They hailed from institutions ranging from small engineering schools to large, research-focused universities. Of all the face-to-face interviews, roughly half occurred at the Engineers for a Sustainable World Conference in October 2005 and the other half at their own institutions or at other conferences. After describing our undergraduate HE program and HEE initiative, we asked participants to respond to our primary research question: What barriers and opportunities might foster or prevent HEE implementation? Interviewees signed approved informed consent forms, and the interviews were audio taped and later transcribed.

To identify emergent categories in the interview data, we enacted the following emergent-design sequence [19,20]:

- Began collecting interview and observation data
- Looked for key issues, recurrent events, which later became categories of focus.
- Collected more data that provided many incidents of the categories of focus to better understand the range of viewpoints on and dimensions in these categories.
- Wrote about the categories, describing and accounting for all the incidents in the data while searching for new incidents.
- Analyzed the data to understand key processes and relationships in the emerging model, the categories of focus combined.
- Engaged in more sampling, coding, and writing as analysis focused on core categories.
- Described the model and its components, categories of focus and subcategories as they have emerged from the data.

We also conducted participant observation in bimonthly meetings for both undergraduate and graduate projects in humanitarian engineering at CSM. This presence, as we collectively wrestle over curricular and programmatic issues, has helped us better understand how the organization, value, and ranking of knowledge in matters when new programs try to make incursion into well-disciplines.

FINDINGS: BARRIERS

Prior to conducting the research, our working hypothesis was that barriers preventing the implementation of HE go beyond expected normative (e.g., "involvement in HE does not count high for tenure and promotion"), curricular (e.g., "there is no room for HE courses in the curriculum"), and institutional (e.g., "we don't have a culture of community involvement among our students") factors to include epistemic (e.g., engineering problem solving), and communicative (e.g., writing research articles and proposals) practices in engineering education and research.

Our interview data indicate that the problem of knowledge is probably the most significant barrier in the implementation of HEE. The problem of knowledge can be defined as a resistance to non-quantitative solutions that emanates from the way in which knowledge is organized, characterized, and/or valued. After analyzing the interview data, we identified three categories of focus in the problem of knowledge: 1) organization of knowledge, 2) content of knowledge, and 3) hierarchy of knowledge.

ORGANIZATION OF KNOWLEDGE AROUND DISCIPLINES

The organization of knowledge in engineering education into the categories including basic sciences, engineering sciences, applied engineering, design, and humanities and social sciences goes back to the Cold War [21]. The engineering sciences were formally institutionalized as research categories at NSF and as curricular categories in most US engineering schools in the 1960s. According to the 1968 Goals Report of the American Society of Engineering Education (ASEE), "all courses that displace engineering science should be scrutinized. *The most important engineering background of the student lies in the basic sciences and engineering sciences*" (italics added) [22]. ABET accreditation criteria quickly and decisively came to reflect this emphasis on science [23]. As international work, customer satisfaction trends, and organizational change in corporations raise the profile of design and the humanities and social sciences, some new curricular categories have gained the attention of engineering educators. Yet, for the most part, the engineering sciences remain unquestioned and in most cases the engineering disciplines remain intact.

Multiple interviewees told us that HEE's inter- and multidisciplinary collaboration would collide with the "stove pipe" mentality, wherein disciplinary specialists remain in separate niches; interviewees indicated this mentality, a phenomenon not limited to engineering, would persist despite the growing acceptance for the notion that many complex problems are best solved by drawing from multidisciplinary expertise. As one participant put it, "the barrier is the way we educate at the present time" [24]. Our interviewees noted that because faculty are often wary of leaving disciplinary comfort zones, they need to be informed of the scope and objectives of HEE and be convinced that the benefits of a more holistic, interdisciplinary approach to teaching and learning engineering outweigh the time and complexity involved.

Some interviewees suggested that the organization of disciplinary knowledge presents a barrier for faculty who were educated in the engineering sciences. Even with the necessity for interdisciplinary collaboration between engineering and humanities/social sciences foregrounded by ABET's current accreditation criteria [25], most faculty have a difficult time knowing how to integrate social and cultural dimensions into technical work, particularly in the ways required by community-service and service-learning projects. Referring to the double challenge of integration brought by ABET and community-service projects, one program coordinator said, "They [faculty] just don't know how to implement it and assess it. We have the same problem, I think, all over the US—it was nice of [ABET] to give us the reigns, but then the burden on figuring out how [to meet our and ABET outcomes] … might actually be harder than them saying, 'you need to teach them these sixteen courses, and this is three credits, and this is three credits.' … Now [students] are supposed to have global awareness and be able to work in teams. The teamwork we're getting really good at because that has been part of this curricular change for the past ten or fifteen years …. But how are they are supposed to look at the social and economic and political aspects of what they are doing [in community-service projects]? I don't think faculty necessarily know how to do that. … The faculty who have been here for a really long time think that 'you can teach [students] that in a humanities course. [Students] have the option to take five humanities courses, and you can teach them sociology there. But it does not make sense to teach them the sociological aspects of designing a new wind turbine. …'" [26]

Additionally, the organization of knowledge around disciplines also conditions how students in multidisciplinary community-service projects view each other. Describing how students organized themselves in service-learning teams, one faculty member involved in a large-scale program told us that "... students will stereotype each other, from day one. When they introduce themselves, they will peg them, based on their major, their year, gender, race, other things, too. So if you've got a female liberal arts student, she's not going to be looked at to deal with [anything technical]. The computer science student will probably be the webmaster ... There are language differences. So [we need] to get students to start to talk about the different languages. The engineering students use words, acronyms and other stuff the other folks won't get ... A lot of the faculty really didn't understand how to facilitate multi-disciplinary activities. They would stereotype students from different disciplines. Or they would ask the liberal arts students to do engineering work. They'd grade them—in one extreme, we had a faculty member who gave the liberal arts students bad grades because there wasn't any engineering design in their work. They were sociologists, [so] there shouldn't have been" [27].

CONTENT OF KNOWLEDGE

The core content of engineering sciences is the engineering problem-solving (EPS) method. Beginning usually in the Introduction to Engineering course and continuing throughout the engineering science courses, students commonly learn EPS in a particular sequence. First, they are given representative textbook problems that tell them what to find and learn to extract the technical, relevant information necessary to solve those problems. Second, students learn to create idealized visual abstractions, such as free-body diagrams, of the problems. Third, students learn to make assumptions to simplify the problems (e.g., fluid is non-compressible) so it can be solved more rapidly and effectively. Fourth, students learn to identify and apply specific scientific principles to the problems; these principles, often in the form of equations, generally come exclusively from the engineering sciences. Fifth, once the equations are in place, students learn to deploy mathematical strategies to solve these equations. Sixth, students learn to produce a single solution, for which they do or do not receive full credit. Finally, most students, but not all, learn to reflect back on the answer and ask whether it makes sense in the physical world. Frequently, students are not taught how to consider non-technical issues throughout this process or are taught that such issues are extraneous variables. Hence, EPS draws a sharp boundary between what is to be considered as engineering (what stays in the problem) and what is not (what stays out the problem) [23]. Although EPS is a powerful analytical tool, it conditions students to not factor in social, cultural, and other non-technical issues and to remain passive problem-solvers who come to expect pre-defined problems to be given to them.

Study participants identified EPS in engineering sciences as a barrier. When asked how students respond to community-service design projects after three years of predominately engineering sciences, one interviewee, for example, told us, "when you put students with a community partner somewhere in the world, local or wherever, they freak out.... The phenomenon is freezing. ... We're giving them tools to muddle around problem solving, but, yeah, they freeze" [27]. By contrast, participants at universities with significant pre-senior design experiences stated that seniors were more accustomed to the kind of problem-based learning characterized by open-ended problem-solving methods.

HIERARCHY OF KNOWLEDGE

Within this dimension, participants reported three specific knowledge value differentiations.

Engineering Science Valued over Engineering Design

A now-tenured interviewee related that a former dean of engineering at his university did not think involvement with design helped a faculty member's tenure case; when he received attention for student design projects, the dean sent the then-untenured faculty design advisor a letter both thanking him and telling him not to be involved again. After that, the young faculty member's department head, who by contrast was supportive of his efforts, was listed as the advisor of all design competitions to shield the untenured faculty member [29].

The hierarchy of knowledge even influences the ways in which people value grants and funding sources. Often grants for education reform, such as those that fund most service-learning initiatives, are not perceived as having the same status as research grants, regardless of their financial size. Rather, the purpose of the grant matters; as one program coordinator explained, "With a hundred-thousand dollars [in] curriculum development that we did get from NSF, the peers didn't see that as [equivalent to] a hundred-thousand dollar EPA, studying this biological contamination of water" [26].

High-Tech Design Valued over Low-Tech Design

Within design, many engineering faculty value high-tech, sophisticated solutions over low-tech, simple solutions, even when the latter might be more appropriate to HE problems. One interviewee told us that in community service projects "you let [local] needs drive the sophistication of the project" yet faculty "look at how elegant something is ... [even] when a group in need may be in need of an efficient solution, not necessarily the most high tech" [27].

During a recent review of our HE projects, a senior design group presented a simple LED circuit as a low-energy solution to the need for portable nightlights in an Ecuadorian village. Reacting to the design, one engineering science faculty member said, "That is not high tech enough. I am concerned that we are not teaching our students to be high-tech, and then they will not be competitive in the job market" [28].

Engineering Valued over Non-Engineering

Some interviewees noted that many engineers consider ethics easy, soft, or just common sense, so it is disregarded or trivialized. Ethics is not alone in its marginalization. In engineering education, some study participants indicated, service learning is also frequently seen as lacking in academic rigor. The difference in the levels of available funding for engineering research vis-à-vis humanities/social sciences research also reinforces a hierarchy of knowledge where engineering is more highly valued because it can bring in significant funding, discouraging engineering faculty from collaborating with humanities/social science faculty. As one program coordinator puts it, "... it's not that [humanities social science faculty] are not doing research on the other side of campus; it's just that they don't get the [same levels of] dollar signs attached to it" [26].

An engineering faculty participant noted that he was able to launch several HE projects at his university primarily because he had first built a strong reputation doing traditional engineering science research, which "gave us cover to do this sort of [humanitarian engineering] work ... but I think it would be hard to start from ground zero if we weren't doing it under the cover of something else" [29].

FINDINGS: OPPORTUNITIES

For each of the dimensions of the problem of knowledge above, respondents provided insights into potential opportunities to address the stated barriers.

ORGANIZATION OF KNOWLEDGE AROUND DISCIPLINES

Several interviewees suggested opportunities to help address the rigidity of disciplinary thinking. For instance, they suggested we locate and recruit big-picture, holistic faculty, especially senior faculty who have tenure and some campus clout. Because of their established reputations, such faculty can build new initiatives in existing disciplinary programs.

Interviewees also highlighted the importance of effective interdisciplinary collaboration. Some interviewees advised us to draw on expertise in colleagues from across the disciplines, such as the economics and business faculty for economic sustainability plans, and so on. They encouraged us especially to teach and learn from each other: liberal arts faculty need to learn about engineering issues, and engineers need to learn about how knowledge from non-technical fields, can make them better engineers and instructors [24,29,30]. Recognizing that faculty support what they help to create, interviewees also recommended directly involving all affected faculty in HEE pedagogical content development [24,31].

CONTENT OF KNOWLEDGE

Traveling to developing communities was identified as an opportunity to question the content of knowledge, particularly EPS. When faculty and students travel, often they find that the immediacy of the problems, the lack of financial and technical resources, and the complexity of social and cultural issues require an approach that differs significantly from EPS.

Another opportunity to alter the content of knowledge came from authoritative challenges for community development such as those stated in the UN's Millennium Development Goals. Even if these challenges are viewed as a set of given problems to be solved, and hence do not challenge the passivity reinforced by EPS, the challenges at least change the focus of traditional engineering problems [26]. However, these challenges did not represent a new challenge to engineering education. After all, many such reports have been published before. When asked why the UN report made such a difference now, even if other reports have been released before, the program coordinator said "I think [it is] the student response to the hands-on [experiences]. [Our university] has been focusing on changing their curriculum to more hands-on engineering ... [and] that philosophical curricular change is now more than a decade old. ... So these students that are here now have had a taste of what it's like to create a project from the ground up ... [our lead faculty] said it was the first time in fifteen years of teaching, or twenty years of teaching, that the students had a learning experience and came back and were demanding more" [26]. At the same time, these desires are supported by institutionalized service-learning programs.

Some students involved in HE projects have broken the rigid boundaries of EPS by involving non-engineering perspectives in solving HE problems. Presenting at the 2006 Engineers without Borders International Conference, students at Tufts University, for example, included on their project team students from the following six disciplines after recognizing that their water project in Tibet involved issues related to engineering, art, economics, international relations, physics, and public health. The engineering students had high praise for students from all other majors, especially the international relations students, who helped educate them and facilitated communication with locals [32].

HIERARCHY OF KNOWLEDGE

Since many of the interviewees were interested in the success of HE projects, they provided case studies as a plausible inquiry-based pedagogy that could foster reconceptualizations of the hierarchies of knowledge outlined above. Case studies from the World Bank and USAID, for example, can help us learn from mistakes and successes regarding issues such as the efficacy and complexity of single-sector vs. multi-sector approaches; planning and installing a clean drinking water system, for instance, involves more than engineering science. Health, community development, language, cultural, social, and political issues also play key roles, and dealing with these issues also opens new yet realistic constraints and complexities. According to some interviewees, hopelessness in developing communities is fostered or perpetuated when outsiders solve problems for people instead of creating an atmosphere in which people have the resources, training, and ownership to solve their own problems [24,29]. Thus, HEE gives the opportunity to teach engineering students and some faculty about the efficacy of less hierarchical and stratified approaches to problem solving, ones that conceptualize vast arrays of knowledge as potential problem-solving tools.

As another opportunity for addressing disciplinary hierarchies, we were also advised to bring in influential outside speakers who are respected for their engineering and/or science work yet who explicitly support the mission and goals of HEE [24]. Such a move could help some in the academic community transcend the simplistic either/or dichotomy in which researchers either work on traditional research or on progressive service-learning initiatives.

DISCUSSION

Participants described both barriers and opportunities to HEE implementation in terms of the problem of knowledge, specifically regarding three specific dimensions: organization, content, and hierarchy of knowledge. On the latter category, interviewees' responses signify three knowledge tensions, between engineering science and engineering design, high and low-tech solutions, and engineering and non-engineering concepts and/or disciplines.

One recurring theme among study participants focused on what garners respect in academic engineering disciplines. Since research in engineering science brings the lion's share of funding and has more outlets for publication, it brings with it greater prestige than less traditional forms of research and pedagogy, perhaps especially those that involve ethics, design, and service learning. Thus, participants' responses suggest that our task involves addressing deeper, systemic issues of disciplinary knowledge structures.

Further, the aforementioned hierarchy runs counter to the purpose of HEE. HEE solutions usually need to be low-tech, low-cost, and high impact; however, many engineering solutions are relatively high-tech, high-cost, and low impact.

RECOMMENDATIONS

These findings suggest areas for further consideration:

- A need to reconsider the effects of existing knowledge priorities. The focus on engineering sciences may work against locating students in positions of power, ones in which they are capable of solving a vast array of problems by drawing from diverse yet appropriate knowledge bases, including engineering science, engineering design, social sciences, humanities, communications, and others.
- The success of an engineering solution should explicitly consider human welfare and well being.
- Some barriers can become opportunities if committed faculty frame HE courses, until now undervalued within the curriculum, in terms of ABET criteria and clearly address the calls for designing systems within realistic constraints, making HE courses critical for accreditation.
- A multidisciplinary HE team can be made into an opportunity instead of becoming the source of stereotypical labeling. For example, some researchers have proposed a method of problem-solving to include human perspectives that allows engineering students to understand, analyze, and value perspectives of others who think differently than engineers do [33].

Among others, our research has provided insight into the importance of exploring whether the way knowledge is valued, organized, created, and disseminated most constrains a graduate program in humanitarian engineering ethics and possibly engineering ethics education in general.

As we look to the future, we see signs of change:

- That NSF funded our HEE initiative.
- The aforementioned shift in ABET accreditation criteria, which now place greater emphasis on non-technical, professional aspects of engineering [25].
- The awarding of the prestigious Bernard M. Gordon Prize by the National Academy of Engineering (NAE) to the aforementioned EPICS Program at Purdue [34].
- The NAE's current definition of engineering emphasizes that "engineering is about design under constraint" which includes "constraints provided by technical, economic, business, political, social, and ethical issues." [35].
- The NAE co-sponsored the 2006 National Conference on Service Learning in Engineering.
- These signs and others suggest the problem of knowledge will play a role in discussions regarding engineering education reform for years to come.

ACKNOWLEDGMENTS

We would like to acknowledge the support of the National Science Foundation for award EEC-0529777. For her contributions to this research, we would also like to recognize E. Heidi Bauer, graduate student at Colorado School of Mines.

REFERENCES

1. Nash, G., "The Life of Herbert Hoover," *Humanitarianism*, vol. 3, 1983.
2. American Society of Mechanical Engineers, "Hoover Medal," http://www.asme.org/Governance/Honors/UnitAwards/Hoover_Medal.cfm. Accessed May 17, 2006.
3. Mitcham, C., Lucena, J., and Moon S., "Humanitarian Science and Engineering," in *Encyclopedia of Science, Technology, and Ethics*, C. Mitcham, Ed. New York: MacMillan, vol. 2, pp. 947–50, 2005.
4. Shallcross, L., "A Human Touch," *Prism*, vol. 15, no. 2 October 2005, pp. 48–49.
5. Mulraine, A., "To the Rescue," *Prism*, vol. 15, n. 7 March 2006, pp. 28–33.
6. Selingo, J., "May I help You?" *Prism*, vol. 15, no. 9 summer 2006, pp. 40–45.
7. "Engineering Projects in Community Service," EPICS Program, Purdue University. http://epics.ecn.purdue.edu/. Accessed March 16, 2006.
8. "Engineering for Developing Communities," EDC Program, The University of Colorado-Boulder. http://www.edc-cu.org/. Accessed March 16, 2006.
9. "Master's Degree in International Peace Corps," Civil and Environmental Engineering, Michigan Technological University. http://www.cee.mtu.edu/peacecorps/index.html. Accessed March 16, 2006.
10. Herkert, J.R., "Engineering ethics education in the USA: Content, pedagogy, and curriculum," *European Journal of Engineering Education*, vol. 25, pp. 303–13, 2000.
11. Herkert, J.R., *Social, Ethical and Policy Implications in Engineering*. New York: IEEE Press, 2000.
12. "Welcome to service learning," The National Service-Learning Clearinghouse. http://www.servicelearning.org/welcome_to_service-learning/index.php. Accessed May 25, 2006.
13. "Humanitarian Engineering," Division of Engineering and Division of Liberal Arts and International Studies, Colorado School of Mines. http://humanitarian.mines.edu/. Accessed March 16, 2006.
14. Newberry, B. "The Dilemma of ethics in engineering education," *Science and Engineering Ethics*, vol. 10, 2004, pp. 343–52.
15. Meyers, C. "Institutional culture and individual behavior: Creating an ethical environment," *Science and Engineering Ethics*, vol. 10, 2004, pp. 269–76.
16. Foundation Coalition, "Curricular and Organizational Change in Engineering Education," 2005. http://www.foundationcoalition.org/publications/brochures/change_one_pager.pdf," Accessed March 16, 2006.
17. Patton, M. Q., *Qualitative Evaluation and Research Methods*, 2nd Ed., Newbury Park, CA: Sage, 1990.
18. Creswell, J.W., *Educational Research: Planning, Conducting, and Evaluating Quantitative and Qualitative Research*, Upper Saddle River, NJ: Pearson, 2002.

19. Glaser, B. *Theoretical Sensitivity: Advances in the Methodology of Grounded Theory*, Mill Valley, CA: Sociology Press, 1978.
20. Bogdan, R.C. and Biklen, S.K., *Qualitative Research for Education: An Introduction to Theories and Methods*, 4th Ed., Boston: Allyn and Bacon, 2003.
21. Seely, B. "The other re-engineering of engineering education, 1900–1965." *Journal of Engineering Education*, vol. 88, no. 33, 1999, pp. 285–94.
22. American Society of Engineering Education. 1968 Goals Report, p. 82, 1968. Washington, DC: ASEE.
23. Lucena, J.C. "Flexible engineers: History, challenges, and opportunities for engineering education," *Bulletin of Science, Technology, and Society*, vol. 23, 2003, pp. 419–435,
24. Interview with Participant A, 2006.
25. Accreditation Board for Engineering and Technology, Criteria for Accrediting Engineering Programs, 2004, http://www.abet.org Accessed October 20, 2005.
26. Interview with Participant B, 2005.
27. Interview with Participant C, 2005.
28. Participant Observation by both authors in fall 2005.
29. Interview with Participant D, 2006.
30. Interview with Participant E, 2005.
31. Interview with Participant F, 2005.
32. Engineers Without Borders International Conference, Hosted by Rice University, Houston, TX, February 16–18, 2006.
33. Downey, G.L., Lucena, J.C, Moskal, B.M., et al., "The Globally Competent Engineer: Working Effectively with People Who Define Problems Differently," *Journal of Engineering Education*, vol. 95, no. 2, 2006, pp. 107–122.
34. "2005's Highest Engineering Honors," The Bernard M. Gordon Prize, http://www4.nationalacademies.org/news.nsf/isbn/02222005?OpenDocument. National Academy of Engineering, Accessed March 16, 2006.
35. National Academy of Engineering, "Technological context of engineering practice," in *The Engineer of 2020: Visions of Engineering in the New Century*, p. 7, Washington, DC: National Academy of Engineering, National Academies Press, 2004.

22 The New Engineer
Between Employability and Social Responsibility

Edward Conlon

CONTENTS

INTRODUCTION

Given the demands for a broader education for future engineers 'it seems justifiable to speak of a general crisis in engineering education calling for "a new engineer"', ([1]: 13–4). The 'New Engineer' will be a broad based professional who is socially and environmentally responsible ([2]: x).

The demand for the 'New Engineer' is reflected in changing approaches to the accreditation of professional engineering programmes. Like professional bodies in other countries, Engineers Ireland (EI), previously known as the Institution of Engineers (IEI), has changed accreditation criteria to include outcomes focused on ethical standards, responsibilities towards people and the environment, teamwork and communication. Programmes are required to develop an awareness of the social and commercial context of engineers' work and the constraints that arise from that context. ([3]: 11–12,15–16).

This article critically examines how engineering education can adequately address the demands that are to be imposed on future engineers. It argues for the importance of the social sciences in helping engineers understand the context in which they will work and how it both constrains and enables their capacity for social responsibility. In this paper social responsibility will be understood as involving a commitment to a socially just, equitable and sustainable world [4]. I will argue that a focus on employability alone will not equip engineers to be socially responsible because it fails to problematise the current structure of work and society.

I write as a sociologist concerned to highlight the role of sociological insights in helping engineering students achieve an understanding of the norms and intuitions required for social responsibility. Societal barriers to social responsibility appear to have grown in recent times with a key feature of current trends being the 'privatisation of everything' ([5]: 263) including basic resources such as water. In an unequal world where interests conflict (between the developed and developing worlds, between the rich and the poor and between workers and employers) educators need to ask themselves who are they producing knowledge for and what will be done with this knowledge [6] and engineers need to reflect on the kinds of problems they choose to solve and the criteria used to solve them ([7]: 315).

These points will be elaborated by examining the reasons proposed for creating the 'New Engineer'. These reasons include concerns about the social skills of engineers (Section 2) and, secondly, about the role and perception of engineering in society (Section 3). In Section 4 I will state my conclusions and make some proposals.

ENGINEERING SKILLS AND EMPLOYABILITY[1]

Under this heading two interrelated reasons for change can be identified [8–12]. Firstly new skills are required to make engineers more effective as engineers and secondly, many engineers become managers or spend much of their working lives on management and supervisory tasks. This literature highlights the importance of acquiring non-technical generic competencies in areas such as communications, project management, leadership and teamwork, rather than the acquisition of theoretical knowledge in a range of 'socio-economic' subjects.

This emphasis on generic professional practice skills is not surprising given the changes in the organisation of work resulting from increased global competition. Features of the 'new industrial paradigm' ([1]: 19), include: total quality management; new forms of work organisation with an emphasis on team work; flexible production systems and employment contracts and a focus on customer needs.

This new orientation requires new skills. In a sense engineering educators are being asked to prepare graduates to insert themselves in the flexible globalised workplace:

> the focus of training must increasingly be on *employability* and that there is an urgent need for a concerted effort … to ensure that a well-trained flexible workforce is available as a means of sustaining a national competitive advantage in a world of megacompetition. ([13]: 179, emphasis added).[2]

It is also the case that engineers are being asked to be agents of globalisation and advocates of new organisational forms. For example, Devon [14] argues that lean production should be viewed in 'a positive ethical light'. Others have argued that 'modern approaches to quality management are starting to improve the quality of work for many employees' ([15]: 138).

This optimistic view of modern work systems is problematic given engineers' responsibilities for the way work is organised for others, their role as originators of ideas about work organisation and the substantial evidence that the experience of work has not improved for those working in systems such as lean production (e.g Bradley *et al.* [16]). Ciulla [17] argues that Reengineering, the last major management theory of the twentieth century, shares with scientific management a concern 'with the speed of production. Time is still money, only now it moves faster and costs more' (p. 147) and while fairness is the central moral issue in the workplace, income inequality is increasing 'across virtually all the developed economies of the world' ([18]: 472).[3] A review of Irish evidence on workplace change in the 1990s concludes:

> the period since the 1980s has witnessed a regression in the quality of work life as many workers are expected to undertake increased workloads and experience intensification in the pace of work (without) an increase in their influence over day to day activities [19].

The significance of this for social responsibility is that it raises questions about whose problems engineers are trying to solve and on what basis. In most cases engineers tend to be absorbed in management hierarchies and values and tend to use business considerations as appropriate criteria for engineering decision-making (see Ref [20]: 9). Johnston [21] have argued that the discourse of business (and science) has dominated engineering. They argue that while engineers are keenly focused on productivity they do not see the fair distribution of the benefits of economic activity as their concern. This focus on productivity is now to the fore and finds expression in the employability discourse.

Downey and Lucena [22] note the rise, during the 1980s in the USA, of concerns about competitiveness, which elevated 'engineering to the status of a national problem' (183). With globalisation this focus on competitiveness has grown worldwide and has led to a focus on the employability skills of engineers. There are increasing demands for education to be more responsive to the needs of industry but there must be concern that that this will lead to a narrow focus in engineering education whereby graduates are trained to insert themselves into the 'runaway world' [23] of globalisation. Employability denotes the requirement to adapt to the demands of employment and for individuals to remain competitive in the labour market ([24]: 2). In this context students may end up believing that they 'are only responsible for themselves' [25].

The employability agenda is about getting graduates to adapt to the new flexible workplace. Crucially, graduates skills are problematised while the employment practices of employers are left untouched: 'The skills identified as core produce the type of knowledge and understanding that is required to maintain dominant cultural and political arrangements' ([26]: 137). But a critical focus on these arrangements is required if the quality of working lives is to be enhanced.

A focus on employability skills alone will not give engineers the capacities required to reflect critically on the structure of work and the manner in which the rewards of productive activity are distributed. It will not force them to ask questions about whether the work they design for others provides 'opportunities for workers to realise their human potential through creative, meaningful, and productive work' ([23]: 310).[4]

A sociological approach to work emphasises the inequality inherent in the employment contract, the conflicting interests and asymmetry of power of the parties [27]. It focuses not just on the workplace but also on the wider social arrangements, which constrain or enable the power of workers and employers. This focus helps us understand why the quality of work may vary across different societies.

For example, the comparative study of engineers by Meiksins and Smith [20] considers the experiments in work humanisation in Sweden. They argue that:

> Conditions in Swedish society have imposed on engineers more constraints, and created the conditions for a dual agenda for production efficiency and work humanization' (265).

What can be noted is the manner in which constraints become enablers for work humanisation. Important factors in the Swedish case included societal commitments to full employment, which led to tight labour markets, and egalitarianism, a comprehensive welfare system and the strong position of trade unions with a wide membership base (including many engineers) and comprehensive legal rights at the workplace. This means we have to consider the regulatory environment, and the way it can shape the balance of power at the workplace, at a time when globalisation is promoting lightly regulated labour markets.[5]

Meiksins and Smith also argue that work humanisation was facilitated because Swedish engineers were closely aligned with manual workers and were engaged in a dialogue with social scientists 'exposing engineers in their training and practice to the benefits of work humanization' (1996: 265).

What the above example suggests is the importance of engineers being exposed in their education to criteria other than narrowly conceived productivity, efficiency and flexibility, and the importance of them understanding the wider social context of their work, including the regulatory environment, and how it enables or constrains the possibilities for designing meaningful work for others. A focus on the wider social context is also required if engineers are to contribute to creating a sustainable society.

ENGINEERING AND SOCIETY

The second set of reasons behind the demand for the 'New Engineer' focus on the relationship between engineers and society. There is concern that the status of engineering is being undermined as engineers are identified with environmentally damaging technologies. There is particular concern about the failure to attract women into the profession. To attract women, the humanitarian role of engineering should be highlighted including the role of engineering in promoting sustainable development.[6]

SUSTAINABLE DEVELOPMENT (SD)

There is increasing pressure to practice engineering more sustainably. The mission statement of Engineers Ireland, along with the Code of Ethics, contains a commitment to the promotion of SD. Engineers are seen to have a key role in making economic and technological activities sustainable and some have argued that engineers are uniquely placed to take a lead in moving towards sustainability but only if 'they have a broad understanding of their own discipline and an awareness of how it fits with other disciplines and into the social fabric of their society' ([15]: 316). The evidence suggests that engineers tend to have a narrow understanding of the concept.

SD focuses on the relationship between 'three pillars': the ecological, social and economic, yet many commentators have highlighted the failure of engineers to grapple with the social dimension of SD [28–31]. Turek and Mistina claim that 'engineering education takes a prevailing technocratic approach, aimed especially at maximising production and economic efficiency' (2007: 397). Again the failure to address issues of distribution can be noted [30].

This approach to SD seems to be reflected in engineering students' understanding of SD. Research with students [32,33] suggests there are substantial knowledge gaps across all stages of engineering programmes. Students appear to be relatively knowledgeable about environmental issues but significant knowledge gaps exist with respect to the other two components (social and economic) of SD. There is clearly a need for students to embrace a fuller understanding of SD including the social and economic dimensions.

Even then problems remain in moving towards sustainability (see Ref. [34]). Taylor [35] has suggested that Irish environmental policy is constrained by the concern not to compromise the market ethos and the inward investment upon which recent economic growth was predicated. There are powerful vested interests opposed to the kind of radical change required to move towards a sustainable and just society. The very operation of free market systems encourages growth for growth's sake [36] and overconsumption [37]. This means that more fundamental questions have to be asked. Sustainability 'implies cultural, social and economic restructuring simultaneously with technological restructuring' ([34]: 150). Here again we see a focus on the wider context in which engineers work. Donnelly and Boyle highlight the importance of changing that context by changing the regulatory environment to favour sustainable solutions and outcomes. Current trends towards deregulation are contributing towards ecological devastation because the role of governments in correcting environmental externalities is reduced (see Ref. [36,38]).

Assuming the goal of sustainability and given the obstacles to moving towards it an (exclusive) approach to engineering ethics which focuses on the ethics of individual engineers must be questioned. This can be justified by looking at how sociology understands human action.

ENGINEERING ETHICS AND SOCIAL THEORY

Sociology is concerned with the relationship between social structures and human action (agency). The structure-agency debate is at the heart of social theorising [39] and has increasingly focused on how social structures *both* constrain and facilitate agency.[7] Social structures can be seen as the rules of society but also the sets of social relations which provide differential access to material and cultural resources. A focus on social structures requires that we examine how social activity is organised, the manner in which social relations provide differential access to power, and the legitimating values used to maintain or modify these relations.

Engineering ethics is concerned with the values of engineers. The focus is often on the ethical behaviour of the individual engineer. (Bucciarelli [40] provides some examples.) But in reality while engineers *may* be committed to ethical practices it is not always possible to behave ethically. Social theory suggests that the capacity to be socially responsible is not solely a feature of the values of actors. To exercise agency, commitment to particular outcomes *is* necessary, but so is the power to achieve these outcomes. To exercise agency actors must have choices, but these are constrained by the physical world, the social structure and the power of other agents.

A key constraint identified in relation to engineering ethics is that most engineers are employees they do not have control over the projects on which they work. They tend to solve problems 'framed and formulated by others' [21]. If engineers are to solve or diminish the ethical dilemmas they face and increase their capacity for social responsibility they have to understand the broader context from which ethical dilemmas originate and they have to play an active role in helping to reshape that context wherever that may be necessary ([41]: 297, see also Ref. [42]). The engineering profession

> must start working to influence the restructuring of current social, political, economic, and institutional paradigms ... thus increasing the diversity of acceptable options and our ability to move in more sustainable directions ([34]: 153).

Two issues are crucial here. Firstly, engineers need to address the 'contradiction', highlighted 30 years ago by Mike Cooley [43], of the gap between what technology could provide for society and what it actually does provide. Rather than simply promote globalised competition and conspicuous consumption engineers should promote economic activity which meets vital social needs.

Secondly, if we are to refocus engineering activity, and diminish some of the ethical dilemmas that engineers face in their daily activity, then it is vital that engineers and engineering educators move beyond seeing rules, such as laws and other regulatory devices, just as constraints but also as enablers that *may* facilitate socially responsible action. It is the case that engineers need good laws [44][8] and need to engage in debate about the nature of these laws. They also and need to make alliances with those seeking regulations requiring sustainable and socially just practices.

Engineers need to consider how they intervene in the public policy arena and whether these interventions enable or constrain the move towards a sustainable and just world. For example in Ireland EI supported the government's Critical Infrastructural Act which aims to fast track planning processes and which 'may constrain future collective action on environmental issues' ([45]: 238). This is significant in the context of Leonard's assessment that it has been the combination of top-down EU legislation and bottom-up grassroots agitation that has shaped Irish environmental policy.

This suggests that engineering ethics must focus on more than the individual ethical dilemmas faced by engineers. The requirement to widen the scope of engineering ethics, including a greater engagement with STS scholarship, is increasingly recognised and should be encouraged and developed.[9] A focus on the agency/structure relationship will help in integrating micro and macro issues in engineering ethics teaching by giving a focus to a number of key questions:

- What meaning does social responsibility have for engineers both individually and as a profession? Whose problems do they choose to solve?
- What criteria (e.g profit or need) do they use in solving engineering problems and whose interests do these solutions serve?

- What constraints stop them acting in a socially responsible manner? Do they have the power to act or does the power of others stop them?
- How can these constraints be changed to facilitate social responsibility? What changes in public policy, including laws, or social practices are needed and what resources and allies can they call on to help them seek these changes?

CONCLUSIONS AND IMPLICATIONS FOR THE CURRICULUM

It has been argued here that engineering education needs to widen its focus if students are to be educated as socially responsible engineers. A narrow focus on the skills and values of individual students related to employability is not adequate to prepare them for the challenge of delivering sustainable and just engineering solutions. Students need to develop the capacity to situate their individual practice as engineers in its wider social context. How is this to be done?

In his article Bucciarelli suggest a wholesale reorganisation of engineering education to broaden its focus and embed the social dimension in a multidisciplinary approach to engineering education. This can be endorsed but from an Irish perspective the extent of reform proposed is wide and unlikely to be realised in the short term,[10] although discussions on moving to a full three plus two model to comply with the Bologna Accord will open up the possibility of broadening the early years of engineering courses. It also opens up the possibility of attracting more women to engineering (see [46]).

In the interim a number of priorities can be identified:

1. Engineers and engineering educators need to more fully embrace a commitment to social justice, equality, work humanisation and the principles of SD. These should provide the underpinning for all engineering programmes. Students should be introduced to these principles in the first year of their studies so that they come to see them as inherent to engineering and come to see engineering as a social as well as a technical process;
2. All project work undertaking by students should address explicit social criteria on which they should be assessed. Students should undertake project work with an explicit public policy dimension;
3. Engineering students could be offered modules in the social sciences including STS studies to help them understand the manner in which technology is socially shaped;
4. Given that many engineers study management these modules should address principles of work humanisation and the importance of redressing the imbalances of power inherent in the employment relationship;
5. Ethics modules should specifically deal with the obstacles inhibiting an ethical engineering practice and the public policy role of the engineering profession. Students should be given the opportunity to critically evaluate the public policy positions of the profession.

ENDNOTES

1. Employability has been defined as 'a set of achievements – skills, understandings and personal attributes – that make graduates more likely to gain employment and be successful in their chosen occupations' by the UK Higher Education Academy, Engineering Subject Centre. See http://www.engsc.ac.uk/er/employability.
2. This purely economic view of globalisation is reflected in the accreditation documentation of Engineers Ireland which requires students 'To understand and contend with globalisation and its impacts on the marketing and manufacture of product' (IEI 2003:16).
3. There is also evidence of rising global inequality (see Ref. [38]).
4. It is hard to find treatment of issues related to work design in engineering ethics textbooks. Even Goujan and Dubreuil [47], which is broad ranging in content, does not deal with the

topic in any depth. In a SEFI document from 1995 on *Educating the Whole Engineer* it is suggested that industrial sociology is a 'vital technical subject' (p. 3). This is problematic in that it suggests that work design is a matter of the application of technology. But a broader view is necessary if we are to develop work systems in which the humanity of the people who work in them is recognised.

5. Further evidence for the importance of the regulatory environment can be found in Lorenz and Valeyre [48]. They examine different work regimes across the EU 15 and show that deregulated labour markets, especially in Ireland and the UK, do not provide the necessary institutional support for establishing substantial forms of autonomy at work.

6. It is true that there have always been engineers who are concerned about the social impact of engineering irrespective of concerns about the image of the profession. But recent research on the image of the profession does highlight the importance of the societal impact of engineering. For example a recent report from the UK Royal Academy of Engineering on *Public Attitudes to and Perceptions of Engineering* states: 'The social responsibility of engineering is an important issue underpinning attitudes towards the profession' (p. 38). See http://www.raeng.org.uk/news/publications for the full report.

7. Most introductory text books in sociology deal with the structure/agency debate. A good introduction can be found in various editions of George Ritzer, *Sociological Theory* (Mc Graw Hill), which is now in its 7*th* edition.

8. Zandvoort [44] has highlighted important areas for reform including the need for change in the rules of liability. He has argued for the need of strict legal liability in view of the responsible management of the environment, technological risks, and sustainability. (p. 25) He is also concerned to promote the idea that laws can be solutions to prisoner's dilemmas. A law is a solution to a prisoners dilemma in Zandvoort's terms if the law makes each individual better off, at least in the long run, than would be the case without the law. Such a law could in principle be adopted with unanimity. Zandvoort appears to suggest (but does not explicitly state) that a transition from the current limited or fault liability laws to strict liability laws may represent solutions to prisoner's dilemmas, and hence might proceed on the basis of unanimity, as the effects of the transition may ultimately be beneficial for all. However, it can be doubted whether such legal change could be expected on the basis of unanimity. This latter requirement would seem to allow vested interests, mainly corporate actors, the opportunity to effectively block legal change which they oppose.

9. Rather than provide a long list of references I refer readers to the: *European Journal of Engineering Education*, **25** (4); *IEEE Technology and Society Magazine* Fall 2001 and Winter 2001/2 [49]; Goujan and Dubreuil [47] and Herkert [42]. See also Bucciarelli in this issue.

10. In a recent address the President of one of Ireland's main universities has highlighted the emphasis on depth, rather than breadth, in Irish higher education which makes it hard for students to take options outside their chosen specialisation. See http://www.ul.ie/presoff /Inaugural%20Speech.htm.

REFERENCES

1. Christensen, H., Meganck, M. and Delahousse, B., 2007. Introduction. Occupational building in engineering education. *In*: H. Christensen *et al.*, eds. *Philosophy in Engineering*. Aarus: Academia, pp. 13–22.
2. Beder, S., 1998. *The New Engineer*. Sydney: Macmillan.
3. Institution of Engineers of Ireland, 2003. *Accreditation Criteria for Engineering Education Programmes*, Dublin, November.
4. Pritchard, J. and Baillie, C., 2006. How can engineering education contribute to a sustainable future? *European Journal of Engineering Education*, 31 (5), 555–565.
5. Burawoy, M. 2005. American Sociological Association Presidential address: For public sociology. *The British Journal of Sociology*, 56 (2), 259–294.

6. Burawoy, M., 2004. American Sociological Association Presidential address: For public sociology. *The British Journal of Sociology*, 2005, 56 (2), 259–294.

7. Johnston, S., McGregor, H. and Taylor, E., 2000b. Practice-focused ethics in Australian engineering education. *European Journal of Engineering Education*, 25 (4), 315–324.

8. Batley, T., 1998. Management training of professional engineers in New Zealand. *Journal of European Industrial Training*, 22 (7), 309–312.

9. Bodmer, C., *et al.*, 2002. *SPINE (Successful Practices in International Engineering Education) Benchmarking Study*. Final Report, May.

10. Palmer, S.R., 2003. Framework for undergraduate engineering management studies. *Journal of Professional Issues in Engineering Education and Practice*, 129 (2), 92–99.

11. Markes, I., 2006. A review of the literature on employability skill needs in engineering. *European Journal of Engineering Education*, 31 (6), 673–650.

12. Scott, G. and Yates, K.W., 2002. Using successful graduates to improve the quality of undergraduate engineering programmes. *European Journal of Engineering Education*, 27 (4), 363–378.

13. Richardson, P., 2000. Employability in a globalising economy: Implications for engineering training. *Engineering Science and Education Journal*, 9 (4), 179–184.

14. Devon, R., 1999. Towards a social ethics of engineering: The norms of engagement. *Journal of Engineering Education*, 88 (1), 87–92.

15. Johnston S.F., Gostelow, J.P. and King, W.J., 2000a. *Engineering and society*. New Jersey: Prentice Hall.

16. Bradley, H., *et al.*, 2000. *Myths at work*. Cambridge: Polity.

17. Ciulla, J.B., 2000. *The working life*. New York: Three Rivers Press.

18. Cappelli, P., 2006. Conclusions. *In*: M. Korczynski *et al.*, eds. *Social theory at work*. Oxford: Oxford University Press, pp. 464–486.

19. Gunnigle, P., 1999. *Involvement, participation and partnership: A Review of the debate and some reflections on the Irish contex*. The 24th Countess Markievecz Lecture, Dublin.

20. Meiksins, P. and Smith, C., 1996. *Engineering labour*. London and New York: Verso.

21. Johnston, S.F., Lee, A. and Mc Gregor, H., 1996. Engineering as captive discourse. *Society for Philosophy and Technology Quarterly Electronic Journal*, 1 (3/4) [Online] Available: http://scholar.lib.vt.edu/ejournals/SPT/spt.html

22. Downey, G.L. and Lucena, J.C., 1995. Engineering studies. *In*: S. Jasonoff *et al.*, eds. *Handbook of Science and Technology Studies*. California: Sage, pp. 167–188.

23. Legge, K., 2006. Ethics and work. In: M. Korczynski *et al.*, eds. *Social Theory at Work*, Oxford: Oxford University Press, pp. 299–324.

24. Garsten, C. and Jacobsen, K., 2003. Learning to be employable. Basingstoke: Palgrave.

25. Winner, L., 1998. Technology as 'Big Magic' and other myths. *IEEE Technology and Society Magazine*, Fall, 4–16.

26. Morley, L., 2001. Producing new workers: Quality, equality and employability in higher education. *Quality in Higher Education*, 7 (2), 131–138.

27. Brown, R.K., 1998. The employment relationship in sociological theory. *In*: D. Gaille, ed. *Employment in Britain*. Oxford: Basil Blackwell, pp. 33–66.

28. Herkert, J.R., 1997. Sustainable development and engineering: Ethical and public policy implications. *IEEE Technology and Society International Symposium on Technology and Society at a Time of Sweeping Change*, 1997.

29. Herkert, J.R., 1998. Sustainable development, engineering and multinational corporations: Ethical and public policy implications. *Science and Engineering Ethics*, 4 (3), 333–346.

30. Johnston, S., 1997. Sustainability, engineering and Australian Academe. *Society for Philosophy and Technology Quarterly Electronic Journal*, 2 (3/4) [Online] Available: http://scholar.lib.vt.edu/ejournals/SPT/spt.html

31. Szymkowiak, S., 2003. Why build a network about introduction of sustainable development into scientific education? *European Journal of Engineering Education*, 28 (2), 179–186.

32. Carew, A.L. and Mitchell, C.A., 2002. Characterising undergraduate engineering students' understanding of sustainability. *European Journal of Engineering Education*, 27 (4), 349–361.

33. Azapagic, A., Perdan, S. and Shallcross, D., 2005. How much do engineering students know about sustainable development? The findings of an international survey and possible implications for the engineering curriculum. *European Journal of Engineering Education*, 30 (1), 1–19.

34. Donnelly, R. and Boyle, C., 2006. The Catch-22 of engineering sustainable development. *Journal of Environmental Engineering*, 132 (2) 149–155.

35. Taylor, G., 2005. *Negotiated governance and public policy in Ireland*. Manchester: Manchester University Press.
36. Smith, T., 1997. The case against free market environmentalism. *Journal of Agricultural and Environmental Ethics*, 8 (2), 126–144.
37. Woodhouse, E.J., 2001. Curbing overconsumption: Challenge for ethically responsible engineering. *IEEE Technology and Society Magazine*, Fall, 23–30.
38. Riley, D., 2007. Resisting neoliberalism in global development engineering, 114*th ASEE Annual Conference*, Hawai [online]. Available http://www.icee.usm.edu/icee/conferences/asee2007/papers/2072
39. Carter, B. and New, C., 2004. Introduction: Realist social theory and empirical research. *In*: B. Carter and C. New, eds. *Making realism work*. Abingdon: Routledge, pp. 1–20.
40. Bucciarelli, L.L., 2008. Ethics and engineering education. *European Journal of Engineering Education*, 33 (2), 139–147.
41. Zandvoort, H., Van de Poel, I. and Brumsen, M., 2000. Ethics in the engineering curricula: Topics, trends and challenges for the future. *European Journal of Engineering Education*, 25 (4), 291–302.
42. Herkert, J.R., 2006. Confession of a shoveler. *Bulletin of Science, Technology and Society*, 26 (5), 410–418.
43. Cooley, M., 1978. Design, technology and production for social needs. *In*: K. Coates, ed. *The right to useful work*, Nottingham: Spokesman, pp. 195–211.
44. Zandvoort, H., 2005. Good engineers need good laws. *European Journal of Engineering Education*, 30 (1), 21–36.
45. Leonard, L., 2006. *Green Nation*. Drogheda: Greenhouse/Choice Publishing.
46. Beraud, A., 2003. Potentials of interdisciplinary courses in engineering, information technology, natural and socioeconomic sciences in a changing society. *European Journal of Engineering Education*, 28 (4), 435–451.
47. Goujan, P. and Dubreuil, B.H., 2001. *Technology and Ethics*. Leuven: Peeters.
48. Lorenz, E. and Valeyre, A., 2004. *Organisational Change in Europe*. DRUID Working Paper No.04-04.
49. IEEE Technology and Society Magazine Fall 2001 and Winter 2001/2.
50. Turek, I and Mistina, J., 2007. Globalisation and its impact on engineering education. *In*: H. Christensen *et al.*, eds. *Philosophy in engineering*. Aarhus: Academia, pp. 391–408.

ABOUT THE AUTHOR

Eddie Conlon is a lecturer at the School of Multidiciplinary Technology and the Assistant Head of the Department of Engineering Science and General Studies at the Dublin Institute of Technology. His current research interest include engineers' education for sustainability, the sociology of work, industrial relations and the general education of engineers. He has recently acquired funding to examine Irish engineers' understandings of sustainability.

Appendix I: Editors' Note

WHY CARE ABOUT ENGINEERING ETHICS?

Engineering ethics is the moral struggle faced by engineering educators and practicing engineers alike, to construct general, allowable metrics of what it means to act as a socially responsible practitioner. In the broadest sense, codes of ethics in engineering are meant to regulate the relationship between (a) engineers and their clients and (b) engineers and fellow professionals. Following the philosopher of engineering Michael Davis, "ethics" in "engineering ethics" regards "special standards" of conduct, which, it must be stressed, "do not *automatically* carry a normative change [1]." Precisely, it is their membership to a specific professional society that makes such standards applicable to the engineers. Thus, apart from choice in ethical deliberations over engineering typically following a utilitarian framework, codes of ethics "reflect *voluntary* standards of professional and personal practice *recommended* for . . . engineers" [2] (emphasis ours).

Specifically, though, engineering codes of ethics are used to decide what makes a "better" or "worse" professional and can thus greatly influence individual, corporate, and institutional decision-making. By debating engineering ethics, engineers who are part of professional bodies are more likely to act morally responsible in their work. Moreover, engineering codes provide evidence as to how professional societies change over time and are a reflection of the broader engineering landscape.

Hence, as metrics of professional status, the articulation and acceptance/rejection of such codes challenges the professional to encompass technical and social complexities. As a result, the moral struggle in engineering ethics is also a struggle for reflexivity. Engineers have expressed, in a way, which perhaps, resembles behavioral patterns observed in the medical profession, the desire to reassess their practices in an uncertain and changing world [3].

Commentators have remarked that the content of codes of ethics in engineering makes apparent the dominant identities of engineering professionals [4]. Professional insecurity and a "dissatisfaction with their status" have always been a major source of concern for practitioners, manifested in the creation of a ". . . philosophy of professionalism [which] carried engineers' ambitions beyond technology, to politics and policy-making generally [5]." The strategy of examining codes of ethics by putting their content *and* the process of writing a code into historical perspective is key, for once codes are seen as products of their time and place, the locally different conceptualizations of ethical values are better appreciated [6]. This is not the same with, say, the National Society of Professional Engineers (NSPE) claim that previous versions of codes and canons be viewed "*solely* for historic purposes [7]." Thinking about past codes of ethics as museum objects involves the risk of downplaying the value judgments playing out in the code-making process.

Importantly, then, provided that engineering ethics is a polemic issue, codes of ethics deserve our attention for they are used as means to safeguard controversial determination in the management of technology. An examination of the ways by which technology policy comes to entrench moral assumptions prompts several questions, which once taken into consideration, may help the reader frame his or her evaluation of the recent codes of ethics in American engineering:

- Who decides what constitutes an ethically desirable outcome within a specific engineering context?
- How, where, and by whom does the teaching of professional ethics for engineers take place?

- How do the mandates of codes of ethics acquire meaning, if any, in the field, the lab, or the corporate environment?
- Which factors determine the ways practicing engineers conceptualize their commitment to, or resistance against, codes of professional ethics?
- How does one go about improving the effectiveness of such codes, given that they are tools, which measure, and simultaneously dictate the social, political, and environmental accountability of engineers?

Questions of this sort, however, must be thought of in relation to their relevant national and institutional settings. The analysis of the type of engineering considerations, which acquire moral value, would necessarily follow the examination of the individuals (in our case the *professional* engineers in America) and the institutional configurations to which those codes of ethics are designed to apply. With that being said, what follows is a compilation of several codes of ethics from the major engineering organizations in the United States, including the World Federation of Engineering Societies.

REFERENCES

1. Davis, Michael. *Thinking like an Engineer: Essays in the Ethics of a Profession.* New York: Oxford University Press, 1998; Davis, Michael. Engineering ethics, individuals, and organizations. *Science and Engineering Ethics* 12 (2006): 223–231; Davis, Michael. What's philosophically interesting about engineering ethics? *Science and Engineering Ethics* 9 (2003): 353–361.
2. Vesilind, Aarne. Why do engineers wear black hats? *Journal of Professional Issues in Engineering Education and Practice* 119, no. 1 (1993): 1–7; Biomedical Engineering Society. *Code of Ethics.* Landover, MD: Biomedical Engineering Society, 2004.
3. Groopman, Gerome. *How Doctors Think.* Boston, MA: Houghton Mifflin, 2007, p. 151.
4. Downey, Gary Lee, Lucena, Juan C., and Mitcham, Carl. Engineering ethics and identity: Emerging initiatives in comparative perspective. *Science and Engineering Ethics* 13, no. 4 (2007): 463–487.
5. Layton, Edwin. *The Revolt of the Engineers: Social Responsibility and the American Engineering Profession.* Cleveland, OH: Case Western Reserve University Press, 1971, p. 61.
6. Henderson, Kathryn. Ethics, culture, and structure in the negotiation of straw bale building codes. *Science, Technology, and Human Values* 31, no. 3 (2006): 261–288.
7. National Society of Professional Engineers. NSPE Webpage: History of the Code of Ethics for engineers. http://www.nspe.org/Ethics/CodeofEthics/CodeHistory/historyofcode.html. Accessed July 13, 2018.

Appendix II: Codes of Engineering Ethics

Gail Palmer School of Electrical and Computer Engineering Georgia Institute of Technology

<u>ACCREDITATION BOARD FOR ENGINEERING AND TECHNOLOGY</u>

CODE OF ETHICS OF ENGINEERS

The Fundamental Principles

Engineers uphold and advance the integrity, honor and dignity of the engineering profession by:

1. Using their knowledge and skill for the enhancement of human welfare
2. Being honest and impartial, and serving with fidelity the public, their employers and clients
3. Striving to increase the competence and prestige of the engineering profession
4. Supporting the professional technical societies of their disciplines.

The Fundamental Canons

1. Engineers shall hold paramount the safety, health and welfare of the public in the performance of their professional duties.
2. Engineers shall perform services only in the areas of their competence.
3. Engineers shall issue public statements only in an objective and truthful manner.
4. Engineers shall act in professional matters for each employer or client as faithful agents or trustees, and shall avoid conflicts of interest.
5. Engineers shall build their professional reputation on the merit of their services and shall not compete unfairly with others.
6. Engineers shall act in such a manner as to uphold and enhance the honor, integrity and dignity of the profession.
7. Engineers shall continue their professional development throughout their careers and shall provide opportunities for the professional development of those engineers under their supervision.

ABET
345 East 47th St., New York, NY 10017
1987
For additional information, please contact Gail Palmer at gpalmer@ece.gatech.edu.

<u>AMERICAN ASSOCIATION OF ENGINEERING SOCIETIES</u>

MODEL GUIDE FOR PROFESSIONAL CONDUCT

Preamble

Engineers recognize that the practice of engineering has a direct and vital influence on the quality of life for all people. Therefore, engineers should exhibit high standards of competency, honesty and impartiality; be fair and equitable; and accept a personal responsibility for adherence to applicable laws, the protection of the public health, and maintenance of safety in their professional actions and behavior. These principles govern professional conduct in serving the interests of the public, clients, employers, colleagues and the profession.

The Fundamental Principle

The engineer as a professional is dedicated to improving competence, service, fairness and the exercise of well-founded judgment in the practice of engineering for the public, employers and clients with fundamental concern for the public health and safety in the pursuit of this practice.

Canons of Professional Conduct

1. Engineers offer services in the areas of their competence and experience, affording full disclosure of their qualifications.
2. Engineers consider the consequences of their work and societal issues pertinent to it and seek to extend public understanding of those relationships.
3. Engineers are honest, truthful and fair in presenting information and in making public statements reflecting on professional matters and their professional role.
4. Engineers engage in professional relationships without bias because of race, religion, sex, age, national origin or handicap.
5. Engineers act in professional matters for each employer or client as faithful agents or trustees, disclosing nothing of a proprietary nature concerning the business affairs of technical processes of any present or former client of employer without specific consent.
6. Engineers disclose to affected parties known or potential conflicts of interest or other circumstances which might influence—or appear to influence—judgment of impair the fairness of quality of their performance.
7. Engineers are responsible for enhancing their professional competence throughout their careers and for encouraging similar actions by their colleagues.
8. Engineers accept responsibility for their action; seek and acknowledge criticism of their work; offer honest criticism of the work of others; properly credit the contributions of others; and do not accept credit for work not theirs.
9. Engineers perceiving a consequence of their professional duties to adversely affect the present or future public health and safety shall formally advise their employers or clients and, if warranted, consider further disclosure.
10. Engineers act in accordance with all applicable laws and the _____[1] rules of conduct, and lent support to others who strive to do likewise.

NATIONAL SOCIETY OF PROFESSIONAL ENGINEERS

CODE OF ETHICS FOR ENGINEERS

Preamble

Engineering is an important and learned profession. As members of this profession, engineers are expected to exhibit the highest standards of honesty and integrity. Engineering has a direct and vital impact on the quality of life for all people. Accordingly, the services provided by engineers require honesty, impartiality, fairness, and equity, and must be dedicated to the protection of the public health, safety, and welfare. Engineers must perform under a standard of professional behavior that requires adherence to the highest principles of ethical conduct.

I. Fundamental Canons

Engineers, in the fulfillment of their professional duties, shall:

1. Hold paramount the safety, health, and welfare of the public.
2. Perform services only in areas of their competence.
3. Issue public statements only in an objective and truthful manner.

4. Act for each employer or client as faithful agents or trustees.
5. Avoid deceptive acts.
6. Conduct themselves honorably, responsibly, ethically, and lawfully so as to enhance the honor, reputation, and usefulness of the profession.

II. Rules of Practice

1. Engineers shall hold paramount the safety, health, and welfare of the public.
   a. If engineers' judgment is overruled under circumstances that endanger life or property, they shall notify their employer or client and such other authority as may be appropriate.
   b. Engineers shall approve only those engineering documents that are in conformity with applicable standards.
   c. Engineers shall not reveal facts, data, or information without the prior consent of the client or employer except as authorized or required by law or this Code.
   d. Engineers shall not permit the use of their name or associate in business ventures with any person or firm that they believe is engaged in fraudulent or dishonest enterprise.
   e. Engineers shall not aid or abet the unlawful practice of engineering by a person or firm.
   f. Engineers having knowledge of any alleged violation of this Code shall report thereon to appropriate professional bodies and, when relevant, also to public authorities, and cooperate with the proper authorities in furnishing such information or assistance as may be required.
2. Engineers shall perform services only in the areas of their competence.
   a. Engineers shall undertake assignments only when qualified by education or experience in the specific technical fields involved.
   b. Engineers shall not affix their signatures to any plans or documents dealing with subject matter in which they lack competence, nor to any plan or document not prepared under their direction and control.
   c. Engineers may accept assignments and assume responsibility for coordination of an entire project and sign and seal the engineering documents for the entire project, provided that each technical segment is signed and sealed only by the qualified engineers who prepared the segment.
3. Engineers shall issue public statements only in an objective and truthful manner.
   a. Engineers shall be objective and truthful in professional reports, statements, or testimony. They shall include all relevant and pertinent information in such reports, statements, or testimony, which should bear the date indicating when it was current.
   b. Engineers may express publicly technical opinions that are founded upon knowledge of the facts and competence in the subject matter.
   c. Engineers shall issue no statements, criticisms, or arguments on technical matters that are inspired or paid for by interested parties, unless they have prefaced their comments by explicitly identifying the interested parties on whose behalf they are speaking, and by revealing the existence of any interest the engineers may have in the matters.
4. Engineers shall act for each employer or client as faithful agents or trustees.
   a. Engineers shall disclose all known or potential conflicts of interest that could influence or appear to influence their judgment or the quality of their services.
   b. Engineers shall not accept compensation, financial or otherwise, from more than one party for services on the same project, or for services pertaining to the same project, unless the circumstances are fully disclosed and agreed to by all interested parties.

   c. Engineers shall not solicit or accept financial or other valuable consideration, directly or indirectly, from outside agents in connection with the work for which they are responsible.

   d. Engineers in public service as members, advisors, or employees of a governmental or quasi-governmental body or department shall not participate in decisions with respect to services solicited or provided by them or their organizations in private or public engineering practice.

   e. Engineers shall not solicit or accept a contract from a governmental body on which a principal or officer of their organization serves as a member.

5. Engineers shall avoid deceptive acts.

   a. Engineers shall not falsify their qualifications or permit misrepresentation of their or their associates' qualifications. They shall not misrepresent or exaggerate their responsibility in or for the subject matter of prior assignments. Brochures or other presentations incident to the solicitation of employment shall not misrepresent pertinent facts concerning employers, employees, associates, joint venturers, or past accomplishments.

   b. Engineers shall not offer, give, solicit, or receive, either directly or indirectly, any contribution to influence the award of a contract by public authority, or which may be reasonably construed by the public as having the effect or intent of influencing the awarding of a contract. They shall not offer any gift or other valuable consideration in order to secure work. They shall not pay a commission, percentage, or brokerage fee in order to secure work, except to a bona fide employee or bona fide established commercial or marketing agencies retained by them.

III. Professional Obligations

1. Engineers shall be guided in all their relations by the highest standards of honesty and integrity.

   a. Engineers shall acknowledge their errors and shall not distort or alter the facts.

   b. Engineers shall advise their clients or employers when they believe a project will not be successful.

   c. Engineers shall not accept outside employment to the detriment of their regular work or interest. Before accepting any outside engineering employment, they will notify their employers.

   d. Engineers shall not attempt to attract an engineer from another employer by false or misleading pretenses.

   e. Engineers shall not promote their own interest at the expense of the dignity and integrity of the profession.

2. Engineers shall at all times strive to serve the public interest.

   a. Engineers are encouraged to participate in civic affairs; career guidance for youths; and work for the advancement of the safety, health, and well-being of their community.

   b. Engineers shall not complete, sign, or seal plans and/or specifications that are not in conformity with applicable engineering standards. If the client or employer insists on such unprofessional conduct, they shall notify the proper authorities and withdraw from further service on the project.

   c. Engineers are encouraged to extend public knowledge and appreciation of engineering and its achievements.

   d. Engineers are encouraged to adhere to the principles of sustainable development in order to protect the environment for future generations.

3. Engineers shall avoid all conduct or practice that deceives the public.
   a. Engineers shall avoid the use of statements containing a material misrepresentation of fact or omitting a material fact.
   b. Consistent with the foregoing, engineers may advertise for recruitment of personnel.
   c. Consistent with the foregoing, engineers may prepare articles for the lay or technical press, but such articles shall not imply credit to the author for work performed by others.
4. Engineers shall not disclose, without consent, confidential information concerning the business affairs or technical processes of any present or former client or employer, or public body on which they serve.
   a. Engineers shall not, without the consent of all interested parties, promote or arrange for new employment or practice in connection with a specific project for which the engineer has gained particular and specialized knowledge.
   b. Engineers shall not, without the consent of all interested parties, participate in or represent an adversary interest in connection with a specific project or proceeding in which the engineer has gained particular specialized knowledge on behalf of a former client or employer.
5. Engineers shall not be influenced in their professional duties by conflicting interests.
   a. Engineers shall not accept financial or other considerations, including free engineering designs, from material or equipment suppliers for specifying their product.
   b. Engineers shall not accept commissions or allowances, directly or indirectly, from contractors or other parties dealing with clients or employers of the engineer in connection with work for which the engineer is responsible.
6. Engineers shall not attempt to obtain employment or advancement or professional engagements by untruthfully criticizing other engineers, or by other improper or questionable methods.
   a. Engineers shall not request, propose, or accept a commission on a contingent basis under circumstances in which their judgment may be compromised.
   b. Engineers in salaried positions shall accept part-time engineering work only to the extent consistent with policies of the employer and in accordance with ethical considerations.
   c. Engineers shall not, without consent, use equipment, supplies, laboratory, or office facilities of an employer to carry on outside private practice.
7. Engineers shall not attempt to injure, maliciously or falsely, directly or indirectly, the professional reputation, prospects, practice, or employment of other engineers. Engineers who believe others are guilty of unethical or illegal practice shall present such information to the proper authority for action.
   a. Engineers in private practice shall not review the work of another engineer for the same client, except with the knowledge of such engineer, or unless the connection of such engineer with the work has been terminated.
   b. Engineers in governmental, industrial, or educational employ are entitled to review and evaluate the work of other engineers when so required by their employment duties.
   c. Engineers in sales or industrial employ are entitled to make engineering comparisons of represented products with products of other suppliers.
8. Engineers shall accept personal responsibility for their professional activities, provided, however, that engineers may seek indemnification for services arising out of their practice for other than gross negligence, where the engineer's interests cannot otherwise be protected.
   a. Engineers shall conform with state registration laws in the practice of engineering.
   b. Engineers shall not use association with a nonengineer, a corporation, or partnership as a "cloak" for unethical acts.

9. Engineers shall give credit for engineering work to those to whom credit is due, and will recognize the proprietary interests of others.

   a. Engineers shall, whenever possible, name the person or persons who may be individually responsible for designs, inventions, writings, or other accomplishments.

   b. Engineers using designs supplied by a client recognize that the designs remain the property of the client and may not be duplicated by the engineer for others without express permission.

   c. Engineers, before undertaking work for others in connection with which the engineer may make improvements, plans, designs, inventions, or other records that may justify copyrights or patents, should enter into a positive agreement regarding ownership.

   d. Engineers' designs, data, records, and notes referring exclusively to an employer's work are the employer's property. The employer should indemnify the engineer for use of the information for any purpose other than the original purpose.

   e. Engineers shall continue their professional development throughout their careers and should keep current in their specialty fields by engaging in professional practice, participating in continuing education courses, reading in the technical literature, and attending professional meetings and seminars.

Footnote 1 "Sustainable development" is the challenge of meeting human needs for natural resources, industrial products, energy, food, transportation, shelter, and effective waste management while conserving and protecting environmental quality and the natural resource base essential for future development.

As Revised July 2007

"By order of the United States District Court for the District of Columbia, former Section 11(c) of the NSPE Code of Ethics prohibiting competitive bidding, and all policy statements, opinions, rulings or other guidelines interpreting its scope, have been rescinded as unlawfully interfering with the legal right of engineers, protected under the antitrust laws, to provide price information to prospective clients; accordingly, nothing contained in the NSPE Code of Ethics, policy statements, opinions, rulings or other guidelines prohibits the submission of price quotations or competitive bids for engineering services at any time or in any amount."

Statement by NSPE Executive Committee

In order to correct misunderstandings which have been indicated in some instances since the issuance of the Supreme Court decision and the entry of the Final Judgment, it is noted that in its decision of April 25, 1978, the Supreme Court of the United States declared: "The Sherman Act does not require competitive bidding."

It is further noted that as made clear in the Supreme Court decision:

1. Engineers and firms may individually refuse to bid for engineering services.
2. Clients are not required to seek bids for engineering services.
3. Federal, state, and local laws governing procedures to procure engineering services are not affected, and remain in full force and effect.
4. State societies and local chapters are free to actively and aggressively seek legislation for professional selection and negotiation procedures by public agencies.
5. State registration board rules of professional conduct, including rules prohibiting competitive bidding for engineering services, are not affected and remain in full force and effect. State registration boards with authority to adopt rules of professional conduct may adopt rules governing procedures to obtain engineering services.
6. As noted by the Supreme Court, "nothing in the judgment prevents NSPE and its members from attempting to influence governmental action ...

Note: In regard to the question of application of the Code to corporations vis-à-vis real persons, business form or type should not negate nor influence conformance of individuals to the Code. The Code deals with professional services, which services must be performed by real persons. Real persons in turn establish and implement policies within business structures. The Code is clearly written to apply to the Engineer, and it is incumbent on members of NSPE to endeavor to live up to its provisions. This applies to all pertinent sections of the Code.

1420 King Street
Alexandria, Virginia 22314-2794
703/684-2800 • Fax:703/836-4875
www.nspe.org
Publication date as revised: July 2007 • Publication #1102

WFEO/FMOI (UNESCO)

WORLD FEDERATION OF ENGINEERING ORGANIZATIONS

FEDERATION MONDIALE DES ORGANISATIONS D'INGENIEURS

CODE OF ETHICS

Since 1990, WFEO has worked to prepare a Code of Ethics under the supervision of Donald Laplante (Canada), David Thom (New Zealand), Bud Carroll (USA), and others. It is expected that this model code will be used to define and support the creation of codes in member institutions, which will be adopted in the near future.

WFEO MODEL CODE OF ETHICS
   I. BROAD PRINCIPLES
   II. PRACTICE PROVISION ETHICS
   III. ENVIRONMENTAL ENGINEERING ETHICS
   IV. CONCLUSION

INTERPRETATION OF THE CODE
   Sustainable Development & Environment
   Protection of the Public and the Environment
   Faithful Agent of Clients and Employers
   Competence & Knowledge
   Fairness and Integrity in the Workplace
   Professional Accountability & Leadership

WFEO MODEL CODE OF ETHICS

I. BROAD PRINCIPLES

Ethics is generally understood as the discipline or field of study dealing with moral duty or obligation. This typically gives rise to a set of governing principles or values, which in turn are used to judge the appropriateness of a particular conduct or behaviour. These principles are usually presented either as broad guiding principles of an idealistic or inspirational nature or, alternatively, as a detailed and specific set of rules couched in legalistic or imperative terms to make them more enforceable. Professions which have been given the privilege and responsibility of self regulation, including the engineering profession, have tended to opt for the first alternative, espousing sets of underlying principles as codes of professional ethics which form the basis and framework for

responsible professional practice. Arising from this context, professional codes of ethics have sometimes been incorrectly interpreted as a set of 'rules' of conduct intended for passive observance.
A more appropriate use by practicing professionals is to interpret the essence of the underlying
principles within their daily decision-making situations in a dynamic manner, responsive to the
needs of the situation.

As a consequence, a code of professional ethics is more than a minimum standard of conduct;
rather, it is a set of principles which should guide professionals in their daily work.

In summary, the model Code presented herein expresses the expectations of engineers and society
in discriminating engineers' professional responsibilities. The Code is based on broad principles of
truth, honesty and trustworthiness, respect for human life and welfare, fairness, openness, competence and accountability. Some of these broader ethical principles or issues deemed more universally
applicable are not specifically defined in the Code, although they are understood to be applicable as
well. Only those tenets deemed to be particularly applicable to the practice of professional engineering
are specified. Nevertheless, certain ethical principles or issues not commonly considered to be part
of professional ethics should be implicitly accepted to judge the engineer's professional performance.

Issues regarding the environment and sustainable development know no geographical boundar-
ies. The engineers and citizens of all nations should know and respect the environmental ethic.
It is desirable, therefore, that engineers in each nation continue to observe the philosophy of the
Principles of Environmental Ethics delineated in Section III of this code.

II. PRACTICE PROVISION ETHICS

Professional engineers shall:

- Hold paramount the safety, health and welfare of the public, including people with activity
  limitations, and the protection of both the natural and the built environments in accordance
  with the Principles of Sustainable Development;
- Promote health and safety within the workplace;
- Offer services, advise on or undertake engineering assignments only in areas of their competence, and practice in a careful and diligent manner;
- Act as faithful agents of their clients or employers, maintain confidentially and disclose
  conflicts of interest;
- Keep themselves informed in order to maintain their competence, strive to advance the
  body of knowledge within which they practice and provide opportunities for the professional development of their subordinates and fellow practitioners;
- Conduct themselves with fairness, and good faith towards clients, colleagues and others,
  give credit where it is due and accept, as well as give, honest and fair professional criticism;
- Be aware of and ensure that clients and employers are made aware of the societal and
  environmental consequences of actions or projects, and endeavour to interpret engineering
  issues to the public in an objective and truthful manner;
- Present clearly to employers and clients the possible consequences of overruling or disre-
  garding engineering decisions or judgment;
- Report to their association and/or appropriate agencies any illegal or unethical engineering
  decisions or practices of engineers or others.

III. ENVIRONMENTAL ENGINEERING ETHICS

Engineers, as they develop any professional activity, shall:

- Try with the best of their ability, courage, enthusiasm and dedication, to obtain a superior
  technical achievement, which will contribute to and promote a healthy and agreeable surrounding for all people, including people with activity limitations, in open spaces as well
  as indoors;

- Strive to accomplish the beneficial objectives of their work with the lowest possible consumption of raw materials and energy and the lowest production of wastes and any kind of pollution;
- Discuss in particular the consequences of their proposals and actions, direct or indirect, immediate or long term, upon the health of people (including people with activity limitations), social equity and the local system of values;
- Study thoroughly the environment that will be affected, assess all the impacts that might arise in the structure, dynamics and aesthetics of the eco-systems involved, urbanized or natural, as well as in the pertinent socio-economic systems, and select the best alternative for development that is both environmentally sound and sustainable;
- Promote a clear understanding of the actions required to restore and, if possible, to improve the environment that may be disturbed, and include them in their proposals;
- Reject any kind of commitment that involves unfair damages for human surroundings and nature, and aim for the best possible technical, social, and political solution;
- Be aware that the principles of eco-systemic interdependence, diversity maintenance, resource recovery and inter-relational harmony form the basis of humankind's continued existence and that each of these bases poses a threshold of sustainability that should not be exceeded.

IV. CONCLUSION

Always remember that war, greed, misery and ignorance, plus natural disasters and human induced pollution and destruction of resources, are the main causes of the progressive impairment of the environment and that engineers, as active members of society, deeply involved in the promotion of development, must use our talent, knowledge and imagination to assist society in removing those evils and improving the quality of life for all people, including people with activity limitations.

INTERPRETATION OF THE CODE

The interpretive articles which follow expand on and discuss some of the more difficult and inter-related components of the Code, especially with regard to the Practice Provisions. No attempt is made to expand on all clauses of the Code, nor is the elaboration presented on a clause-by-clause basis. The objective of this approach is to broaden the interpretation, rather than narrow its focus. The ethics of professional engineering is an integrated whole and cannot be reduced to fixed 'rules'. Therefore, the issues and questions arising from the Code are discussed in a general framework, drawing on any and all portions of the Code to demonstrate their inter-relationship and to expand on the basic intent of the Code.

SUSTAINABLE DEVELOPMENT AND ENVIRONMENT

Engineers shall strive to enhance the quality of the biophysical and socioeconomic urban environment and of buildings and spaces, and to promote the principles of sustainable development.

Engineers shall seek opportunities to work for the enhancement of safety, health, and the social welfare of both their local community and the global community through the practice of sustainable development.

Engineers whose recommendations are overruled or ignored on issues of safety, health, welfare, or sustainable development, shall inform their contractor or employer of the possible consequences.

PROTECTION OF THE PUBLIC AND THE ENVIRONMENT

Professional Engineers shall hold paramount the safety, health and welfare of the public, including people with activity limitations, and the protection of the environment. This obligation to the safety, health and welfare of the general public, which includes one's own work environment, is often dependent upon engineering judgments, risk assessments, decisions and practices incorporated into structures, machines, products, processes and devices. Therefore, engineers must control and ensure that what they are involved with is in conformity with accepted engineering practices, standards and applicable codes, and would be considered safe based on peer adjudication. This responsibility extends to include all and any situations which an engineer encounters, and includes an obligation to advise the appropriate authority if there is reason to believe that any engineering activity, or its products, processes, etc., do not conform with the above stated conditions.

The meaning of paramount in this basic tenet is that all other requirements of the Code are subordinate, if protection of public safety, the environment or other substantive public interests are involved.

FAITHFUL AGENT OF CLIENTS AND EMPLOYERS

Engineers shall act as faithful agents or trustees of their clients and employers with objectivity, fairness and justice to all parties. With respect to the handling of confidential or proprietary information, the concept of ownership of the information and protecting that party's rights is appropriate. Engineers shall not reveal facts, data or information obtained in a professional capacity without the prior consent of its owner. The only exception to respecting confidentially and maintaining a trustee's position is in instances where the public interest or the environment is at risk, as discussed in the preceding section; but even in these circumstances, the engineer should endeavour to have the client and/or employer appropriately redress the situation, or at least, in the absence of a compelling reason to the contrary, should make every reasonable effort to contact them and explain clearly the potential risks, prior to informing the appropriate authority.

Professional Engineers shall avoid conflict of interest situations with employers and clients but, should such conflict arise, it is the engineer's responsibility to fully disclose, without delay, the nature of the conflict to the party/parties with whom the conflict exists. In those circumstances where full disclosure is insufficient, or seen to be insufficient, to protect all parties' interests, as well as the public, the engineer shall withdraw totally from the issue or use extraordinary means, involving independent parties if possible, to monitor the situation. For example, it is inappropriate to act simultaneously as agent for both the provider and the recipient of professional services. If a client's and an employer's interests are at odds, the engineer shall attempt to deal fairly with both. If the conflict of interest is between the intent of a corporate employer and a regulatory standard, the engineer must attempt to reconcile the difference, and if that is unsuccessful, it may become necessary to inform his/her association and the appropriate regulatory agency.

Being a faithful agent or trustee includes the obligation of engaging, or advising to engage, experts or specialists when such services are deemed to be in the client's or employer's best interests. It also means being accurate, objective and truthful in making public statements on behalf of the client or employer when required to do so, while respecting the client's and employer's rights of confidentiality and proprietary information.

Being a faithful agent includes not using a previous employer's or client's specific privileged or proprietary information and trade practices or process information, without the owner's knowledge and consent. However, general technical knowledge, experience and expertise gained by the engineer through involvement with the previous work may be freely used without consent or subsequent undertakings.

COMPETENCE AND KNOWLEDGE

Professional Engineers shall offer services, advise on or undertake engineering assignments only in areas of their competence by virtue of their training and experience. This includes exercising care and communicating clearly in accepting or interpreting assignments, and in setting expected outcomes. It also includes the responsibility to obtain the services of an expert if required or, if the knowledge is unknown, to proceed only with full disclosure of the circumstances and, if necessary, of the experimental nature of the activity to all parties involved. Hence, this requirement is more than simply duty to a standard of care, it also involves acting with honesty and integrity with one's client or employer, and one's self. Professional Engineers have the responsibility to remain abreast of developments and knowledge in their area of expertise, that is, to maintain their own competence. Should there be a technologically driven or individually motivated shift in the area of technical activity, it is the engineer's duty to attain and maintain competence in all areas of involvement including being knowledgeable with the technical and legal framework and regulations governing their work. In effect, it requires a personal commitment to ongoing professional development, continuing education and self-testing.

In addition to maintaining their own competence, Professional Engineers have an obligation to strive to contribute to the advancement of the body of knowledge within which they practice, and to the profession in general. Moreover, within the framework of the practice of their profession, they are expected to participate in providing opportunities to further the professional development of their colleagues.

This competence requirement of the Code extends to include an obligation to the public, the profession and one's peers, that opinions on engineering issues are expressed honestly and only in areas of one's competence. It applies equally to reporting or advising on professional matters and to issuing public statements. This requires honesty with one's self to present issues fairly, accurately and with appropriate qualifiers and disclaimers, and to avoid personal, political and other non-technical biases. The latter is particularly important for public statements or when involved in a technical forum.

FAIRNESS AND INTEGRITY IN THE WORKPLACE

Honesty, integrity, continuously updated competence, devotion to service and dedication to enhancing the life quality of society are cornerstones of professional responsibility. Within this framework, engineers shall be objective and truthful and include all known and pertinent information in professional reports, statements and testimony. They shall accurately and objectively represent their clients, employers, associates and themselves, consistent with their academic experience and professional qualifications. This tenet is more than 'not misrepresenting'; it also implies disclosure of all relevant information and issues, especially when serving in an advisory capacity or as an expert witness.

Similarly, fairness, honesty and accuracy in advertising are expected. If called upon to verify another engineer's work, there is an obligation to inform (or make every effort to inform) the other engineer, whether the other engineer is still actively involved or not. In this situation, and in any circumstance, engineers shall give proper recognition and credit where credit is due and accept, as well as give, honest and fair criticism on professional matters, all the while maintaining dignity and respect for everyone involved.

Engineers shall not accept, nor offer covert payment or other considerations for the purpose of securing, or as remuneration for, engineering assignments. Engineers should prevent their personal or political involvement from influencing or compromising their professional role or responsibility.

Consistent with the Code, and having attempted to remedy any situation within their organization, engineers are obligated to report to their association or other appropriate agency any illegal or unethical engineering decisions by engineers or others. Care must be taken not to enter into legal arrangements which compromise this obligation.

PROFESSIONAL ACCOUNTABILITY AND LEADERSHIP

Engineers have a duty to practice in a careful and diligent manner, and accept responsibility and be accountable for their actions. This duty is not limited to design, or its supervision and management, but applies to all areas of practice. For example, it includes construction supervision and management, preparation of shop drawings, engineering reports, feasibility studies, environmental impact assessments, engineering developmental work, etc.

The signing and sealing of engineering documents indicates the taking of responsibility for the work. This practice is required for all types of engineering endeavour, regardless of where or for whom the work is done, including but not limited to, privately and publicly owned firms, crown corporations, and government agencies or departments. There are no exceptions; signing and sealing documents is appropriate whenever engineering principles have been used and public welfare may be at risk.

Taking responsibility for engineering activity includes being accountable for one's own work and, in the case of a senior engineer, accepting responsibility for the work of a team. The latter implies responsible supervision where the engineer is actually in a position to review, modify and direct the entirety of the engineering work. This concept requires setting reasonable limits on the extent of activities, and the number of engineers and others, whose work can be supervised by the responsible engineer. The practice of a 'symbolic' responsibility or supervision is the situation where an engineer, say with the title of *Chief Engineer*, takes full responsibility for all engineering on behalf of a large corporation, utility or government agency/department, even though the engineer may not be aware of many of the engineering activities or decisions being made daily throughout the firm or department. The essence of this approach is that the firm is taking the responsibility by default, whether engineering supervision or direction is applied or not.

Engineers have a duty to advise their employer and, if necessary, their clients and even their professional association, in that order, in situations when the overturning of an engineering decision may result in breaching their duty to safeguard the public, including people with activity limitations. The initial action is to discuss the problem with the supervisor/employer. If the employer does not adequately respond to the engineer's concern, then the client must be advised in the case of a consultancy situation, or the most senior officer should be informed in the case of a manufacturing process plant or government agency. Failing this attempt to rectify the situation, the engineer must advise in confidence his/her professional association of his/her concerns.

In the same order as mentioned above, the engineer must report unethical engineering activity undertaken by other engineers, or by non-engineers. This extends to include, for example, situations in which senior officials of a firm make 'executive' decisions which clearly and substantially alter the engineering aspects of the work, or protection of the public welfare or the environment arising from that work.

Because of developments in technology and the increasing ability of engineering activities to impact on the environment, engineers have an obligation to be mindful of the effect that their decisions will have on the environment and the wellbeing of society, and to report any concerns of this nature in the same manner as previously mentioned. Further to the above, with the rapid advancement of technology in today's world and the possible social impacts on large populations of people, engineers must endeavour to foster the public's understanding of technical issues and the role of Engineering more than ever before.

Sustainable development is the challenge of meeting current human needs for natural resources, industrial products, energy, food, transportation, shelter, and effective waste management while conserving and, if possible enhancing, the Earth's environmental quality, natural resources, ethical, intellectual, working and affectionate capabilities of people and the socio-economic bases essential for the human needs of future generations. The proper observance of these principles will considerably help to eradicate world poverty.

Final Version, Adopted 2001.

1. **Member's Code of Conduct**—The Membership of SWE commits itself to ethical, businesslike, and lawful conduct, including proper use of authority and decorum at the highest level when acting on behalf of SWE. The Membership of SWE will consistently fulfill the purposes set forth in SWE's Bylaws, Policies, Code of Conduct, Core Values, and Strategic Vision.

2. **Policy against Harassment**—Please see SWE's "Policy Against Harassment" document to see the zero tolerance policy that shall be followed by all members.

3. **Disposition of Complaints and Disputes involving SWE Members**—Complaints or disputes should be discussed immediately with your SWE Leader or skip level officer or to SWE's Ethics Committee. All issues will be handled based on the "Procedures for Review of SWE Member Conduct."

AMERICAN SOCIETY OF MECHANICAL ENGINEERS

SOCIETY POLICY

Ethics

ASME requires ethical practice by each of its members and has adopted the following Code of Ethics of Engineers as referenced in the ASME Constitution, Article C2.1.1.

Code of Ethics of Engineers

Engineers uphold and advance the integrity, honor and dignity of the engineering profession by:

I. using their knowledge and skill for the enhancement of human welfare;
II. being honest and impartial, and serving with fidelity their clients (including their employers) and the public; and
III. striving to increase the competence and prestige of the engineering profession.

The Fundamental Canons

1. Engineers shall hold paramount the safety, health and welfare of the public in the performance of their professional duties.
2. Engineers shall perform services only in the areas of their competence; they shall build their professional reputation on the merit of their services and shall not compete unfairly with others.
3. Engineers shall continue their professional development throughout their careers and shall provide opportunities for the professional and ethical development of those engineers under their supervision.
4. Engineers shall act in professional matters for each employer or client as faithful agents or trustees, and shall avoid conflicts of interest or the appearance of conflicts of interest.
5. Engineers shall respect the proprietary information and intellectual property rights of others, including charitable organizations and professional societies in the engineering field.
6. Engineers shall associate only with reputable persons or organizations.
7. Engineers shall issue public statements only in an objective and truthful manner and shall avoid any conduct which brings discredit upon the profession.
8. Engineers shall consider environmental impact and sustainable development in the performance of their professional duties.
9. Engineers shall not seek ethical sanction against another engineer unless there is good reason to do so under the relevant codes, policies and procedures governing that engineer's ethical conduct.

Engineers who are members of the Society shall endeavor to abide by the Constitution, By-Laws and Policies of the Society and they shall disclose knowledge of any matter involving another member's alleged violation of this Code of Ethics or the Society's Conflicts of Interest Policy in a prompt, complete and truthful manner to the chair of the Committee on Ethical Standards and Review.

The Committee on Ethical Standards and Review maintains an archive of interpretations to the ASME Code of Ethics (P-15.7). These interpretations shall serve as guidance to the user of the ASME Code of Ethics and are available on the Committee's website or upon request.

Responsibility:

Centers Board of Directors/Center for Professional Career and Professional Advancement/ Committee on Ethical Standards and Review

Reassigned from Centers Board of Directors/Center for Professional Development, Practice and Ethics/Committee on Ethical Standards and Review 4/23/09

Reassigned from Council and Member Affairs/Board on Professional Practice & Ethics 6/1/05

Adopted:

March 7, 1976

Revised:

December 9, 1976 December 7, 1979 November 19, 1982 June 15, 1984 (editorial changes 7/84) June 16, 1988 September 12, 1991 September 11, 1994 June 10, 1998 September 21, 2002 September 13, 2003 (editorial changes 6/1/05) November 5, 2006 (editorial changes to the responsible unit 4/09)

AMERICAN SOCIETY OF SAFETY ENGINEERS

CODE OF PROFESSIONAL CONDUCT

American Society of Safety Engineers Code of Professional Conduct Membership in the American Society of Safety Engineers evokes a duty to serve and protect people, property and the environment. This duty is to be exercised with integrity, honor and dignity. Members are accountable for following the Code of Professional Conduct.

Fundamental Principles

1. Protect people, property and the environment through the application of state-of-the-art knowledge.
2. Serve the public, employees, employers, clients and the Society with fidelity, honesty and impartiality.
3. Achieve and maintain competency in the practice of the profession.
4. Avoid conflicts of interest and compromise of professional conduct.
5. Maintain confidentiality of privileged information.

Fundamental Canons

In the fulfillment of my duties as a safety professional and as a member of the Society, I shall:

1. Inform the public, employers, employees, clients and appropriate authorities when professional judgment indicates that there is an unacceptable level of risk.
2. Improve knowledge and skills through training, education and networking.
3. Perform professional services only in the area of competence.
4. Issue public statements in a truthful manner, and only within the parameters of authority granted.
5. Serve as an agent and trustee, avoiding any appearance of conflict of interest.
6. Assure equal opportunity to all.

Approved by House of Delegates June 9, 2002

CODE OF ETHICS[2]

Fundamental Principles[3]

Engineers uphold and advance the integrity, honor and dignity of the engineering profession by:

1. Using their knowledge and skill for the enhancement of human welfare and the environment;
2. Being honest and impartial and serving with fidelity the public, their employers and clients;

3. Striving to increase the competence and prestige of the engineering profession; and

4. Supporting the professional and technical societies of their disciplines.

Fundamental Canons

1. Engineers shall hold paramount the safety, health and welfare of the public and shall strive to comply with the principles of sustainable development[4] in the performance of their professional duties.

2. Engineers shall perform services only in areas of their competence.

3. Engineers shall issue public statements only in an objective and truthful manner.

4. Engineers shall act in professional matters for each employer or client as faithful agents or trustees, and shall avoid conflicts of interest.

5. Engineers shall build their professional reputation on the merit of their services and shall not compete unfairly with others.

6. Engineers shall act in such a manner as to uphold and enhance the honor, integrity, and dignity of the engineering profession and shall act with zero-tolerance for bribery, fraud, and corruption.

7. Engineers shall continue their professional development throughout their careers, and shall provide opportunities for the professional development of those engineers under their supervision.

GUIDELINES TO PRACTICE UNDER THE FUNDAMENTAL CANONS OF ETHICS

CANON 1.

Engineers shall hold paramount the safety, health and welfare of the public and shall strive to comply with the principles of sustainable development in the performance of their professional duties.

a. Engineers shall recognize that the lives, safety, health and welfare of the general public are dependent upon engineering judgments, decisions and practices incorporated into structures, machines, products, processes and devices.

b. Engineers shall approve or seal only those design documents, reviewed or prepared by them, which are determined to be safe for public health and welfare in conformity with accepted engineering standards.

c. Engineers whose professional judgment is overruled under circumstances where the safety, health and welfare of the public are endangered, or the principles of sustainable development ignored, shall inform their clients or employers of the possible consequences.

d. Engineers who have knowledge or reason to believe that another person or firm may be in violation of any of the provisions of Canon 1 shall present such information to the proper authority in writing and shall cooperate with the proper authority in furnishing such further information or assistance as may be required.

e. Engineers should seek opportunities to be of constructive service in civic affairs and work for the advancement of the safety, health and well-being of their communities, and the protection of the environment through the practice of sustainable development.

f. Engineers should be committed to improving the environment by adherence to the principles of sustainable development so as to enhance the quality of life of the general public.

CANON 2.

Engineers shall perform services only in areas of their competence.

a. Engineers shall undertake to perform engineering assignments only when qualified by education or experience in the technical field of engineering involved.
b. Engineers may accept an assignment requiring education or experience outside of their own fields of competence, provided their services are restricted to those phases of the project in which they are qualified. All other phases of such project shall be performed by qualified associates, consultants, or employees.
c. Engineers shall not affix their signatures or seals to any engineering plan or document dealing with subject matter in which they lack competence by virtue of education or experience or to any such plan or document not reviewed or prepared under their supervisory control.

CANON 3.

Engineers shall issue public statements only in an objective and truthful manner.

a. Engineers should endeavor to extend the public knowledge of engineering and sustainable development, and shall not participate in the dissemination of untrue, unfair or exaggerated statements regarding engineering.
b. Engineers shall be objective and truthful in professional reports, statements, or testimony. They shall include all relevant and pertinent information in such reports, statements, or testimony.
c. Engineers, when serving as expert witnesses, shall express an engineering opinion only when it is founded upon adequate knowledge of the facts, upon a background of technical competence, and upon honest conviction.
d. Engineers shall issue no statements, criticisms, or arguments on engineering matters which are inspired or paid for by interested parties, unless they indicate on whose behalf the statements are made.
e. Engineers shall be dignified and modest in explaining their work and merit, and will avoid any act tending to promote their own interests at the expense of the integrity, honor and dignity of the profession.

CANON 4.

Engineers shall act in professional matters for each employer or client as faithful agents or trustees, and shall avoid conflicts of interest.

a. Engineers shall avoid all known or potential conflicts of interest with their employers or clients and shall promptly inform their employers or clients of any business association, interests, or circumstances which could influence their judgment or the quality of their services.
b. Engineers shall not accept compensation from more than one party for services on the same project, or for services pertaining to the same project, unless the circumstances are fully disclosed to and agreed to, by all interested parties.
c. Engineers shall not solicit or accept gratuities, directly or indirectly, from contractors, their agents, or other parties dealing with their clients or employers in connection with work for which they are responsible.
d. Engineers in public service as members, advisors, or employees of a governmental body or department shall not participate in considerations or actions with respect to services solicited or provided by them or their organization in private or public engineering practice.

e. Engineers shall advise their employers or clients when, as a result of their studies, they believe a project will not be successful.

f. Engineers shall not use confidential information coming to them in the course of their assignments as a means of making personal profit if such action is adverse to the interests of their clients, employers or the public.

g. Engineers shall not accept professional employment outside of their regular work or interest without the knowledge of their employers.

CANON 5.

Engineers shall build their professional reputation on the merit of their services and shall not compete unfairly with others.

a. Engineers shall not give, solicit or receive either directly or indirectly, any political contribution, gratuity, or unlawful consideration in order to secure work, exclusive of securing salaried positions through employment agencies.

b. Engineers should negotiate contracts for professional services fairly and on the basis of demonstrated competence and qualifications for the type of professional service required.

c. Engineers may request, propose or accept professional commissions on a contingent basis only under circumstances in which their professional judgments would not be compromised.

d. Engineers shall not falsify or permit misrepresentation of their academic or professional qualifications or experience.

e. Engineers shall give proper credit for engineering work to those to whom credit is due, and shall recognize the proprietary interests of others. Whenever possible, they shall name the person or persons who may be responsible for designs, inventions, writings or other accomplishments.

f. Engineers may advertise professional services in a way that does not contain misleading language or is in any other manner derogatory to the dignity of the profession. Examples of permissible advertising are as follows:

- Professional cards in recognized, dignified publications, and listings in rosters or directories published by responsible organizations, provided that the cards or listings are consistent in size and content and are in a section of the publication regularly devoted to such professional cards.
- Brochures which factually describe experience, facilities, personnel and capacity to render service, providing they are not misleading with respect to the engineer's participation in projects described.
- Display advertising in recognized dignified business and professional publications, providing it is factual and is not misleading with respect to the engineer's extent of participation in projects described.
- A statement of the engineers' names or the name of the firm and statement of the type of service posted on projects for which they render services.
- Preparation or authorization of descriptive articles for the lay or technical press, which are factual and dignified. Such articles shall not imply anything more than direct participation in the project described.
- Permission by engineers for their names to be used in commercial advertisements, such as may be published by contractors, material suppliers, etc., only by means of a modest, dignified notation acknowledging the engineers' participation in the project described. Such permission shall not include public endorsement of proprietary products.

g. Engineers shall not maliciously or falsely, directly or indirectly, injure the professional reputation, prospects, practice or employment of another engineer or indiscriminately criticize another's work.

h. Engineers shall not use equipment, supplies, laboratory or office facilities of their employers to carry on outside private practice without the consent of their employers.

CANON 6.

Engineers shall act in such a manner as to uphold and enhance the honor, integrity, and dignity of the engineering profession and shall act with zero tolerance for bribery, fraud, and corruption.

a. Engineers shall not knowingly engage in business or professional practices of a fraudulent, dishonest or unethical nature.
b. Engineers shall be scrupulously honest in their control and spending of monies, and promote effective use of resources through open, honest and impartial service with fidelity to the public, employers, associates and clients.
c. Engineers shall act with zero-tolerance for bribery, fraud, and corruption in all engineering or construction activities in which they are engaged.
d. Engineers should be especially vigilant to maintain appropriate ethical behavior where payments of gratuities or bribes are institutionalized practices.
e. Engineers should strive for transparency in the procurement and execution of projects. Transparency includes disclosure of names, addresses, purposes, and fees or commissions paid for all agents facilitating projects.
f. Engineers should encourage the use of certifications specifying zero tolerance for bribery, fraud, and corruption in all contracts.

CANON 7.

Engineers shall continue their professional development throughout their careers, and shall provide opportunities for the professional development of those engineers under their supervision.

a. Engineers should keep current in their specialty fields by engaging in professional practice, participating in continuing education courses, reading in the technical literature, and attending professional meetings and seminars.
b. Engineers should encourage their engineering employees to become registered at the earliest possible date.
c. Engineers should encourage engineering employees to attend and present papers at professional and technical society meetings.
d. Engineers shall uphold the principle of mutually satisfying relationships between employers and employees with respect to terms of employment including professional grade descriptions, salary ranges, and fringe benefits.

AMERICAN INSTITUTE OF CHEMICAL ENGINEERS

CODE OF ETHICS

Members of the American Institute of Chemical Engineers shall uphold and advance the integrity, honor and dignity of the engineering profession by: being honest and impartial and serving with fidelity their employers, their clients, and the public; striving to increase the competence and prestige of the engineering profession; and using their knowledge and skill for the enhancement of human welfare. To achieve these goals, members shall

Hold paramount the safety, health and welfare of the public and protect the environment in performance of their professional duties.

Formally advise their employers or clients (and consider further disclosure, if warranted) if they perceive that a consequence of their duties will adversely affect the present or future health or safety of their colleagues or the public.

Accept responsibility for their actions, seek and heed critical review of their work and offer objective criticism of the work of others.

Issue statements or present information only in an objective and truthful manner.

Act in professional matters for each employer or client as faithful agents or trustees, avoiding conflicts of interest and never breaching confidentiality.

Treat faiSrly and respectfully all colleagues and co-workers, recognizing their unique contributions and capabilities.

Perform professional services only in areas of their competence.

Build their professional reputations on the merits of their services.

Continue their professional development throughout their careers, and provide opportunities for the professional development of those under their supervision.

Never tolerate harassment.

Conduct themselves in a fair, honorable and respectful manner.

(Revised January 17, 2003)

AMERICAN NUCLEAR SOCIETY

CODE OF ETHICS

Preamble

Recognizing the profound importance of nuclear science and technology in affecting the quality of life throughout the world, members of the American Nuclear Society (ANS) are committed to the highest ethical and professional conduct.

Fundamental Principle

ANS members as professionals are dedicated to improving the understanding of nuclear science and technology, appropriate applications, and potential consequences of their use.

To that end, ANS members uphold and advance the integrity and honor of their professions by using their knowledge and skill for the enhancement of human welfare and the environment; being honest and impartial; serving with fidelity the public, their employers, and their clients; and striving to continuously improve the competence and prestige of their various professions.

ANS members shall subscribe to the following practices of professional conduct:

Practices of Professional Conduct

We hold paramount the safety, health, and welfare of the public and fellow workers, work to protect the environment, and strive to comply with the principles of sustainable development in the performance of our professional duties.

1. We will formally advise our employers, clients, or any appropriate authority and, if warranted, consider further disclosure, if and when we perceive that pursuit of our professional duties might have adverse consequences for the present or future public and fellow worker health and safety or the environment.
2. We act in accordance with all applicable laws and these practices, lend support to others who strive to do likewise, and report violations to appropriate authorities.
3. We perform only those services that we are qualified by training or experience to perform, and provide full disclosure of our qualifications.
4. We present all data and claims, with their bases, truthfully, and are honest and truthful in all aspects of our professional activities. We issue public statements and make presentations on professional matters in an objective and truthful manner.
5. We continue our professional development and maintain an ethical commitment throughout our careers, encourage similar actions by our colleagues, and provide opportunities for the professional and ethical training of those persons under our supervision.
6. We act in a professional and ethical manner towards each employer or client and act as faithful agents or trustees, disclosing nothing of a proprietary nature concerning the

business affairs or technical processes of any present or former client or employer without specific consent, unless necessary to abide by other provisions of this Code or applicable laws.

7. We disclose to affected parties, known or potential conflicts of interest or other circumstances, which might influence, or appear to influence, our judgment or impair the fairness or quality of our performance.

8. We treat all persons fairly.

9. We build our professional reputation on the merit of our services, do not compete unfairly with others, and avoid injuring others, their property, reputation, or employment.

10. We reject bribery and coercion in all their forms.

11. We accept responsibility for our actions; are open to and acknowledge criticism of our work; offer honest criticism of the work of others; properly credit the contributions of others; and do not accept credit for work not our own.

Approved by ANS Special Committee on Ethics November 17, 2002.
Approved by ANS Board of Directors June 5, 2003.

SOCIETY OF PETROLEUM ENGINEERS

GUIDE FOR PROFESSIONAL CONDUCT

Preamble

Engineers recognize that the practice of engineering has a vital influence on the quality of life for all people. Engineers should exhibit high standards of competency, honesty, integrity, and impartiality; be fair and equitable; and accept a personal responsibility for adherence to applicable laws, the protection of the environment, and safeguarding the public welfare in their professional actions and behavior. These principles govern professional conduct in serving the interests of the public, clients, employers, colleagues, and the profession.

The Fundamental Principle

The engineer as a professional is dedicated to improving competence, service, fairness, and the exercise of well-founded judgment in the ethical practice of engineering for all who use engineering services with fundamental concern for protecting the environment and safeguarding the health, safety and well-being of the public in the pursuit of this practice.

Canons of Professional Conduct

1. Engineers offer services in the areas of their competence and experience, affording full disclosure of their qualifications.

2. Engineers consider the consequences of their work and societal issues pertinent to it and seek to extend public understanding of those relationships.

3. Engineers are honest, truthful, ethical, and fair in presenting information and in making public statements, which reflect on professional matters and their professional role.

4. Engineers engage in professional relationships without bias because of race, religion, gender, age, ethnic or national origin, attire, or disability.

5. Engineers act in professional matters for each employer or client as faithful agents or trustees disclosing nothing of a proprietary or confidential nature concerning the business affairs or technical processes of any present or former client or employer without the necessary consent.

6. Engineers disclose to affected parties any known or potential conflicts of interest or other circumstances, which might influence, or appear to influence, judgment or impair the fairness or quality of their performance.
7. Engineers are responsible for enhancing their professional competence throughout their careers and for encouraging similar actions by their colleagues.
8. Engineers accept responsibility for their actions; seek and acknowledge criticism of their work; offer honest and constructive criticism of the work of others; properly credit the contributions of others; and do not accept credit for work not their own.
9. Engineers, perceiving a consequence of their professional duties to adversely affect the present or future public health and safety, shall formally advise their employers or clients, and, if warranted, consider further disclosure.
10. Engineers seek to adopt technical and economical measures to minimize environmental impact.
11. Engineers participate with other professionals in multi-discipline teams to create synergy and to add value to their work product.
12. Engineers act in accordance with all applicable laws and the canons of ethics as applicable to the practice of engineering as stated in the laws and regulations governing the practice of engineering in their country, territory, or state, and lend support to others who strive to do likewise.

—Approved by the Board of Directors 26 September 2004

SPE.ORG

©2003–2011 Society of Petroleum Engineers Student Ethics Competition

IEEE ETHICS AND MEMBER CONDUCT COMMITTEE (EMCC)

STUDENT ETHICS COMPETITION

Guidelines and Requirements

Prepared and Sponsored by the IEEE Ethics and Member Conduct Committee (EMCC)
March 2008
Copying and distributing this material is permitted with proper attribution to the IEEE Ethics & Member Conduct Committee.
STUDENT ETHICS COMPETITION (SEC)

INTRODUCTION

The IEEE Student Ethics Competition is sponsored by the IEEE Ethics and Member Conduct Committee (EMCC). It was developed for use at IEEE student events to encourage the study and awareness of professional ethics by IEEE Student and Graduate Student Members. The competition includes a presentation and defense of a case analysis by teams of students. Specific objectives of the competition program are:

1. To foster familiarity with the IEEE Code of Ethics and ethical concepts,
2. To promote a model for discussing and analyzing ethical questions, and
3. To provide experience in applying ethical concepts to typical professional situations.

The use of the competition as part of curricula assignments or extra credit activities is encouraged.

OVERVIEW OF THE STUDENT ETHICS COMPETITION

Activities

Participants will be tested on their knowledge and application of the IEEE Code of Ethics as demonstrated by student analysis and findings resulting from a study of a theoretical ethics case.

Materials provided by the Ethics and Member Conduct Committee

The EMCC provides the following:

! Guidelines for structuring a competition are all provided in the SEC package.
Competition Guidelines, which includes: Overview of competition, Judging Forms, Presentation Guidelines, IEEE Code of Ethics and related Bylaw, competition Resources (including brief Ethics Glossary), Student Analysis Format, and Sample Case.
! Posters and/or CD-ROM with customizable poster files for publicity of the event.

Additionally, for competitions that receive funding approval through the EMCC, the following is provided:

- Access to an EMCC case study for use at event. The competition organizer may develop their own case study but review by the EMCC of any self writing case study is required.
- Funding for prizes.
- Certificates for winning team signed by the EMCC Chair and the IEEE President.
- Certificate of participation for all participants signed by the EMCC Chair and the IEEE President. Certificates of participation will be sent to the competition organizer prior to the competition in order to be presented at the conclusion of the event.

Participants

Based on time considerations, see ("organization of competition"), it is recommended that between four and six teams of two or three IEEE Student or Graduate Student members each be selected for the competition. For events receiving funding by the EMCC, the individual team members must belong to the same IEEE Student Branch as the prize money is distributed to the winning Student Branch and not to individuals.

Eligibility

Participants must be IEEE Student or Graduate Student Members in good standing.

Location of Competition

It is recommended that the competition be part of a major IEEE Student or Graduate Student event that attracts a cross-section of IEEE Student and Graduate Student Members. The competition must be a visible part of the event's public/advance agenda/program.

Prizes

Based on the judging criteria, the competition judges will determine winner and runner-up winning teams. Prize amounts are based on the event type described in the "funding" section below. Upon completion of the event, the competition organizer shall notify the EMCC of the winning teams so that the prize money can be disbursed.

Certificates of participation are available to all participants.

All certificates will be signed by the President of the IEEE and by the Chair of the IEEE Ethics and Member Conduct Committee.

Funding

The IEEE Ethics and Member Conduct Committee provides funding for prizes for up to ten events each year. Each IEEE Region is eligible to receive funding every two years for competitions held at the Region level.

Events, if approved by the EMCC, may be funded as follows:

An event involving two or more IEEE Student Branches is provided $600 US ($400 first place & $200 runner up)

An event involving a single IEEE Student Branch is provided $300 US.

An event involving teams consisting of participants from multiple IEEE Student Branches may receive funding up to $800 US. Teams may consist of a maximum of 4 participants, each from a different IEEE Student Branch. Distribution of funds shall be ($125 US for each first place participant's Student Branch and $75 US for each runner up participant's Student Branch.

All funding is to be used for future activities of those Student Branches.

Judges

Competition organizers will select no less than three, but no more than six judges. Judges should be selected based upon their knowledge of ethics.

Selection of Winners

Judges will evaluate participants based on the following:

1. Analysis of a hypothetical case study presented orally (70%)
2. Question and answers with the judges in defense of the case study conclusions (30%)

Judging Forms for both two and three member teams are attached.

Each judge will rank the teams using the attached Judging Forms. The points awarded by the judges will be tallied and the winners determined by the totaled scores. Ties will be resolved by majority vote of the judges. All other judge decisions will be by majority vote and will be final with regard to adherence to rules, disputes, team eligibility, disqualifications, and other competition conduct.

Notification of Winners

Winners will be notified at the Competition Award Ceremony. A list of winners will be made available on the IEEE Ethics & Member Conduct Committee webpage based on official notice to the IEEE Ethics & Member Conduct Committee at email ethics@ieee.org, or mail, IEEE, 445 Hoes Lane, Piscataway, NJ 08854, Attention: IEEE Ethics & Member Conduct Committee.

Hosting a Student Ethics Competition

Individuals wishing to host a competition should first contact their Student Branch Chair, or Regional Student Branch Committee Chair to coordinate the activity.

Once a decision to host a competition has been made, the competition organizer must provide the following information to the EMCC in order to be considered for funding:

- Proposed date
- Proposed location
- Type of event the competition at which the competition will be held (i.e. Student Branch Congress, Student Activities Committee meeting, etc.)
- Approximate number of attendees at event
- Participants names, IEEE member number and their student branches
- List of judges

- Indication if a case study is needed. The case study will be released to the competition organizer only upon approval of the event. If the competition organizer has developed the case study it must be provided to the EMCC for review. In this instance the organizer should also indicate whether the EMCC may use the case study for other competitions in the future.

Such requests for funding should be received at least 45 days in advance in order to be considered.

Responsibility of Competition Organizers

1. Set date/location for competition
2. Provide competition details to EMCC if requesting funding. (refer to "Hosting a Student Ethics Competition" section above)
3. Organize & advertise the competition
4. Select judges
5. Notify thze EMCC of competition results

For Further Information

Competition information is available at (www.ieee.org/ethics).
For questions contact the IEEE Ethics & Member Conduct Committee at ethics@ieee.org

ORGANIZATION OF COMPETITION

Competition Length

As a guideline, an average competition should take approximately four hours total time. The competition should not be broken with meal or other unsupervised breaks to assure that contestants do not have an opportunity to discuss their decision outside of the group and time allocation. Once teams have presented their decision for judging, they may join the audience for further presentations. Teams which have not yet made their presentation must be isolated from the presentation venue during presentations of other teams.

Registration—Distribution of Competition Case and Draw for Speaking Order

Upon arrival, teams should register with the competition organizer. Once all teams have registered they will draw for speaking order. The competition organizer will then distribute the competition case study when the teams have all assembled and are ready to start the competition.

Materials and Sample Cases Provided Prior to the Competition

The competition materials and case studies are provided to the organizer, by the EMCC, prior to the competition. The organizer should distribute the materials, except for the competition's case study, in advance to all participants.

Preparation of Presentation

Two hours is the recommended time for teams to prepare their presentations. The teams are limited to written competition material distributed at the beginning of the competition, i.e. no extra or outside resources. Teams will work in isolated teams and will require tools to develop presentation materials, for example MS PowerPoint and CD burning or memory stick capabilities.

Preparation by Judges

While teams prepare their presentations, judges will work together to set case expectations and Q&A Procedure.

Collect CDs and Sequester Teams

After two hours, the teams will save their presentations to a CD. The CDs will be collected and the teams will continue to be sequestered.

Presentations Sequence

A random method, or draw from a hat, of determining presentation sequence is recommended.

Open Session for Presentations and Defense of Case Analyses

The teams will present their case recommendations and decisions according to the speaking order. Each team shall have 20 minutes for their analysis according to the following breakdown:

8 to 12-minute team presentation
5-minute judge Q&A and team defenses,
3 minutes for completing judging forms

Close Competition, Count Ballets, and Tabulate Results

After all teams have presented, the competition will be closed, ballots counted and results tabulated.

Present Participation Certificates, Announce Results, and Present Awards

Participation certificates can be awarded at the competition site. However, it is recommended that the certificates for the winner and runner-up teams be awarded at a central or plenary activity of the Regional event.

IEEE STUDENT ETHICS COMPETITION JUDGING FORM (2 MEMBER TEAM OPTION)

CATEGORIES SCORES

DEDUCTIONS

Time Adherence (deduction of 5 points for every 30 seconds outside of time limits) (Timing Lights or Signals at 8, 10, and 12 minute points)
Lack of significant involvement of all team members in presentation (10 point maximum deduction)
TEAM PRESENTATION (70 points)
Case Facts—restatement of relevant facts (5 points)
Question(s)—summary of ethical questions (10 points)
References—identification of relevant sections from IEEE code (5 points)
Discussion—complete analysis of case with logic/reasons (20 points)
Organization and Clear Conclusion—(5 points)
Knowledge and Mastery of Content—(5 points)
Communication Effectiveness—delivery and PowerPoint quality (includes terminology, appearance, voice, physical, use of visuals, etc.)
Team Member #1 (10 points)
Team Member #2 (10 points)
ORAL DEFENSE (30 points)
Team Member #1 (15 points)
Team Member #2 (15 points)
TOTAL SCORE (expressed as a percentage of 100)

NAMES OF TEAM MEMBERS

NAME OF JUDGE SIGNATURE OF JUDGE

TEAM RANK (Circle Choice)

First (5 points) Second (4 points) Third (3 points)
Fourth (2 points) Fifth (1 point) Other (No points)

TEAM POINTS

Each judge will rank the teams and award 5 points to first, 4 points to second, 3 points to third, 2 points to fourth, and 1 point to fifth. The judge's points awarded will be tallied and the winners determined by the scores. Ties will be resolved by majority vote of the judges. All questions of eligibility, adherence to rules, etc. will be resolved by majority vote of the judges.

IEEE STUDENT ETHICS COMPETITION JUDGING
FORM (3 MEMBER TEAM OPTION)

CATEGORIES SCORES

DEDUCTIONS

Time Adherence (deduction of 5 points for every 30 seconds outside of time limits) (Timing Lights or Signals at 8, 10, and 12 minute points)
Lack of significant involvement of all team members in presentation (10 point maximum deduction)
TEAM PRESENTATION (70 points)
Case Facts—restatement of relevant facts (5 points)
Question(s)—summary of ethical questions (10 points)
References—identification of relevant sections from IEEE code (5 points)
Discussion—complete analysis of case with logic/reasons (20 points)
Organization and Clear Conclusion—(5 points)
Knowledge and Mastery of Content—(5 points)
Communication Effectiveness—delivery and PowerPoint quality (includes terminology, appearance, voice, physical, use of visuals, etc.)
Team Member #1 with summarizing role (8 points)
Team Member #2 (6 points)
Team Member #3 (6 points)
ORAL DEFENSE (30 points)
Team Member #1 (10 points)
Team Member #2 (10 points)
Team Member #3 (10 points)
TOTAL SCORE /100

NAMES OF TEAM MEMBERS

NAME OF JUDGE SIGNATURE OF JUDGE

TEAM RANK (Circle Choice)

First (5 points) Second (4 points) Third (3 points)
Fourth (2 points) Fifth (1 point) Other (No points)

TEAM POINTS

Each judge will rank the teams and award 5 points to first, 4 points to second, 3 points to third, 2 points to fourth, and 1 point to fifth. The judge's points awarded will be tallied and the winners determined by the scores. Ties will be resolved by majority vote of the judges. All questions of eligibility, adherence to rules, etc. will be resolved by majority vote of the judges.

IEEE STUDENT ETHICS COMPETITION PRESENTATION GUIDELINE

I. **Purpose**: To present and defend an analysis of a situation in professional ethics.

II. **Topic**: A hypothetical case generally dealing with

1. Public Safety and Welfare,
2. Conflict of Interest,
3. Engineering Practice, or
4. Research Ethics. The selected case will have two or more ethical questions or components.

III. **Preparation**:

a. Three hours to analyze a selected case and prepare a PowerPoint presentation (access will be provided to a computer with no internet connection)
b. Collaboration is limited to members of individual teams.
c. Resources are limited to written competition materials. (Internet access, books, etc. are not allowed)
d. All teams will receive the same case.
e. Teams will not be allowed to collaborate, practice, modify presentation, etc. after the CDs are collected. Teams may observe other presentations after their presentation.

IV. **Requirements**:

a. PowerPoint presentation with significant speaking involvement of all team members
b. Presentation Time 8–12 minutes
c. The order of presentation among the teams will be randomly chosen.
d. Required Components (see example case studies)
    Case Facts—restatement of relevant facts
    Question(s)—summary of ethical questions
    References—identification of relevant sections from IEEE code
    Discussion—analysis of case. The analysis of the case should be performed using the IEEE Code of Ethics.
    Conclusion—position statement on each of the identified ethical questions and recommendation for action

V. **Oral Defense**

a. The judges will ask questions relating to the selected case and the presented analysis.
b. Each team member must respond to at least one question.
c. Time for the defense period will be approximately 5 minutes.

VI. **Comments**

a. Timing lights or other indications will be provided.

INSTITUTE OF ELECTRICAL AND ELECTRONICS ENGINEERS (IEEE)

CODE OF ETHICS

Approved by the IEEE Board of Directors, February 2006
www.ieee.org/about/whatis/code.html
We, the members of the IEEE, in recognition of the importance of our technologies in affecting the quality of life throughout the world, and in accepting a personal obligation to our profession, its members and the communities we serve, do hereby commit ourselves to the highest ethical and professional conduct and agree:

1. to accept responsibility in making decisions consistent with the safety, health and welfare of the public, and to disclose promptly factors that might endanger the public or the environment;
2. to avoid real or perceived conflicts of interest whenever possible, and to disclose them to affected parties when they do exist;
3. to be honest and realistic in stating claims or estimates based on available data;
4. to reject bribery in all its forms;
5. to improve the understanding of technology, its appropriate application, and potential consequences;
6. to maintain and improve our technical competence and to undertake technological tasks for others only if qualified by training or experience, or after full disclosure of pertinent limitations;
7. to seek, accept, and offer honest criticism of technical work, to acknowledge and correct errors, and to credit properly the contributions of others;
8. to treat fairly all persons regardless of such factors as race, religion, gender, disability, age, or national origin;
9. to avoid injuring others, their property, reputation, or employment by false or malicious action;
10. to assist colleagues and co-workers in their professional development and to support them in following this code of ethics.

IEEE Bylaws I-110.4. Member Discipline and Support—Requests for Support

www.ieee.org/bylaws
The IEEE may offer support to engineers and scientists involved in matters of ethical principle that stem in whole or in part from adherence to the principles embodied in the IEEE Code of Ethics, and that can jeopardize a person's livelihood, can compromise the discharge of the person's professional responsibilities, or that can be detrimental to the interests of IEEE or of the engineering profession. All requests for support containing allegations against persons not members of IEEE or against employers or others, requests for advice, and matters of information considered to be relevant to the ethical principles or ethical conduct supported by IEEE shall be submitted initially to the Ethics and Member Conduct Committee. Requests for support shall not include requests that the Ethics and Member Conduct Committee support a member who is the subject of a complaint as set forth in Bylaw I-110.2. IEEE support of persons requesting intervention or amicus curiae participation in legal proceedings shall be limited to issues of ethical principle.

The Ethics and Member Conduct Committee, following a preliminary investigation of any requests for support received, shall submit a report to the Executive Committee, which shall include findings and recommendations for consideration by the Executive Committee. The Executive Committee may, if it deems it appropriate to do so, appoint an advisory board to assist it in considering such report. On the basis of information available, the Executive Committee may thereafter offer support to the person making the request as appropriate to the circumstances and consistent with Sections 7.9 and 7.10 of the current IEEE Policies. The Executive Committee shall make the final decision as to supporting the person, unless the Executive Committee or the Board of Directors determines that the Board of Directors should make such final decision.

The Board of Directors, or the Executive Committee upon approval of the Board of Directors, may publish findings, opinions, or comments in support of the person and take such further action as may be in the interests of that person, the IEEE, or the engineering profession.

Guidelines for Engineers Dissenting on Ethical Grounds
IEEE Ethics Committee 11/11/96
Used with permission from the IEEE

Introduction
The goal of these guidelines is to provide general advice to engineers, including engineering managers, who find themselves in conflicts with management over matters with ethical implications. Much of this advice is pertinent to more general conflicts within organizations. For example, it is not unusual in technical organizations for there to be hard fought battles regarding purely technical decisions that do not necessarily have any ethical implications—but do have impacts on the probabilities of success of products. The assumption here is that the engineer's objective is to prevent some serious harm, while minimizing career damage.

Many ethics related disputes are caused by attempts to satisfy irreconcilable constraints. For example, suppose it is impossible to test a product adequately in time to meet a delivery date.

Missing the delivery date constitutes a highly visible failure, with clearly defined penalties.

There may be no obvious indication that an important set of tests has been omitted, even if this leads to a substantial increase in the probability of a life threatening system failure. Under such conditions, there is a temptation to meet the deadline by skipping or shortening the tests. Such decisions might or might not be in accordance with company policy. If not, then an engineer or manager objecting on ethical grounds usually has an easier, but usually not easy, problem. The chances of resolving the problem within the organization may be quite good. If the decision is consistent with the views of upper management, then the problem is far more serious for the dissenter. The following guidelines, based on the experiences of many people, are designed to maximize the chances of a favorable outcome for the ethically concerned manager or engineer.

   1. ESTABLISH A CLEAR TECHNICAL FOUNDATION

      One should check out the alleged facts and technical arguments as thoroughly as possible. If feasible, get the advice of colleagues that you respect. Carefully consider counterarguments made by others. A good way to ensure that you understand someone else's position, is to restate it to the satisfaction of that person. At any stage, if convinced that the other person's arguments are valid, do not hesitate to change your position accordingly.

      This does NOT mean that you must be able to validate your position with near mathematical certainty. This is seldom possible in the real world. In most engineering work, we must operate with incomplete information and make reasonable engineering judgments. For example, the engineers in the Challenger case could not PROVE that a launch would lead to a disaster. But, in such a situation it was sufficient to show that the likelihood of failure of the O-ring joints was clearly too great with respect to established safety standards. The burden was on the other side to justify the launch—a burden that was not met.

   2. KEEP YOUR ARGUMENTS ON A HIGH PROFESSIONAL PLANE, AS IMPERSONAL AND OBJECTIVE AS POSSIBLE, AVOIDING EXTRANEOUS ISSUES AND EMOTIONAL OUTBURSTS

      For example, do not mix personal grievances into an argument about whether further testing is necessary for some critical subsystem. Keep calm and avoid impugning the motives of an opponent. (Of course, there might be a situation in which the central issue is that an incompetent person has been given critical responsibilities. In that case, it may be necessary to attack that person's qualifications. But this should be done without malice.) Try to structure the situation so that accepting your position will involve as little embarrassment as possible to those being asked to change a decision. For example, you might be able to allow a manager to take credit for realizing that a course reversal was called for. Avoid overstating your case. Your credibility can be seriously undermined by exaggerated, invalid figures—even on matters not central to the main issue.

If the matter turns into a serious conflict, efforts will be made to portray you as some sort of crackpot. Avoid behavior that could be used to support such an attack. In both written and oral arguments be cool, clear, concise and accurate. At all times behave as a competent, ethical professional.

3. TRY TO CATCH PROBLEMS EARLY, AND KEEP THE ARGUMENT AT THE LOWEST MANAGERIAL LEVEL POSSIBLE

Calling attention to a problem at an early stage makes a satisfactory solution much more likely.

As time goes on, personal commitments to a particular course of action become deeper, and making changes becomes increasingly expensive. It is always less costly to resolve the dispute at the lowest organizational level possible. Move up the chain of command only when it is clear that this is necessary.

4. BEFORE GOING OUT ON A LIMB, MAKE SURE THAT THE ISSUE IS SUFFICIENTLY IMPORTANT

If a situation reaches the point where further protest may be costly, consider whether the stakes are sufficiently high. For example, if the issue involves only financial risks for the employer, then, if managers are acting unreasonably, it is probably not worth risking your career.

5. USE ORGANIZATIONAL DISPUTE RESOLUTION MECHANISMS

Good organizations have procedures, not always formal, for resolving disputes. After having exhausted informal efforts to persuade your manager, then you must consider using these mechanisms. Since this will almost certainly damage relations with your manager, this step should be taken only after a careful review along the lines discussed in guidelines 1 and 2. If you have an ally higher up in the management chain, you might appeal to that person for advice and possibly to intervene as a mediator.

If your organization lacks such a dispute resolution procedure, consider championing the creation of one. This could be invaluable in minimizing future problems.

6. KEEP RECORDS AND COLLECT PAPER

As soon as you realize that you are getting into a situation that may become serious, you should initiate a log, recording, with times and dates, the various steps that you take (e.g., conversations, email messages, etc.) Keep copies of pertinent documents or computer files at home, or in the office of a trusted friend—to guard against the possibility of a sudden discharge and sealing off of your office. But be careful not to violate any laws!

7. RESIGNING

If efforts to resolve the conflict within your organization fail, then a decision must be made as to whether to go further. It should be realized that there will almost certainly be a significant personal cost involved if you proceed. It is very unlikely that you would be able to remain with the organization, unless your job is governmental in nature, protected by civil service regulations or the like. One obvious choice is to resign. The advantages are: (1) This adds credibility to your position—makes it obvious you are a serious person. (2) Arguments that you are being disloyal to your employer are disarmed. (3) Since you are likely to be fired, resigning may look better on your record.

The drawbacks are: (1) Once you are gone, it may be easier for the organization to ignore the issues you raised, as others in the organization may be unwilling to carry on the fight. (2) The right to dissent from within the organization may be one of the points you wish to make. (3) You might thereby lose pension rights, unemployment compensation, and the right to sue for improper discharge. It would be wise to consult an attorney before making this decision.

8. ANONYMITY

In some situations, engineers may see serious harm being done within their organizations, but recognize that publicly calling attention to it may cause personal repercussions beyond what they are willing to accept. It might be possible to report the problem anonymously to others who may be able to take action, e.g. a regulatory agency, a senator, or a reporter. One problem is that an anonymous report may not be taken seriously. Providing enough information to make the report more credible may make it easy for the organization to identify its source. Being exposed as a purveyor of an anonymous report may be even more damaging to the engineer than the effect of openly making the report would have been. A reporter might distort the facts to make the case more "newsworthy." Nevertheless, this route is sometimes taken in preference to doing nothing at all. In such a case, one should be particularly careful not to malign any individuals and one should convey in the message means for verifying the claims made.

9. OUTSIDE RESOURCES

If, after the failure of internal conflict resolution measures, you decide to take the matter outside the organization, whether or not you decide to resign, care must be taken in choosing where to go. In many cases, an obvious place is a cognizant regulatory or law enforcement agency. Other possibilities include Members of Congress (from one's own district or state, or the head of a relevant committee), state or local government officials or legislators, or public interest organizations. Of course some combination of these might be chosen. Although it is usually not a good idea to take one's case directly to the news media, they generally become involved eventually, usually in reporting actions taken by whatever entity the engineer has contacted. One must take special pains to be accurate and clear when dealing with journalists so as to minimize sensationalism and distortion. When given a choice among media organizations, choose those with reputations for fairness and accuracy.

Guidance and support from one's professional society is potentially a powerful aid to engineers in the kinds of situations considered here. Efforts are under way within the IEEE to improve the machinery for providing such support. Regardless of whether one obtains professional society support, it would be useful to engage an attorney to advise on the many legal aspects of the situation. But in considering their advice, one must take into account the tendency of attorneys to discourage any acts accompanied by legal risks.

Conclusions

Following the above guidelines will often lead to a satisfactory resolution of the problem at issue. However the situations treated here are inherently difficult. No tactics or strategies can guarantee a happy outcome. It takes courage and dedication to risk one's job, or even career, on ethical grounds. Many who have done so have suffered severe consequences, at least in the short run. It is not uncommon for the engineer's position to prevail, while the engineer is fired.

Sometimes, the immediate battle is lost, but the result of the battle is that fewer such bad decisions are made in the future. Finally, one should also consider the personal consequences of yielding on a matter of principle when the result may be severe harm to others. This can cause a lifelong loss of self esteem.

INSTITUTE OF ELECTRICAL AND ELECTRONICS ENGINEERS (IEEE)

IEEE STUDENT ETHICS COMPETITION RESOURCES

Glossary of Selected Terms

Many expert resources exist, see for example: The Online Ethics Center for Engineering and Science at Case Western Reserve University glossary at: http://www.onlineethics.org/glossary.html [1].

Complainant

As used in an IEEE ethics investigation, anyone who files an official complaint concerning the action or actions of another person who is a member of the IEEE. In general, any person who provides witness to a wrongdoing or problem.

Confidential

Information that must have its access limited to only those who have a need-to-know is considered confidential. Confidential information may be personal, financial, trade-secret technical, or other information that could cause unnecessary embarrassment or negative financial impact if disclosed beyond the control group. Confidential information that must be shared with another person must be shared only when they understand its confidential nature and agree to handle the information accordingly.

Conflict of interest

When a person or group is involved in a decision making process on behalf of others and they have or appear to have a personal or financial interest in the outcome, they could be considered to have a conflict of interest, in making said decision(s). The issue of conflict of interest may be mitigated by full disclosure of any such conflict(s) to the affected group, who may determine its interests are best served by allowing the person or group to retain the decision making responsibility.

Fabrication

Information concerning or gained by any event that is untrue or unfounded by fact or other witness may be considered a fabrication. All information concerning ethical behavior must be founded on physical facts and/or on an oath of truth when provided by an eye witness. As used herein, fabrications does not refer to the assembly of a product.

Falsification

Testimony or other official information provided to facilitate an ethics investigation that is not true and accurate, by design or accident, is a falsification. Any act by an individual or group of individuals that represents or portrays as fact information that is not known to be true and accurate may be perpetrating an act of falsification.

Negligence

An act that in which a responsibility is not discharged because of lack of prudent discharge of one's responsibilities and authorities, whether through ignorance or by intention, is an act of negligence.

Plagiarism

IEEE defines plagiarism as the reuse of someone else's prior ideas, processes, results or words without explicitly acknowledging the original author and source. Plagiarism in any IEEE publication is unacceptable and considered a serious breach of professional conduct, with potentially severe legal consequences.

Profession

A service or action offered by an individual for pay that requires a high degree of competence in a complex field normally established through advanced education and extensive experience.

Professional Engineer

An engineer who is certified by license by an authority, such an authorized agency of a state government in the United States, as having met a set of qualifying requirements as demonstrated by education, experience and satisfactory performance on a written examination.

Responsibility, Official

What your job requirements are or what your corporation says you are supposed to do as it relates to them. The set of standards that are required by a particular assignment i.e. it is the responsibility of a U.S. Ambassador to represent the U.S.

Responsibility, Professional

What is expected of me as defined by my profession. For example, a nurse has the professional responsibility to help someone who may be in need of services only they can render i.e. CPR.

SAFETY

Making sure your working environment and work practices ensure that nobody (company employees, contract employees, and visitors) is injured while performing any type of task.

WHISTLE-BLOWER

Someone who exposes, to those outside the organization, any type of unsafe, unethical, or unlawful activities going on within an organization. The person who releases the information does so regardless of the ramifications (positive and/or negative) of their actions.

   For further information on these terms see the Online Ethics Center's Glossary of Terms
   http://onlineethics.org/glossary.html

IEEE STUDENT ETHICS COMPETITION CASE AND ANALYSIS FORMAT

CASE CRITERIA

The competition cases should meet the following criteria.

- Each case must contain multiple ethical questions that student can identify.
- Cases should not depend on specialized technical knowledge to make a determination.
- The ethical issues should not be intentionally vague, i.e. the results of the analysis should not require significant assumptions.
- Preferably, the anticipated analysis would not result in findings that all of the ethical questions have a negative or a positive result.

   The recommended length of the case descriptions should not exceed one page. Also, the cases must contain all needed information to make a determination as no outside references are allowed in the competition.

ANALYSIS FORMAT

CASE FACTS: Restatement of Relevant Facts
QUESTIONS: Summary of Ethical Questions
REFERENCES: Relevant Sections of the IEEE Code of Ethics
DISCUSSION: Analysis of Case. Any assumptions or special perspectives must be explicitly stated
CONCLUSION: Position Statement on Each Identified Ethical Question

IEEE STUDENT ETHICS COMPETITION SAMPLE CASES

CASE DESCRIPTION

A graduating engineering student is interviewing with several companies for an entry-level position. He receives an attractive offer from company A. Since the job market is very competitive, he feels it unlikely that another company will give an offer, much less an attractive one. The student accepts company A's offer and returns a signed letter of acceptance which documents the terms of the position. However, he receives an offer from company B one week afterwards. This new opportunity has a higher salary, more benefits, better advancement prospects, and a more desirable location. It is significantly better in all respects. Since only one week has past since the first acceptance was returned and the new opportunity is clearly in his professional and financial interests, he tells company A that he has changed his mind and accepts the offer of company B. Company A does not express any criticism of the student's actions. Did the student act unethically?

ETHICAL QUESTIONS TO BE IDENTIFIED BY STUDENTS

Is the student ethically bound to honor the signed letter of acceptance with company A?
Has company A been harmed by the student's action?

EXAMPLE ANALYSIS AND PRESENTATION OUTLINE FOR SAMPLE CASE

CASE FACTS: RESTATEMENT OF RELEVANT FACTS

The student formally accepted a position in which all significant terms of employment were specified. The student backed out of this agreement to accept a second, more desirable offer.
QUESTIONS: Summary of ethical questions
Is the student ethically bound to honor the signed letter of acceptance with company A?
Has company A been harmed by the student's action?
REFERENCES: Relevant sections of the IEEE code
Preamble: … to the highest ethical and professional conduct. …
9. to avoid injuring others, their property, reputation, or employment by false or malicious action.
DISCUSSION: Analysis of case
The student did not act in good faith with the highest standards of conduct. He made a commitment to company A, which presumably was acted on by the company. The professional and financial self-interest of the student was no excuse. While the company probably has a legal case against the student, it has little to gain by pursuing litigation. Despite the short (one week) length of time, company A invested time and resources in processing employment paperwork and may have turned away other applicants for the position. The student thereby injured both the company and other potential employees.
CONCLUSION: Position statement on the identified ethical questions The student was ethically bound to honor the first acceptance. He had formally completed an agreement. Company A gave no cause for a change in this agreement. Company A potentially suffered harm in that other applicants for the position were turned away or found other employment.

IEEE STUDENT ETHICS COMPETITION RESOURCES

Internet Resources:
Materials Available from the Online Ethics Center for Engineering and Science at Case
Western Reserve University at http://www.onlineethics.org/index.html
"Moral Exemplars" at http://www.onlineethics.org/moral/index.html
"Roger Boisoly on the *Challenger* Disaster"
"William LeMessurier and the Fifty-Nine-Story Crisis:
A Lesson in Professional Behavior"
"Professional Ethics in Engineering Practice: Discussion Cases" at http://www.onlineethics
.org/cases/nspe/index.html#safety
(Note that these case studies use the format expected in the case analysis.)

"Public Safety and Welfare" 15 Discussion Cases
"Conflicting Interests and Conflict of Interest" 12 Discussion Cases
"Ethical Engineering/Fair Trade" 23 Discussion Cases
"Research Ethics" 4 Discussion Cases

BIOMEDICAL ENGINEERING SOCIETY

CODE OF ETHICS

Approved February 2004
Biomedical engineering is a learned profession that combines expertise and responsibilities in engineering, science, technology, and medicine. Since public health and welfare are paramount considerations in each of these areas, biomedical engineers must uphold those principles of ethical conduct embodied in this Code in professional practice, research, patient care, and training. This Code reflects voluntary standards of professional and personal practice recommended for biomedical engineers.

BIOMEDICAL ENGINEERING PROFESSIONAL OBLIGATIONS

Biomedical engineers in the fulfillment of their professional engineering duties shall:

1. Use their knowledge, skills, and abilities to enhance the safety, health, and welfare of the public.
2. Strive by action, example, and influence to increase the competence, prestige, and honor of the biomedical engineering profession.

BIOMEDICAL ENGINEERING HEALTH CARE OBLIGATIONS

Biomedical engineers involved in health care activities shall:

1. Regard responsibility toward and rights of patients, including those of confidentiality and privacy, as their primary concern.
2. Consider the larger consequences of their work in regard to cost, availability, and delivery of health care.

BIOMEDICAL ENGINEERING RESEARCH OBLIGATIONS

Biomedical engineers involved in research shall:

1. Comply fully with legal, ethical, institutional, governmental, and other applicable research guidelines, respecting the rights of and exercising the responsibilities to colleagues, human and animal subjects, and the scientific and general public.
2. Publish and/or present properly credited results of research accurately and clearly.

BIOMEDICAL ENGINEERING TRAINING OBLIGATIONS

Biomedical engineers entrusted with the responsibilities of training others shall:

1. Honor the responsibility not only to train biomedical engineering students in proper professional conduct in performing research and publishing results, but also to model such conduct before them.
2. Keep training methods and content free from inappropriate influence from special interests.

COMPUTER SOCIETY CONNECTION

Editor: Mary-Louise G. Piner, Computer,
10662 Los Vaqueros Circle, PO Box 3014,
Los Alamitos, CA 90720-1314; mpiner@computer.org

COMPUTER SOCIETY AND ACM

SOFTWARE ENGINEERING CODE OF ETHICS

Don Gotterbarn, Keith Miller, Simon Rogerson
Executive Committee, IEEE-CS/ACM Joint Task Force
on Software Engineering Ethics and Professional Practices

Software engineering has evolved over the past several years from an activity of computer engineering to a discipline in its own right. With an eye toward formalizing the field, the IEEE Computer Society has engaged in several activities to advance the professionalism of software engineering, such as establishing certification requirements for software developers. To complement this work, a joint task force of the Computer Society and the ACM has recently established another linchpin of professionalism for software engineering: a code of ethics. After an extensive review process, version 5.2 of the Software Engineering Code of Ethics and Professional Practice, recommended last year by the IEEECS/ACM Joint Task Force on Software Engineering Ethics and Professional Practices, was adopted by both the IEEE Computer Society and the ACM.

PURPOSE

The Software Engineering Code of Ethics and Professional Practice, intended as a standard for teaching and practicing software engineering, documents the ethical and professional obligations of software engineers. The code should instruct practitioners about the standards society expects them to meet, about what their peers strive for, and about what to expect of one another. In addition, the code should inform the public about the responsibilities that are important to the profession.

Adopted by the Computer Society and the ACM—two leading international computing societies—the code of ethics is intended as a guide for members of the evolving software engineering profession.

The code was developed by a multinational task force with additional input from other professionals from industry, government posts, military installations, and educational professions.

CHANGES TO THE CODE

Major revisions were made between version 3.0—widely distributed through *Computer* (Don Gotterbarn, Keith Miller, and Simon Rogerson, "Software Engineering Code of Ethics, Version 3.0," November 1997, pp. 88–92) and *Communications of the ACM*—and version 5.2, the recently approved version. The preamble was significantly revised to include specific standards that can help professionals make ethical decisions. To facilitate a quick review of the principles, a shortened version of the code was added to the front of the full version. This shortened version is not intended to be a standalone abbreviated code. The details of the full version are necessary to provide clear guidance for the practical application of these ethical principles.

In addition to these changes, the eight principles were reordered to reflect the order in which software professionals should consider their ethical obligations: Version 3.0's first principle concerned the product, while version 5.2 begins with the public. The primacy of well-being and quality of life of the public in all decisions related to software engineering is emphasized throughout the code. This obligation is the final arbiter in all decisions: "In all these judgements concern for the health, safety and welfare of the public shall be held paramount."

About the Joint Task Force

This Code of Ethics was developed by the IEEE-CS/ACM Joint Task Force on Software Engineering Ethics and Professional Practices. Members are:

Executive Committee

Donald Gotterbarn (Chair)
Keith Miller
Simon Rogerson

Members

Steve Barber
Peter Barnes
Ilene Burnstein
Michael Davis
Amr El-Kadi
N. Ben Fairweather
Milton Fulghum
N. Jayaram
Tom Jewett
Mark Kanko
Ernie Kallman
Duncan Langford
Joyce Currie Little
Ed Mechler
Manuel J. Norman
Douglas Phillips
Peter Ron Prinzivalli
Patrick Sullivan
John Weckert
Vivian Weil
S. Weisband
Laurie Honour Werth

ENDNOTES

1. AAES Member Societies are urged to make reference here to the appropriate code of conduct to which their members will be bound.
   Approved by AAES Board of Governers 12/13/84
2. The Society's Code of Ethics was adopted on September 2, 1914 and was most recently amended on July 23, 2006. Pursuant to the Society's Bylaws, it is the duty of every Society member to report promptly to the Committee on Professional Conduct any observed violation of the Code of Ethics.
3. In April 1975, the ASCE Board of Direction adopted the fundamental principles of the Code of Ethics of Engineers as accepted by the Accreditation Board for Engineering and Technology, Inc. (ABET).
4. In October 2009, the ASCE Board of Direction adopted the following definition of Sustainable Development: "Sustainable Development is the process of applying natural, human, and economic resources to enhance the safety, welfare, and quality of life for all of the society while maintaining the availability of the remaining natural resources."

REFERENCE

1. Case Western Reserve University, *Online Ethics Center for Engineering and Science*, (2003) Available WWW: http://www.onlineethics.org/index.html.

FURTHER READING

1. Institute of Electrical and Electronics Engineers (IEEE), *IEEE Ethics and Member Conduct Committee*, (2003) Available at www.ieee.org/ethics. http://www.ieee.org/organizations/committee/emcc/index.html.
2. National Society for Professional Engineers (NSPE), *Engineering Licensure, Ethics, and the Law*, (2003), Available WWW: http://www.nspe.org/pf-home.asp.
3. Murdough Center for Engineering Professionalism, *National Institute for Engineering Ethics*, (2003) Available WWW: http://www.niee.org/.
4. Caroline Whitbeck, *Ethics in Engineering Practice and Research*, (Cambridge University Press, New York, NY, 1998).
5. Institute for Global Ethics, Ethics Newsline TM. Available WWW: http://globalethics.org (free weekly subscription).

Appendix III: Suggestions for Further Reading

ENGINEERING CRUSADERS

Pacey, Arnold. Engineering, the heroic art, pp. xv–xxii from Ronald Turner, and Steven L. Goulden, eds., *Great Engineers and Pioneers in Technology* (vol. 1). New York: St. Martin's Press (1981).

Petroski, Henry. Images of an engineer, pp. 300–303 from *American Scientist* 79 (1991).

ETHICAL ENGINEERING

Baase, Sara. *A Gift of Fire: Social, Legal, and Ethical Issues for Computing Technology*, 4th Edition. Upper Saddle River, NJ: Pearson/Prentice Hall (2014).

Christensen, Steen Hyldgaard, Bernard Delahouse, and Martin Meganck, eds. *Engineering in Context*: Aarhus: Academica (2009).

Gabbay, Dov M., W. M. Anthonie Meijers, John Woods, and Paul Thagard, eds. *Philosophy of Technology and Engineering Sciences*, Amsterdam: Elsevier (2009).

Goldberg, Rube. Rube Goldberg vs. the Machine Age. pp. 20–21 from Clark Kinnaird, ed., *Rube Goldberg vs. the Machine Age: A Retrospective Exhibition of His Work with Memoirs and Annotations*. New York: Hastings House Publications (1968).

Herkert, Joseph R., ed. *Social, Ethical and Policy Implications of Engineering: Selected Readings*. New York: Wiley/IEEE Press (2000).

Layton, Edwin T. Jr. Engineering ethics and the public interest: A historical view, pp. 26–29 from *American Society of Mechanical Engineers* 76-WA/TS-9 (Technical Paper, 1976).

Nader, Ralph. The engineer's professional role: universities, corporations, and professional societies, pp. 450–454 from *Engineering Education* (Feb. 1967).

Shrader-Frechette, Kristin. Engineering ethics, p. 10 from *IEEE Spectrum* (Jun. 1984).

Vesilind, Aarne P. Vestal virgins and engineering ethics, pp. 92–101 from *Ethics and the Environment* 7:1 (Spring 2002).

Wood, Nicholas and Nael Barakat, Inclusion of bio-engineering into existing codes of ethics, from *American Society of Engineering Education (ASEE) 2009 Spring Conference Proceedings*, Grand Rapids, MI (2009).

LETTERS TO THE EDITOR

Aldridge, Bob. The forging of an engineer's conscience, pp. 2–5 from *Spark!*, 3:2 (1973).

Anonymous. Charles Proteus Steinmetz (father of electrical engineering and social activist), from *Spark!*, 1:1 (1971).

Baker, George Pierce. Control: A pageant for engineering progress, pp. 59–60 from George Pierce Baker, *Control: A Pageant for Engineering Progress*. New York: The American Society of Mechanical Engineers (1930).

Crean, Daniel J. Engineers still needed, but demands have changed, pp. B5 and B7 from *Los Angeles Times* (Washington Edition) (Jul. 21, 1995).

Newell, David. The dream builders, pp. 45–48 from *Bostonian* (Jul. 1984).

Sims, Calvin. From inner-city L.A. to Yale engineering, p. 1232 from *Science* 258 (Nov. 13, 1992).

Various authors, Nuclear reactions, pp. 44–49 from *Mechanical Engineering* (Apr. 1976).

POLITICS AND ENGINEERING

Brody, Herb. The political pleasures of engineering: An interview with John Sununu, pp. 22–28 from *Technology Review* (Aug./Sep. 1992).

ENGINEERING EDUCATION

Bordogna, Joseph, Eli Fromm, and Edward W. Ernst. Engineering education: Innovation through integration, pp. 3–8 from *Journal of Engineering Education*, 82:1 (1993).

Christensen, Steen Hyldgaard, Christelle Didier, Andrew Jamison, Martin Meganck, Carl Mitcham, and Byron Newberry, eds. *International Perspectives on Engineering Education: Engineering Education and Practice in Context, Volume 1*. London: Springer (2015).

Christensen, Steen Hyldgaard, Christelle Didier, Andrew Jamison, Martin Meganck, Carl Mitcham, and Byron Newberry, eds. *Engineering Identities, Epistemologies and Values: Engineering Education and Practice in Context, Volume 2*. London: Springer (2015).

Downey, Gary and Juan C. Lucena. National identities in multinational worlds: Engineers and "engineering culture," pp. 252–260 from *International Journal of Continuous Engineering Education and Lifelong Learning* 15 (2005).

Grasso, Domenico, and Melody Brown Burkins, eds. *Holistic Engineering Education: Beyond Technology*. New York: Springer (2010).

Regen, Chester. Engineers are educated in an "ethical vacuum," p. 4 from *Daily Californian* (Nov. 13, 1988).

Schwartz, John. Re-engineering engineering, from *New York Times* (Sep. 30, 2007).

Segal, Howard P. The third culture: C.P. Snow revisited, pp. 29–32 from *IEEE Technology and Society Magazine*, 15:2 (1996).

PROFESSIONAL RESPONSIBILITY

Boland, Richard J. Jr. Organizational control, organizational power and professional responsibility, pp. 15–25 from *Business & Professional Ethics Journal* 2:1 (Fall 1982).

Corporate Crime Reporter. S. David Freeman's twenty-five year plan to phase out coal, oil, and nuclear, from *Corporate Crime Reporter* 26:9 (Feb. 2012).

Leydens, Jon A., ed. *Sociotechnical Communication in Engineering*. London: Routledge Press (2014).

Morgan Claypool Publishers. *Synthesis Lectures on Engineers, Technology and Society* (various volumes): http://www.morganclaypool.com/toc/ets/5/1. San Rafael, CA: Morgan Claypool Publishers (2006–2014).

ENGINEERING WORKPLACE

Carmichael, Douglas. Do EE careers suffer from poor management?, pp. 50–54 from *IEEE Spectrum* (Jul. 1984).

De George, Richard T. Ethical responsibilities of engineers in large organizations: The Pinto case, pp. 1–14 from *Business & Professional Ethics Journal* 1:1 (Fall 1981).

Rowe, Jonathan. Why the engineer left the shop floor, from *The Washington Monthly* (Jun. 1984).

ENGINEERING FAILURES

Bernstein, Aron M., and Mark Kuchment, Trading disasters, p. 8 from *IEEE Spectrum* (Apr. 1987).

Billington, David P. One bridge does not fit all, p. A23 from *New York Times* (Aug. 18, 2007).

Chang, Chris. BART: How safe is it?, pp. 9–13 from *California Engineer* (Oct. 1985).

Curd, Martin, and Larry May. Professional responsibility for harmful actions, pp. 11–21 from Martin Curd and Larry May, *Professional Responsibility for Harmful Actions*. Dubuque, IA: Kendall/Hunt (1984).

Gladwell, Malcolm. Blowup, pp. 32–36 from *The New Yorker* (Jan. 22, 1996).

Pooley, Eric. Nuclear warriors, pp. 46–54 from *Time* (Mar. 4, 1996).

Sowers, George F. The human factor in failure, pp. 72–73 from *Civil Engineering—ASCE* 61: 3 (Mar. 1991).

Whitebeck, Caroline. Knowledge, foresight, and the responsibility for safety, pp. 111–116 from Caroline Whitebeck, *Ethics in Engineering Practice and Research*. Cambridge, UK: Cambridge University Press (1998).

STANDARDS, REGULATIONS, AND LAW

Feliu, Alfred G. The role of law in protecting scientific and technical dissent, pp. 3–9 from *IEEE Technology and Society Magazine* (Jun. 1985).

Subcommittee on Antitrust and Monopoly of the Committee on the Judiciary of the United States of America, *Questions and Answers about Trade Product Standards: A Primer for Consumers*, pp. 1–9. Washington, DC: US Government Publishing Office (1977).

Surowiecki, James. Turn of the century, pp. 84–89 from *Wired* (Jan. 2002).

DESIGN FOR HUMAN POPULATIONS

Catalano, George. Engineering design in a morally deep world, pp. 43–54 from George Catalano, *Engineering, Poverty, and the Earth*. San Rafael, CA: Morgan & Claypool (2007).

Grasso, Domenico, Melody Brown Burkins, Joseph Helble, and David Martinelli. Dispelling the myths of holistic engineering, pp. 27–29 from *PE Magazine* (Aug./Sep. 2008).

Hotchkiss, Ralf D. Ground swell on wheels: Appropriate technology could bring cheap, sturdy wheelchairs to 20 million disabled people, pp. 14–18 from *The Sciences* (Jul./Aug. 1993).

Johnson, Ann. ABS and risk compensation, pp. 157–165 from Ann Johnson, *Hitting the Breaks: Engineering Design and the Production of Knowledge*. Durham, NC: Duke University Press (2009).

Polak, Paul. Design for the other ninety percent, pp. 82–104 (excerpts) from Paul Polak, *Out of Poverty: What Works When Traditional Methods Fail*. San Francisco, CA: Berrett-Koehler Publishers (2008).

Schaumburg, Frank D. Engineering—"As if people mattered," pp. 56–58 from *Civil Engineering—ASCE* (Oct. 1981).

WHISTLEBLOWING

Nader, Ralph. The engineers, pp. 170–209 from Ralph Nader, *Unsafe at Any Speed*. New York: Grossman Publishers (1965).

SUSTAINABLE ENGINEERING

Murcott, Susan. Engineering education for sustainability: Reflections on "the greening of engineers" (A. Ansari), pp. 137–140 from *Science and Engineering Ethics*, 7:1 (2001).

STUDENT ENGINEERS

Bierman, David. A student's viewpoint, p. 3 from *Reconstruct* 2 (2009).

Celis, William 3rd. Students discover the soul of engineering by serving the disabled, from *New York Times* (Feb. 24, 1993).

Masters, Brooke A. Redesigning first-year engineering: Experiment lets students do more, p. A4 from *Washington Post* (Feb. 16, 1993).

Siedsma, Andrea. Engineering students help San Diego region secure $154 million in solar bonds, from *This Week @ UCSD* (Dec. 13, 2010).

Terhune, Linda, and William Meiners. A cooling solution: An international team of students offers up a special delivery on the healthcare front, pp. 26–27 from *Purdue's Engineering Edge* (Dec. 2008).

Wald, Matthew L. At Clarkson U., students are racing with the sun, p. B8 from *New York Times* (May 17, 1995).

GENDER AND RACE IN ENGINEERING

Claris, Lionel M. and Donna M. Riley. Learning to think critically in and about engineering: A liberative perspective, from *ASEE 2007 Conference Proceedings*, Honolulu, HI (2007).

Faulkner, Wendy. Gender dynamics and engineering—How to attract and retain women in engineering, pp. 196–199 from *Engineering: Issues, Challenges and Opportunities for Development*. Paris: United Nations Educational, Scientific and Cultural Organization (2010).

Frize, Monique. Women in engineering: The next steps, pp. 199–200 from *Engineering: Issues, Challenges and Opportunities for Development*. Paris: United Nations Educational, Scientific and Cultural Organization (2010).

ENGINEERS AND SOCIAL RESPONSIBILITY

Blue, Ethan, Michael Levine, and Dean Nieusma. *Engineering and War: Militarism, Ethics, Institutions, Alternatives*. San Rafael, CA: Morgan & Claypool (2013).

Catalano, George D. *Engineering Ethics: Peace, Justice, and the Earth*, 2nd Edition. San Rafael, CA: Morgan & Claypool (2014).

Douglas, David, Greg Papadopoulos, and John Boutelle. *The Citizen Engineer: A Handbook for Socially Responsible Engineering*. Upper Saddle River, NJ: Prentice Hall (2010).

Johnson, Deborah G. The social/professional responsibility of engineers, pp. 106–114 from *Annals of the New York Academy of Sciences* 577 (1992).

Lass, Kelly. *Engineering and Social Justice Bibliography*: http://www.onlineethics.org/Resources /Bibliographies/SocialJustice.aspx. Accessed July 15, 2018. Washington, DC: Online Ethics Center, National Academy of Engineering (Jul. 19, 2010).

Lubar, Steven. Technological intuition, pp. 1504–1505 from *Science* 258 (Nov. 27, 1992).

Ottinger, Gwen. *Refining Expertise: How Responsible Engineers Subvert Environmental Justice Challenges*. New York: New York University Press, 2013.

Vallero, Daniel A., and P. Aarne Vesilind. *Socially Responsible Engineering: Justice in Risk Management*. Hoboken, NJ: John Wiley & Sons, 2007.

Appendix IV: List of Relevant Television Series and Documentaries for Engineers

1. ***Modern Marvels: Engineering Disasters***, released in DVD in May 2008.
   This is a television series.
   Director: Various
   Producer: The History Channel

In 18 thrilling episodes, History Channel dramatizes the world's most notorious engineering failures, using state-of-the-art special effects to investigate not only what went wrong, but what can be learned from these spectacular disasters. The catastrophes recounted led to the creation of Superfund sites for the cleanup of toxic waste; improved safety standards for cars, planes, and ships; and new techniques for building roads, bridges, and buildings that will withstand nature's most destructive forces.

2. ***The Invisible Machine***, 2007.
   This is a documentary.
   Directors: Barbara Doran and Jon Whalen
   Producers: Morag Productions

On a calm Sunday morning in 1978 residents of Bell Island, Newfoundland hear an odd, high-pitched hum, immediately followed by a sudden and terrifying blast resounding for hundreds of miles. Outbuildings are destroyed, livestock electrocuted, televisions explode and power lines vaporize. *The Invisible Machine* unravels the mystery of the Bell Island boom and in doing so takes a chilling look at the US military's experimentation with electromagnetic pulse (EMP) weapons.

3. "**Upgrade Me**," 2009.
   This is a television episode.
   Producer: Jeremy Monblat
   Executive Producer: Dominic Crossley-Holland
   Presenter: Simon Armitage
   "Upgrade Me" was part of BBC's series called *Electric Revolution*.
   See http://www.bbc.co.uk/programmes/b00n1hwj.

Poet and gadget lover Simon Armitage explores people's obsession with upgrading to the latest technological gadgetry.

Upgrade culture drives millions to purchase the latest phones, flatscreen TVs, laptops and MP3 players. But is it design, functionality, fashion or friends that makes people covet the upgrade, and how far does the choice of gadgets define identity? Simon journeys across Britain and to South Korea in search of answers.

4. *e2 design*, debuted on November 10, 2009 (Sundance Channel).
    This is a television series.
    Director: Tad Fettig
    Series Producer: Elizabeth Westrate
    Executive Producers: Karena Albers and Tad Fettig

Narrated by Brad Pitt, "e2 design" looks at the burgeoning sustainability movement among architects and designers worldwide.

    (Episode 1: "The Green Apple"; Episode 2: "Green for All"; Episode 3: "The Green Machine"; Episode 4: "From Gray to Green"; "China: From Red to Green?"; Episode 6: "Deeper Shades of Green")

5. *Water for Mulobere*, 2010.
    This is a documentary.
    Producers: Beth Anderson, the University of Minnesota's Institute on the Environment, and the Initiative for Renewable Energy and the Environment (IREE)
Watch a trailer at http://environment.umn.edu/multimedia/video_waterformulobere_trailer.html.

During the summer of 2009, a team of up-and-coming engineers helped bring clean drinking water to an African village. As part of an Engineers without Borders project, the University of Minnesota student group installed a solar-powered water supply system for an entire school and its surrounding community in Mulobere, Uganda.

6. *The Green House: Design it. Build it. Live it*, 2010.
    This is a documentary.
    Director: Jason Scadron
    Producers: Liv Violette and Jason Scadron

This illuminating documentary chronicles the building of the first carbon-neutral house and the designing of the first green show house in the Washington, D.C. area. The building of the house in McLean, Virginia was captured from start to finish, from the monumental groundbreaking to the exquisitely furnished show home decorated by eco-conscious designers. Audiences are placed in the middle of the action and behind the scenes, receiving firsthand knowledge of the engineering and technology that drives the house and the principles and methods of designing eco-friendly spaces. Environmentalist Philippe Cousteau Jr. appears in the film.

7. *Here Comes the Sun!*, 2008, filmed for the Dutch television program VPRO Tegenlicht (VPRO Backlight).
    This is a documentary.
    Director: Rob van Hattum
    Producer: Judith van den Berg
    Editors-in-Chief: Jos de Putter and Doke Romeijn
    Research: Gijs Meyer Swantee
    See http://www.youtube.com/watch?v=mLHBFyfvK8A&eurl=http://vodpod.com/watch/1158939-here-comes-the-sun-a-documentary-on-solar-energy&feature=player_embedded.

Even without major breakthroughs we know how to catch the sun's energy. A piece of desert the size of France is enough to fulfill all our current energy needs. So with all the global warming, the political oil dependency, the energy security problem, why don't we convert our oil-society rapidly into a Helio-society? In fact, we are doing so; the only thing is you do not know it … yet. The documentary *Here Comes the Sun!* will show that Solar Stocks are soaring, the desert is our new land of glory, algae in the garden will fuel your car and your roof is a new marketplace for Solar real estate managers.

Oil companies say that it's not efficient, too expensive and that it is not going to happen soon. But after having seen *Here Comes the Sun!* you will know better.

8. *Taken for a Ride*, 1996.
   This is a documentary.
   Director: Jim Klein
   Producer: Independent Television Service (ITVS)
   Narrators: Renee Montagne, Jim Klein, and Bradford Snell

Why Does America Have the Worst Public Transit in the Industrialized World, and the Most Freeways? *Taken for a Ride* reveals the tragic and little known story of an auto and oil industry campaign, led by General Motors, to buy and dismantle streetcar lines. Across the nation, tracks were torn up, sometimes overnight, and diesel buses placed on city streets. The highway lobby then pushed through Congress a vast network of urban freeways that doubled the cost of the Interstates, fueled suburban development, increased auto dependence, and elicited passionate opposition. Seventeen city freeways were stopped by citizens, who would become the leading edge of a new environmental movement. With investigative journalism, vintage archival footage and candid interviews, *Taken for a Ride* presents a revealing history of our cities in the 20th century that is also a meditation on corporate power, city form, citizen protest and the social and environmental implications of transportation.

9. "**Build Green**," 2007.
   This is a short documentary.
   Director: Paula Salvador
   Producer: CBC, in a series called *The Nature of Things*.
   Executive Producer: Michael Allder
   Editor: Ed Balevicius

In Build Green, Canada's best architects hype their green creations. From retro-fitting a hip, old Montreal housing complex with state-of-the-art sustainable energy systems, to pitching hay for straw-bale houses, to building transportable "mini-homes" with their own small power plant, Build Green takes a close look at the materials and means we'd be foolish not to adopt.

10. *The Sustainable City*, 2003.
   This is a documentary.
   Writers: Nicolas Vidal and Jean Vercoutère
   Producers: Mosaïque Films-VOI Sénart and Thomas Schmitt

Today, the way ecology is being incorporated into architecture has evolved considerably. Sustainable architecture, or green architecture, aims to minimize the negative impact of buildings on the environment by enhancing efficiency and moderating the use of materials, energy, and space. Spewing carbon dioxide, generating masses of waste, and consuming alarming quantities of energy and water, our cities place a heavy burden on both the global environment and the local ecosystem.

11. *Who Killed the Electric Car*, 2006.
   This is a documentary.
   Director: Chris Paine
   Producer: Jessie Deeter
   Executive Producers: Tavin Marin Titus, Richard D. Titus (Plinyminor), Dean Devlin, Kearie Peak, Mark Roskin, and Rachel Olshan (Electric Entertainment)

The film investigates the events leading to the quiet destruction of thousands of new, radically efficient electric vehicles. Through interviews and narrative, the film paints a picture of an industrial culture whose aversion to change and reliance on oil may be deeper than its ability to embrace ready solutions.

12. *Tapped*, 2009.
>   This is a documentary.
>   Director: Stephanie Soechtig
>   Producer: Atlas Films

Is access to clean drinking water a basic human right, or a commodity that should be bought and sold like any other article of commerce? Stephanie Soechtig's debut feature is an unflinching examination of the big business of bottled water.

From the producers of *Who Killed the Electric Car* and *I.O.U.S.A.*, this timely documentary is a behind the scenes look into the unregulated and unseen world of an industry that aims to privatize and sell back the one source that ought never to become a commodity: our water.

13. *Garbage Dreams*, 2009.
>   This is a documentary.
>   Producer, Director, and Cinematographer: Mai Iskander
>   Executive Producer: Tiffany Schauer
>   Coexecutive Producer: Claudia Miller

Garbage Dreams follows three teenage boys born into the trash trade and growing up in the world's largest garbage village, on the outskirts of Cairo. It is the home to 60,000 Zaballeen, Arabic for "garbage people." Far ahead of any modern "Green" initiatives, the Zaballeen survive by recycling 80% of the garbage they collect. When their community is suddenly faced with the globalization of its trade, each of the teenage boys is forced to make choices that will impact his future and the survival of his community.

14. *Flow*, 2008.
>   This is a documentary.
>   Director: Irena Salina
>   Producer: Steven Starr
>   Coproducers: Gill Holland, Yvette Tomlinson, Stephen Nemeth, Caroleen Feeney, and Brent Meikle

Irena Salina's award-winning documentary investigation into what experts label the most important political and environmental issue of the 21st Century—the World Water Crisis.

Salina builds a case against the growing privatization of the world's dwindling fresh water supply with an unflinching focus on politics, pollution, human rights, and the emergence of a domineering world water cartel. Interviews with scientists and activists intelligently reveal the rapidly building crisis, at both the global and human scale, and the film introduces many of the governmental and corporate culprits behind the water grab, while begging the question "can anyone really own water?"

Beyond identifying the problem, *Flow* also gives viewers a look at the people and institutions providing practical solutions to the water crisis and those developing new technologies, which are fast becoming blueprints for a successful global and economic turnaround.

15. *Building the Impossible*, 2002.
>   This is a television series.
>   Director: Various
>   Producer: BBC

Caroline Baillie, Deputy Director of the UK Centre for Materials Education, was one of two scientists challenged by the BBC, in this recent four-part series, to re-create a succession of engineering feats from the past.

In each episode of the series Caroline, a materials scientist and Chris Wise, a structural engineer attempted to reconstruct an object that hasn't been built for hundreds of years, if at all. As far as possible they had to use only the knowledge, technology and materials that were available at the time.

16. *A Road Not Taken*, 2011.
> This is a documentary.
> Directors: Christina Hemauer and Roman Keller
> Producers: Christina Hemauer and Roman Keller

In 1979, Jimmy Carter, in a visionary move, installed solar panels on the roof of the White House. This symbolic installation was taken down in 1986 during the Reagan presidency. In 1991, Unity College, an environmentally-minded centre of learning in Maine acquired the panels and later installed them on their cafeteria roof.

In "A Road Not Taken," Swiss artists Christina Hemauer and Roman Keller travel back in time and, following the route the solar panels took, interview those involved in the decisions regarding these panels as well as those involved in the oil crisis of the time. They also look closely at the way this initial installation presaged our own era.

17. *The Atom and Eve*, 1966.
> This is a film.
> Director: Gene Starbecker
> Writer: Gene Starbecker
> Producer: Green Mountain Post Films

Few GMP releases have triggered more outraged response than this vintage (1966) piece of pro-nuclear propaganda. Fusing heavy inputs of male chauvinism with conspicuous consumption, this film must have been made with the stag executive luncheon in mind.

The Atom and Eve is, in essence, a cinematic marriage between the desire to fill every kitchen with electric appliances, and to then lock women in. The product is worthy of the theater of the absurd. As the 1972 Atomic Energy Commission catalogue puts it: "In a light, pleasant and entertaining manner, we are introduced to Eve as a baby, then as a girl, and finally as a woman (who dances through the film) in parallel to the growing needs of millions of Eves for more and more electricity."

Not surprisingly, *The Atom and Eve* has proven popular with womens' and antinuclear groups, as well as in classes where the nuclear propaganda battle is an issue. Because of its unique place in the history of media, GMP continues to make this film available to interested groups.

18. *Winds of Change*, 2010.
> This is a documentary.
> Executive Producers: Peter Wiesner and IEEE Future Directions Group
> Producer-Directors: Pat Corbitt and Peter Wiesner
> Field Producers: Ric Zivic, Anindita Roy, Anne O'Neil, Bichlien Hoang, Bill Sweet, Daniel Traub, and Erica Wissolik

Targeting a non-technical audience, "Winds of Change" provides a global overview of the technology with an emphasis on emerging technologies and public policy. Locations for this documentary include India, China, Denmark, Germany, France, United Kingdom, as well as the United States. Appearances include technical and public policy experts as well as leaders in government and industry. All program content was peer reviewed. This one-hour documentary includes the following segments: A background piece on the various technologies involving wind energy and its history. A focused treatment of recent developments in wind energy, including technical explanations appropriate for non-technical audiences. A discussion of the economics of wind energy, An examination of the public policy environment concerning the development renewable energy throughout the world, including "smart grid" technology to support local generation. Produced by the IEEE Future Directions Group, 'Winds of Change' is the first program of a series on, Smarter Energy: A Video Series on Energy Development, Distribution, and Consumption. Aimed at a general audience, this series will consist of three one-hour documentaries for broadcast, internet, and DVD release: Winds of Change, Catch the Sun, and Future Grid.

Note: *Winds of Change* can be watched through the IEEE Smart Grid Media Center's web page. Featured videos also include the following:

- Accelerating Photovoltaics
- A Smart Grid for Intelligent Energy Use
- Dean Kamen Takes the Island Off the Grid
- Geothermal Energy in the Military
- Ride with the Teams at the World Solar Challenge
- Life on 150 Watts with a Nano-hydroelectric System
- Wind Power: The Technology
- NREL Wind Technology Center
- The Story of Hoover Dam
- Coal Gasification at the Polk Power Station
- Formula Hybrid
- Tufts University Nerd Girls Build a Solar Car
- Renewable Energy for Refugee Camps
- Charging Ahead: The Case for Plug-In Hybrid Cars
- Technology Discourse: Bio Fuels

19. ***Doing the Right Thing: Social Implications of Technology***, 2006.[1]
    This is a documentary.

This IEEE.tv episode intends to encourage working engineers and students to engage in discourse concerning the social, economic, and environmental impact of technology. Engineers from industry and academia, attending the 2005 Symposium on Technology and Society (ISTAS), discuss ethics and social responsibility.
    Interviews with engineers were conducted to gain insights into moral and ethic questions related to their profession. How should loyalty to one's employer be balanced with the obligation of serving society? How should engineers evaluate the uncertainties concerning the impact of technology? To what extent is communicating with the public a professional responsibility?

    Editor's note: The National Society of Professional Engineers offers a list of engineering ethics videos at http://www.nspe.org/Ethics/EthicsResources/Videos/index.html.

20. ***Wind over Water***, 2004.
    This is a documentary.
    Producer-Director: Ole Tangen Jr.
    Associate Producers: Sarah Eberle and Chetin Chabuk

Wind over Water is a documentary chronicling the debate over the Cape Wind Project, an offshore wind farm proposed for off the southern coast of Cape Cod, MA. With similar facilities spreading throughout Northern Europe, many people were excited at the prospect of the first offshore wind project ever to be proposed for American shores.
    However, since its plans were revealed in November 2001, many residents of the Cape have banded together to stop the project and prevent its developers from turning the waters of Nantucket Sound into what they categorize as an industrial energy complex. With a colorful cast of characters that includes Sen. Edward Kennedy and Walter Cronkite, this story has developed into an intriguing representation of people's attitudes toward land, energy, politics and NIMBY (Not in My Back Yard).
    Supporters of the project maintain that the promise of wind energy is that it can produce clean, renewable power while helping to stem some of the 2.5 billion tons of pollution released into the atmosphere by traditional fossil fuel plants in the US. While this facet of wind energy appears appealing, its greatest liability is that exposed hilltops and shallow offshore waters, areas once immune to development, are now being sited as ideal locations for wind energy facilities.

Wind over Water will attempt to address the question: Is the American public willing to grant the wind industry access to these lands in exchange for clean, renewable energy?

21. ***The Power of the Sun***, 2005.
   This is a documentary.
   Executive Producer: Walter Kohn, University of California, Santa Barbara
   Producers: David Kennard (*The Ascent of Man, Cosmos*) (InCA) and Victoria Simpson
   Writers: John Perlin (*From Space to Earth: The Story of Solar Electricity*) and David Kennard

"The Power of the Sun" consists of two films on a single DVD: "The Power of the Sun—The Science of the Silicon Solar Cell" (S), is a 20-minute animated educational film for 12th grade high school students, or freshman college/university students with interests in physics and/or chemistry, materials science, engineering.

"The Power of the Sun" (G) is a 56-minute film, telling the story of photovoltaics—light; history and science; implementation; and future.

The film is a scientific morality tale: how, starting from the most pure and basic science, through stages of brilliant applied science and engineering, there emerges one of the most promising multibillion dollar technologies to help deal with one of the great challenges of our time: energy. That is, finding economically realistic, clean and safe energy sources to replace diminishing cheap fossil fuels, while energy demands of the developing world continue to grow rapidly.

22. ***The Next Frontier: Engineering the Golden Age of Green***, 2010.
   This is a documentary.
   Writers, Producers, and Directors: Filmsight Productions, Brad Marshland, Morgan Schmidt-Feng, Rivkah Beth Medow, and Chris Brown

The Next Frontier: Engineering the Golden Age of Green focuses on the renewable, clean energy technologies that can improve our future and create significant economic opportunities.

This entertaining one-hour documentary takes the viewer around the world in search of technologies and policies that will address the serious problem of excessive carbon dioxide emissions and our dangerous dependence on foreign oil. It features interviews with some of the top energy and economic experts along with educators and high-level government officials, all striving to develop clean energy solutions and alternatives to burning fossil fuels.

From windmills in Denmark to tidal turbines in Ireland to concentrated solar plants in California, The Next Frontier takes you on a global journey in stunning high definition. The film is enhanced by the educational and entertaining animation of Emmy Award-winner Charlie Canfield.

23. ***Windfall***, 2011.
   This is a documentary.
   Director: Laura Israel
   Producers: Laura Israel and Autumn Tarleton
   Coproducer: Stacey Foster

Wind power … it's sustainable … it burns no fossil fuels … it produces no air pollution. What's more, it cuts down dependency on foreign oil. That's what the people of Meredith, in upstate New York first thought when a wind developer looked to supplement the rural farm town's failing economy with a farm of their own—that of 40 industrial wind turbines. *Windfall*, a beautifully photographed feature length film, documents how this proposal divides Meredith's residents as they fight over the future of their community. Attracted at first to the financial incentives that would seemingly boost their dying economy, a group of townspeople grow increasingly alarmed as they discover the impacts that the 400-foot high windmills slated for Meredith could bring to their community as well as the potential for financial scams. With wind development in the United States growing annually at 39 percent, *Windfall* is an eye-opener that should be required viewing for anyone concerned about the environment and the future of renewable energy.

24. **Bag It**, 2010.
   This is a documentary.
   Producer: Michelle Hill
   Executive Producer: Judith Kohin
   Writer: Michelle Curry Wright
   Producer and Director: Suzan Beraza

Bag It follows "everyman" Jeb Berrier as he tries to make sense of our dependence on plastic bags. Although his quest starts out small, Jeb soon learns that the problem extends past landfills to oceans, rivers and ultimately human health.

The average American uses about 500 plastic bags each year, for about twelve minutes each. This single-use mentality has led to the formation of a floating island of plastic debris in the Pacific Ocean more than twice the size of Texas.

The film explores these issues and identifies how our daily reliance on plastic threatens not only waterways and marine life, but human health, too. Two of the most common plastic additives are endocrine disruptors, which have been shown to link to cancer, diabetes, autism, attention deficit disorder, obesity and infertility.

25. **The Bridge So Far—A Suspense Story**, 2006.
   This is a documentary.
   Producer and Director: David L. Brown
   Writer: Stephen Most

"The Bridge So Far—A Suspense Story" is an entertaining one-hour documentary on the often outrageous and always controversial history and status of the San Francisco-Oakland Bay Bridge. Tragic, frustrating, comical, and historic, this entertaining documentary/news special follows the Bridge from its original construction through the 1989 Loma Prieta earthquake up to the present day. It recounts the progress, delays, setbacks, and politics during the design and construction of a new, safe bridge to re-complete the connection across the Bay between San Francisco and Oakland.

This was much more than a huge design and construction project. It was local, regional, state, and even federal politics; dollars and delays; finances and finger pointing; the U.S. Navy vs. Caltrans; northern vs. southern alignments; skyway vs. suspension bridge, with a bikeway; conceptual changes during construction; and monumental cost increases caused by such far-flung factors as the upcoming Olympics in China.

In other words, it was a doozy of a story, just waiting to be told even while chapters were being added.

26. **Amazing: The Rebuilding of the MacArthur Maze**, 2007.
   This is a television special.
   Producer, Director, and Photographer: David L. Brown
   Assistant Editor/Associate Producer: Michael Bolner

Amazing: The Rebuilding of the MacArthur Maze is a half-hour television special which tells the remarkable story of the fiery collapse and rebuilding (in only 26 days) of a key connector in the Bay Area's MacArthur Maze, where three major freeways meet just east of the San Francisco–Oakland Bay Bridge."

In addition to interviews, *Amazing* includes a wide variety of news and archival footage of the entire 26-day process beginning with the fire that melted and collapsed the structure and an animated depiction of the gasoline tanker truck which overturned and created the fireball. The film captures the vivid impressions of the first responders, the overnight creation of new design plans, and C.C. Myers capturing the winning bid, planning on a large bonus for early completion. Viewers will see how Caltrans and the Myers team tracked down enough steel and worked day and night to rebuild the structure in record time.

27. *A Span in Time: The Dramatic Story of the 2007 Labor Day Weekend Bay Bridge Demolition and Replacement Project*, 2008.

> This is a documentary.
> Producer, Director, and Photographer: David L. Brown
> Videograpers: Steven Baigel, Ken Day, Richard Schatzman, and Hal Sloane
> Editors: Steven Baigel, Tal Skloot, and David L. Brown
> Animator: Charlie Canfield

Span in Time tells the saga of the 2007 Labor Day weekend Bay Bridge construction project, with the now-legendary C.C. Myers as the contractor. During a three-day bridge closure, Myers' and Caltrans' teams demolished and removed a football field-size bridge, rolled in a new pre-constructed replacement span, and finished the amazingly challenging job eleven hours ahead of schedule! The film tells the story from the perspectives of the construction contractor, Caltrans engineers and designers, and two of the reporters who covered the story. Hilarious cartoon animation introduces soon-to-be-legendary TV anchor "Max Tabloid," who reports on the story as it unfolds on the screen.

This half hour special won an Emmy Award for graphic arts and was produced and directed by David L. Brown, with animation by Charlie Canfield, both of whom won Emmy Awards for the "The Bridge So Far," their previous entertainment/documentary on the Bay Bridge.

28. *Modern Marvels: Renewable Energy*, 2008.

> This is a documentary.
> Producer: Actuality Productions, Inc. for the History Channel
> Writer and Producer: Anthony Lacques
> Producer: Bruce Nash

In the young 21st Century, two realizations are dawning on the world's population: we are hopelessly dependent on petroleum, which is only going to get more expensive; and global warming, caused mainly by our burning of fossil fuels, will impact civilization in ways that we're only beginning to grasp. Stepping in to fight both of these massive problems are the rapidly evolving technologies that harness renewable energy. We will see how air, water, earth, and fire are transformed into clean, reliable sources of heat, electricity, and even automobile fuel. We'll take an in-depth look at the most proven and reliable sources: solar, wind, geothermal, biofuels, and tidal power. From the experimental to the tried-and-true, renewable energy sources are overflowing with potential ... just waiting to be exploited on a massive scale. And unlike fossil fuels, they'll always be there.

29. *The Last Mountain: A Fight for Our Future*, 2011.

> This is a documentary.
> Executive Producer: Tim Disney and Tim Roockwood
> Producers: Clara Bingham, Eric Grunebaum, and Bill Haney
> Director: Bill Haney
> Cinematographers: Jerry Risius, Stephen McCarthy, and Tim Hotchner
> Editor: Peter Rhodes
> Coproducer: Laura Longsworth

In the valleys of Appalachia, a battle is being fought over a mountain. It is a battle with severe consequences that affect every American, regardless of their social status, economic background or where they live. It is a battle that has taken many lives and continues to do so the longer it is waged. It is a battle over protecting our health and environment from the destructive power of Big Coal.

The mining and burning of coal is at the epicenter of America's struggle to balance its energy needs with environmental concerns. Nowhere is that concern greater than in Coal River Valley, West Virginia, where a small but passionate group of ordinary citizens are trying to stop Big Coal corporations, like Massey Energy, from continuing the devastating practice of Mountain Top Removal.

30. ***E-Waste Tragedy (The)***, 2014.

      This is a documentary.
      Director: Cosima Dannoritzer
      Producers: Yuzu Productions, Media 3.14, and Arte France

The illegal recycling of electronics is a toxic business on a global scale. Why do three quarters of Europe's electronic waste disappear from the official recycling system?

In the suburbs of Accra, the capital of Ghana, children play at dismantling scrapped electronic equipment, surrounded by foul smelling and toxic fumes, in a sadly-famous, uncontrolled rubbish dump. The film takes this site as its point of departure, and is driven forward by Ghanaian journalist Mike Anane, who is an expert on the environment. He wants to know why his country has become the trash can of the developed countries. This lucid and efficient investigation by Cosima Dannoritzer (*The Light Bulb Conspiracy*) into several European countries, Asia and the U.S., reveals the apparatus of large-scale trafficking, as well as a complex chain of responsibility and collusion. Awards: Golden Award—Prix Italia 2015 (Torino, Italy).

31. ***The Human Face of Big Data***, 2014.

      This is a documentary.
      Director: Sandy Smolan
      Screenplay: Sandy Smolan
      Narrator: Joel McHale
      Music Composer: Philip Sheppard

With the rapid emergence of digital devices, an unstoppable, invisible force is changing human lives in ways from the microscopic to the gargantuan: Big Data, a word that was barely used a few years ago but now governs the day for many of us from the moment we awaken to the extinguishing of the final late-evening light bulb.

"The Human Face of Big Data" explores how the visualization of data streaming in from satellites, billions of sensors and smart phones is beginning to enable us, as individuals and collectively as a society, to sense, measure and understand aspects of our existence in ways never possible before.

The premise of the documentary is that all of our devices are creating a planetary nervous system and that the massive gathering and analyzing of data in real time is suddenly allowing us address to some of humanity biggest challenges, including pollution, world hunger, and illness.

But as Edward Snowden and the release of the NSA documents has shown, the accessibility of all this data comes at a steep price. Each of us is now leaving an indelible digital trail that will remain forever in our wake.

Narrated by Joel Mchale, the promise and peril in the growing revolution around big data is the playing field of a new documentary by Sandy Smolan, who's feature "Rachel River" was nominated for The Grand Jury Prize at The Sundance Film Festival and won the Jury Prize for Best Cinematography and a special Jury Prize for Acting.

32. ***Dream Big: Engineering Our World***, 2017.

      This is a documentary.
      Director: Greg MacGillivray
      Producer: Shaun MacGillivray
      Narrator: Jeff Bridges
      Music Composer: John Jennings Boyd

Upon first consideration, it might not seem the stuff of grand cinematic adventure. But could engineering secretly be an exciting, creative, heroic realm where the optimists of today are creating the life-saving, world-altering marvels that will make for a safer, more connected, more equal and even more awe-inspiring tomorrow?

With an eclectic, stereotype-bursting engineer cast, the huge story told by *Dream Big* answers that question with a resoundingly "yes" using a series of surprising human stories to expose the hidden world behind the most exciting inventions and structures across the world. It is not only a journey through engineering's greatest wonders, but equally a tale of human grit, aspiration, compassion and the triumph of human ingenuity over life's greatest challenges.

Review: Riley, Donna. Emphasizing service to people and the planet, a new documentary seeks to inspire the next engineers, *Science Magazine Blog*: Feb. 13, 2017.

ENDNOTE

1. Nicholas Sakellariou would like to thank IEEE's Peter K. Wiesner for providing information regarding *Doing the Right Thing*, as well as for sharing with him a DVD copy of the documentary.

Appendix V: Links to Informative Websites for Practicing Engineers and Engineering Students[1]

ENGINEERING PROJECTS IN COMMUNITY SERVICE (EPICS) AT PURDUE UNIVERSITY: https://engineering.purdue.edu/EPICS

Community service agencies face a future in which they must take advantage of technology to improve, coordinate, account for, and deliver the services they provide. They need the help of people with strong technical backgrounds. Undergraduate students face a future in which they will need more than solid expertise in their discipline to succeed. They will be expected to work with people of many different backgrounds to identify and achieve goals. They need educational experiences that can help them broaden their skills.

The challenge is to bring these two groups together in a mutually beneficial way.

In response to this challenge, Purdue University has created EPICS: Engineering Projects in Community Service.

HUMANITARIAN ENGINEERING AT THE COLORADO SCHOOL OF MINES: http://inside.mines.edu/HE-Humanitarian-Engineering-Home

In the past, engineers may have asked, "How do I generate electricity most efficiently?" The humanitarian engineer asks, "How can I help to reduce poverty?" The answer to this question may include generating electricity, but more importantly, Humanitarian Engineers will try to balance technical excellence, economic feasibility, ethical maturity and cultural sensitivity.

> hu•man•i•tar•i•an: an artifact, process, system, or practice promoting present and future wellbeing for the direct benefit of underserved populations.
>
> en•gi•neer•ing: designing and creating a component, subsystem, or system under physical, political, cultural, ethical, legal, environmental, and economic constraints.
>
> humanitarian engineering: design under constraints to directly improve the wellbeing of underserved populations.

THE ILLINOIS FOUNDRY FOR INNOVATION IN ENGINEERING EDUCATION (IFOUNDRY): http://ifoundry.illinois.edu/

The Illinois Foundry for Innovation in Engineering Education (iFoundry) is transforming engineering education for the 21st century. Specifically, iFoundry is forging widespread educational change resulting in engineers aligned with the challenges of a global, creative era by emphasizing (1) philosophical underpinnings and conceptual clarity, (2) pervasive collaboration through organizational and political change, and (3) modern digital media, sharing technology, and other systems innovations.

ENGINEERGIRL: http://www.engineergirl.org/

The EngineerGirl website is part of an NAE project to bring national attention to the opportunity that engineering represents to all people at any age, but particularly to women and girls.

ENGINEERS WITHOUT BORDERS INTERNATIONAL (EWB): http://www.ewb-international.org/index.htm

Engineers without Borders International (EWB-I) is an international association of national EWB/ISF groups whose mission is to facilitate collaboration, exchange of information, and assistance among its member groups that have applied to become part of the association. EWB-I helps the member groups develop their capacity to assist poor communities in their respective countries and create a new generation of *global* engineers.

ENGINEERS FOR A SUSTAINABLE WORLD (ESW): http://www.esustainableworld.org/displaycommon.cfm?an=2

Engineers for a Sustainable World (ESW) is a national, non-profit network committed to building a better world. Established in 2002, ESW is comprised of students, university faculty and professionals who are dedicated to building a more sustainable world for current and future generations.

We believe engineers can be a vital part of the solutions needed to meet global human needs while providing sustainable access to the world's resources for current and future generations. Developed countries contribute to millions of tons of pollution and waste each year, while every day, people around the world struggle to gain sustainable access to clean water, healthy food, and suitable shelter.

Through collegiate chapters across the United States, ESW mobilizes students and faculty members through new educational programs, sustainability-oriented design projects, and volunteer activities that foster practical and innovative solutions to address the world's most critical challenges.

ENGINEERING CULTURES AT VIRGINIA POLYTECHNIC INSTITUTE AND STATE UNIVERSITY: http://www.engcultures.sts.vt.edu/overview.html

The main goal of this course is to help engineers learn to work with people who define problems differently than they do. The course travels around the world, examining how what counts as an engineer and engineering knowledge has varied over time and from place to place. Students gradually become "global engineers" by coming to recognize and value that they live and work in a world of diverse perspectives. Minimally, participants gain concrete strategies for understanding the cultural differences they will encounter on the job and for engaging in shared problem solving in the midst of those differences. When the course works best, it can help students figure out how and where to locate engineering problem solving in their lives while still holding onto their dreams. The title of the course is a pun: it both compares the cultures of engineers at different times and places and explores how engineers participate in and contribute to everyday cultural life.

Dr. Gary Downey and Dr. Juan Lucena jointly developed Engineering Cultures™ at Virginia Tech in 1995. Dr. Downey has taught the course at Virginia Tech nearly every semester since 1996, and Dr. Lucena has been teaching the course at Embry Riddle Aeronautical University, Prescott Campus since 1997. During summer 2001, Downey and Lucena taught a two-week version of the course at the International Institute for Women in Engineering in Paris, France.

OPEN SOURCE ECOLOGY: http://opensourceecology.org.nyud.net/index.php

Open Source Ecology is a network of farmers, engineers, and supporters that for the last two years has been creating the Global Village Construction Set, an open source, low-cost, high performance technological platform that allows for the easy, DIY fabrication of the 50 different Industrial Machines that it takes to build a sustainable civilization with modern comforts. The GVCS lowers the barriers to entry into farming, building, and manufacturing and can be seen as a life-size lego-like set of modular tools that can create entire economies, whether in rural Missouri, where the project was founded, in urban redevelopment, or in the developing world.

ENGINEERING SOCIAL JUSTICE AND PEACE NETWORK: http://esjp.org/

Engineering, Social Justice, and Peace (ESJP) is a network of academics, practitioners, and students in a range of disciplines related to engineering, social justice, and peace. Our approach works toward engineering practices that enhance gender, racial, class, and cultural equity and are democratic, non-oppressive, and non-violent. We seek to better understand the relationships between engineering practices and the contexts that shape those practices, with the purpose of promoting local-level community empowerment through engineering problem solving, broadly conceived.

ENGINEERING FOR CHANGE: https://www.engineeringforchange.org/home.action

Engineering for Change provides a forum to connect, collaborate, solve challenges and share knowledge among a growing community of engineers, technologists, social scientists, NGOs, local governments and community advocates, who are dedicated to improving the quality of life all over the world.

ANITA BORG INSTITUTE FOR WOMEN AND TECHNOLOGY[2]: http://anitaborg.org/get-involved/systers/

Systers is the world's largest email community of technical women in computing. It was founded by Anita Borg in 1987 as a small electronic mailing list for women in "systems." Today, Systers broadly promotes the interests of women in the computing and technology fields. Anita created Systers to "increas[e] the number of women in computer science and mak[e] the environments in which women work more conducive to their continued participation in the field." (Read Why Systers?) It serves this purpose by providing women a private space to seek advice from their peers, and discuss the challenges they share as women technologists. Today, systers is curated by the current Systers-keeper, Robin Jeffries.

ENGINEERING BLOGS: http://engineerblogs.org/

This is a collection of some of the top engineering bloggers on the internet. Surprisingly, scientists seem to outnumber engineers, though we don't think that will happen for long. Some posts link directly back to the author's web page and some stay right here on EngineerBlogs.org. Either way, we promise you some of the best engineering related content on the web.

ENGINEERS AGAINST POVERTY: http://www.engineersagainstpoverty.org/

Engineers against Poverty (AEP) is a specialist NGO working in the field of engineering and international development.

EAP has developed a reputation for producing cutting edge action research and is rapidly establishing itself as a leading agency in its field. EAP demonstrates a high level of innovation both in terms of its programme content and in the range of partners it has mobilised in support of its programmes.

As outlined in EAP's development perspective, Science, Engineering, Technology and Innovation (SETI) plays a critical role in meeting the challenges of sustainable development and poverty reduction. EAP works with partners in industry, government and civil society to identify innovative ways for SETI policy and practice to enhance its contribution to addressing these global challenges.

ONLINE ETHICS CENTER (OEC): http://www.onlineethics.org/

The Online Ethics Center (OEC) is maintained by the National Academy of Engineering (NAE) and is part of the Center for Engineering, Ethics, and Society (CEES). The CEES started in April 2007 and plans conferences and other research and educational activities under the direction of the CEES advisory group.

The Online Ethics Center at the National Academy of Engineering provides readily accessible literature and information, case studies and references, and discussion groups on ethics in engineering and science. It focuses on problems that arise in and for the work life of engineers and scientists. It serves practitioners, educators and students, and individuals interested in professional and research ethics.

The mission of the Online Ethics Center continues to be:

- to provide engineers and engineering students with resources for understanding and addressing ethically significant problems that arise in their work, and
- to serve those who are promoting learning and advancing the understanding of responsible research and practice in engineering.

COMPUTER PROFESSIONALS FOR SOCIAL RESPONSIBILITY: http://cpsr.org/

CPSR is a global organization promoting the responsible use of computer technology. Founded in 1981, CPSR educates policymakers and the public on a wide range of issues. CPSR has incubated numerous projects such as Privaterra, the Public Sphere Project, EPIC (the Electronic Privacy Information Center), the 21st Century Project, the Civil Society Project, and the CFP (Computers, Freedom & Privacy) Conference. Originally founded by U.S. computer scientists, CPSR now has members in 26 countries on six continents.

CENTER FOR THE STUDY OF ETHICS IN THE PROFESSIONS, ILLINOIS INSTITUTE OF TECHNOLOGY: http://ethics.iit.edu/index1.php/Programs

The Center for the Study of Ethics in the Professions is pleased to support a number of ongoing programs and projects that fit the Center's mission of promoting research and teaching on practical moral problems in the professions.

ENGINEERING AND PUBLIC POLICY: http://www.epp.cmu.edu/

Engineering and Public Policy (EPP) is a unique department in the College of Engineering at Carnegie Mellon University which addresses important problems in technology and policy in which the technical details are of central importance. The department offers a research-oriented Ph.D. program and double-major undergraduate B.S. programs with each of the five traditional engineering departments and Computer Science.

Research in the department focuses on problems in:

- energy and environmental systems
- information and communication technology policy
- risk analysis and communication; and
- technology policy and management (including technological innovation and R&D policy).

Across these four focal areas we also study issues in engineered systems and domestic security, issues in technology and organizations and issues in technology and economic development (focusing in particular on Brazil, China, India, and Mexico). We frequently undertake the development of new software tools for the support of policy analysis and research.

MURDOUGH CENTER FOR ENGINEERING PROFESSIONALISM/NATIONAL INSTITUTE FOR ENGINEERING ETHICS, TEXAS TECH UNIVERSITY: http://www.depts.ttu.edu/coe/centers/murdough.php

The Center's purpose is to "[p]rovide engineering ethics and professionalism education, research, and communications to students, faculty, staff, and engineers in industry, government and private practice, other professionals, and citizens in the community, state, and nation. The goal of the center is to increase the awareness of the professional and ethical obligations and responsibilities entrusted to individuals who practice engineering, and encourage cooperation among individuals, universities, professional and technical societies and business organizations with regard to engineering ethics and professionalism issues.

HUMANITARIAN TECHNOLOGY CHALLENGE (HTC), BY IEEE: http://www.ieeehtc.org/index.php/htc/about/

Even in this day and age, far too many people in underdeveloped countries live in darkness, without reliable electricity . . . far too many doctors are forced to treat patients without accurate information . . . and far too many people die from illnesses easily treatable in the world's developed nations.

Think there's nothing you can do about it? Think again.

The Humanitarian Technology Challenge (HTC) was created by people like you, for people like you: technologists, humanitarians, nonprofit organizations, students, government employees—and citizens of the world—who are coming together to identify, and work to solve, some of the world's most pressing challenges. With a unique, open-source collaborative concept, HTC enables you to make a difference—without making a major commitment of your time or resources.

ENDNOTES

1. This list, which is meant to provide evidence that some change in engineering education and practice is underway, is far from comprehensive. The reader who is interested in the most recent developments in the burgeoning field of sustainability in engineering education would benefit from reading the "Special Issue on Sustainability in Civil and Environmental Engineering Education" of the *Journal of Professional Issues in Engineering Education and Practice* 137, no. 49 (2011). A valuable assessment of the status and future of sustainability in engineering education is offered in David Allen et al.'s *Benchmarking Sustainable Engineering Education: Final Report* (2008).
2. Thanks to Stacy Branham, currently Assistant Professor of Informatics in the Donald Bren School of Information and Computer Science at the University of California, Irvine.

Appendix VI: Philosophical Poem

Blueprints for Babel
Taylor Loy

It's an old truth that all towers fall.
Everything we build, we build from rubble.
Even the first stone was fractured from a child
of a star. With our fingernailed hands
we placed it on the altar of the world.

This is the root of ritual: to build a religion
of work. Babel's foundation is laid first
in the mind. A design translating the world
into terms we understand, into heavy things
we cast out into reason's borderlands.

We built with these stones of the mind,
laid with disciplined precision, with wills bent
toward a distant and unseen goal. Perhaps
our great sin was not in our attempt to reach God
but in foolishly imagining heaven
    to be so close by.

Index

Page numbers followed by f, t, and n indicate figures, tables, and notes, respectively.